T0192464

The Science of Wind Power

The Science of Wind Power

Frank R. Spellman

CRC Press
Taylor & Francis Group
Boca Raton London New York

CRC Press is an imprint of the
Taylor & Francis Group, an **informa** business

Cover image: Photo by F. Spellman

First edition published 2022
by CRC Press
6000 Broken Sound Parkway NW, Suite 300, Boca Raton, FL 33487-2742

and by CRC Press
2 Park Square, Milton Park, Abingdon, Oxon, OX14 4RN

CRC Press is an imprint of Taylor & Francis Group, LLC

ISBN: 978-1-032-26579-7 (hbk)
ISBN: 978-1-032-26580-3 (pbk)
ISBN: 978-1-003-28894-7 (ebk)

DOI: 10.1201/9781003288947

Typeset in Times
by Deanta Global Publishing Services, Chennai, India

Contents

PART II The Nuts and Bolts of Wind Power

Preface

The Science of Wind Power is the sixth volume in the acclaimed series that includes *The Science of Water*, *The Science of Air*, *The Science of Environmental Pollution*, *The Science of Renewable Energy*, and *The Science of Waste* (in production), all of which bring this highly successful series fully into the 21st century. *The Science of Wind Power* continues the series based on good science, not feel-good science. It also continues to be presented in the author's trademark conversational style. This book is about wind power and asks the same basic and pertinent questions related to the topic of discussion: Wind Power, what is it? Wind turbine aerodynamics, what is it? Wind power, what is its impact and potential—should we harness it? Wind power, what are its components? Wind power, what are the mechanics and dynamics? Wind power, what are the electrical aspects? Wind power, what are the environmental impacts? Wind power, what are the maintenance requirements?

In addition to asking the same questions, this standard synthesis has been fashioned to answer all the questions. The text deals with the drivers that make wind power important for the future. The text includes the following sections or chapters:

- Fundamentals of wind turbine aerodynamics
- Wind turbine modeling and testing
- Wind turbine design standards
- Environmental impacts of wind turbines
- Wind turbine maintenance requirements

Concern for the environment and for the impact of environmental pollution has brought about a trend (and need) to shift from the use (and reliance) of hydrocarbons to energy-power sources that are pollution neutral or near pollution neutral and renewable. We are beginning to realize that we are responsible for much of the environmental degradation of the past and present—all of which is readily apparent today. Moreover, the impact of 200 years of industrialization and surging population growth has far exceeded the future supply of hydrocarbon power sources.

Why a text on the science of wind power? Simply put, studying wind power without science is analogous to attempting to cook without being able to read a recipe, or to measure ingredients correctly, or to monitor the progress of that which is cooking. However, studying wind power through science affords us the opportunity to maintain a healthful, life-sustaining environment for ourselves and all of Earth's organisms. Keep in mind that I am not striving to make the study of wind power simplistic; I am striving to make it simple. This is a tall order, because there is nothing simple about wind power or its potential and impact on all of us. The bottom line, when you get right down to it, is that sustaining healthy life on Earth is a goal everyone should strive for.

Many of us have come to realize that a price is paid (sometimes a high price) for what is called "the good life." Our consumption and use of the world's resources makes all of us at least partially responsible for the pollution of our environment, due to our use of conventional energy sources such as oil. Pollution and its ramifications are one of the inevitable products of the good life we all strive to attain, but obviously neither is pollution something caused by any single individual nor can one individual totally prevent or correct the situation. To reduce pollution and its harmful effects, everyone must band together as an informed, knowledgeable group and pressurize the elected decision makers to manage the problem now and in the future. Even as individuals, though, we are not impotent in the battle to preserve and repair our environment. The concepts and guidelines in this text will equip you with the scientific rationale required to make decisions that directly affect the environment around you and the use of wind power. This is important, because the information on environmental issues that is seen through the media frequently portends dire catastrophes. Some sources say that pollution is the direct cause of climate change; others deny that the possibility even exists.

Though it is impossible to ensure personal views and perceptions are nonexistent in this text, I have deliberately avoided the intrusion of media hype and political ideology (to me, ideology is thinking without thought); instead, this text attempts to relate facts. As mentioned, I am a firm believer in good science and detest feel-good science. Political ideology and the media work to temper perception, to fog the facts, which, in turn, fogs both the public perception of the issues and the scientific assessment of environmental quality.

Throughout this text, commonsense approaches and practical examples have been presented. Again, because this is a science text, I have adhered to scientific principles, models, and observations, but you need not be a scientist to understand the principles and concepts presented. What is needed is an open mind, a love for the challenge of wading through all the information, an ability to decipher problems, and the patience to answer the questions relevant to each topic presented. The text follows a pattern that is nontraditional; that is, the paradigm used here is based on real-world experience, not on theoretical gobbledygook. Real-life situations are woven throughout the fabric of this text and presented in straightforward, plain English to give the facts, knowledge, and information needed to make informed decisions.

This text is not an answer book. Instead, it is designed to stimulate thought. Although answers to specific energy and pollution problems are not provided, the framework or principles you can use to understand the complexity of any energy use and pollution problem are provided.

Energy usage and environmental issues are attracting ever-increasing attention at all levels. The problems associated with these issues are compounded and made more difficult by the sheer number of factors involved in handling any phase of a problem. Because the issues affect so many areas of society, the dilemma makes us hunt for strategies that solve the problems for all, while maintaining a safe environment without excessive regulation and cost—Gordian knots that defy easy solutions.

The preceding statement goes to the heart of why this text is needed. Presently, only a limited number of individuals have sufficient background in the science of wind power to make informed decisions on energy production, energy usage, and associated environmental issues.

Finally, *The Science of Wind Power* is designed to reach a wide range of student backgrounds and provide a basic handbook or reference for wind technicians. The text focuses on harnessing the wind to produce energy for transmission and use that is critical to preserving what we call the good life. Modern wind turbines bring together mechanical, electrical, hydraulic, and electronic systems to produce a self-contained power generation plant. Most importantly of all, however, *The Science of Wind Power* aims to generate greater awareness of the importance of renewable energy provided by wind energy. Because energy availability is a real-world issue, it logically follows that we can solve energy problems by using real-world methods. That's what *The Science of Wind Power* is all about. Critical to solving these real-world environmental problems is for all of us to remember the old saying: we should "take nothing but pictures, leave nothing but footprints, and kill nothing but time" and sustain ourselves with the flow of wind power.

Frank R. Spellman
Norfolk, Virginia

Author

Frank R. Spellman, PhD, is a retired full-time adjunct assistant professor of environmental health at Old Dominion University, Norfolk, Virginia, and the author of more than 150 books covering topics ranging from concentrated animal feeding operations (CAFOs) to all areas of environmental science and occupational health. Many of his texts are readily available online at Amazon.com and Barnes and Noble.com, and several have been adopted for classroom use at major universities throughout the United States, Canada, Europe, and Russia; two have been translated into Spanish for South American markets. Dr. Spellman has been cited in more than 950 publications. He serves as a professional expert witness for three law groups and as an incident/accident investigator for the US Department of Justice and a northern Virginia law firm. In addition, he consults on homeland security vulnerability assessments for critical infrastructures including water/wastewater facilities nationwide and conducts pre-Occupational Safety and Health Administration (OSHA)/Environmental Protection Agency (EPA) audits throughout the country. Dr. Spellman receives frequent requests to co-author with well-recognized experts in several scientific fields; for example, he is a contributing author of the prestigious text *The Engineering Handbook*, 2nd ed. (CRC Press). Dr. Spellman lectures on wastewater treatment, water treatment, homeland security, and safety topics throughout the country and teaches water/wastewater operator short courses at Virginia Tech (Blacksburg, Virginia). In 2011–2012, he traced and documented the ancient water distribution system at Machu Pichu, Peru, and surveyed several drinking water resources in Amazonia-Coco, Ecuador. He continues to collect and analyze contaminated sediments in the major river systems in the world. Dr. Spellman also studied and surveyed two separate potable water supplies in the Galapagos Islands; he also studied and researched Darwin's finches while in the Galapagos. He holds a BA in public administration, a BS in business management, an MBA, and an MS and PhD in environmental engineering.

Part I

Physical Aspects of Wind Turbines (aka The Horse)

$$HP = \frac{W}{t} = \frac{Fd}{t} = \frac{180 \text{ lbf} \times 2.4 \times 2\pi \times 12 \text{ ft}}{1 \text{ min}} = 33,573 \text{ ft} \cdot \text{lbf} / \text{min}$$

1 Introduction

Wind is the fate we are facing
Wind is the life we are touching every second
Wind is the love we don't understand but feel
Wind is the bridge we cannot see but feel
Wind is wind is wind is wind

—**Nyein Way,** *Dance with Wind*

In the end, the wind takes everything, doesn't it?
And why not?
Why other?

—**Stephen King,** *The Wind through the Keyhole*

BY ANY OTHER NAME

The epigraphs above give us two different views about wind. Way's view is emotional and heartfelt, and King's view veers or turns or twists or is bent more to the other side of emotion garnished with alarm, anxiety, fear, and reality. Whether we call it a breeze, air current, current of air, gale, hurricane, draft, zephyr, sea breeze, or the all-inclusive general term "wind" or by any other name, one thing is certain: the earth's wind has a good and bad side. In this regard and for the truth be known, wind can be said to have a Dr. Jekyll and Mr. Hyde personality or characteristic or aspect; that is, again, like the two well-known literary personalities, one side is good, and the other side is bad. The bad or negative characteristics of wind are tornadoes, hurricanes, movement of pollen and mold (allergens), and the causation of waves in lakes and oceans that can cause destructive flooding. Wind erosion is another bad aspect and is a main geomorphological influence, especially in arid and semi-arid regions. It is also a major source of land desertification, degradation, harmful airborne dust, and crop damage—especially when being magnified far above natural rates by anthropogenic activities such as agriculture, deforestation, and urbanization. This is the "Mr. Hyde" or bad side of the wind. The "Dr. Jekyll" or good side of wind affects world circulation of weather patterns, pollinates plants, reduces air pollution via air circulation, moves warm breezes that comfort us, and provides energy to produce wind power. It is this providing of energy to produce wind power, one of the "good sides" of wind, which is the focus of this book. The bottom line: it is important to keep in mind that the good and/or bad aspects of wind are essential to

DOI: 10.1201/9781003288947-2

the earth's natural processes, all of which are generated with and by the urges or whims of Mother Nature.

Note: Another Hyde characteristic is that wind is one of Mother Nature's most often used tools, a sculpting/grinding/drilling tool without peer, wind erosion, contained within her grand toolbox and continuously used to reshape the earth's surface features. Figures 1.1–1.3 illustrate a few examples of results of Mother Nature's wind tool (water, freeze, and thaw also contribute). Keep in mind that not all people judge these wind carvings as bad but, instead, are amazed by Mother Nature's handiwork.

WIND POWER

Wind power or wind energy—these terms are used interchangeably throughout the text—is the use of wind to provide mechanical power through prime movers—wind turbines—to rotate or turn electric generators for electrical power. Wind power is one of the fastest-growing energy sources on the globe. This is the case because the use of wind power offers many advantages such as being sustainable and a renewable energy source that has a much smaller impact on the global environment compared to the usage of fossil fuels. It is important to point out, however, that like wind, wind power has a good and bad side—specifically wind power has advantages and disadvantages (challenges)—and is covered in detail in this book.

FIGURE 1.1 Wind sculpting in Monument Valley, Utah. Photo by F.R. Spellman.

FIGURE 1.2 Wind sculpting in Monument Valley, Utah. Photo by F.R. Spellman.

FIGURE 1.3 Wind sculpting of Hoodoos in Bryce Canyon, Utah. Photo by F.R. Spellman.

Wind Power Terminology and Development*

To provide comprehensive coverage of wind power and to better prepare the reader for the material contained in this book, wind turbine and wind turbine plan of development terms are defined here in a comprehensible manner. Simply, it is important early on to identify and define key terms used in the book. Although it is common practice to provide a glossary of terms and their definitions at the end of most technical books, in the *Science of* series, the terms and their definitions are provided right up front so that the reader can become familiar with the terminology used throughout the text and thus increase understanding of the subject matter when it is presented. Terms and definitions not provided herein are provided when presented in the book.

Note: These terms and their definitions apply to modern wind power plant construction in the United States.

Terms and Definitions

Access Road: Provides primary access to large blocks of land and connects with, or is an extension of, the public road system. The Access Roads will bring the large turbine components and construction and operations personnel from the public road system to the initial point of entry and inspection on the project site.

Acreage: An area, measured in acres, which is subject to ownership or control by those holding total or fractional shares of working interests. Acreage is considered developed when development has been completed. A distinction may be made between "gross" acreage and "net" acreage:

- Gross—All acreage covered by any working interest, regardless of the percentage of ownership in the interest
- Net—Gross acreage adjusted to reflect the percentage of ownership in the working interest in the acreage

Active power: The component of electric power that performs work, typically measured in kilowatts (kW) or megawatts (MW). Also known as "real power." The terms "active" or "real" are used to modify the base term "power" to differentiate it from Reactive Power.

Accumulator: Fluid-power component used to maintain a fixed volume of fluid at system pressure.

Active braking: Wind turbine yaw brake system that uses electrical or hydraulic actuators to maintain nacelle position when the yaw drive is not activated.

Actuator: Device used to convert electrical, pneumatic, or hydraulic power to useful work.

Aerodynamic force: Force on an object created by fluid flow around the object.

---

* From F. Spellman 2016. *The Science of Renewable Energy*. Boca Raton, FL: CRC Press.

Aerodynamic pressure: Force applied to the surface area of an object created by its interaction with a fluid. Units include pounds per square inch (psi) and megapascal (MPa—a measure of internal pressure).

Note: A fluid is a substance that has no fixed shapes and yields easily to external pressure; a gas or (principally) a liquid.

Adaptation: Adjustment to environmental conditions.

Adaptive management: Focuses on learning and adapting, through partnerships of managers, scientists, and other stakeholders who learn together how to create and maintain sustainable ecosystems. Adaptive management helps science managers maintain flexibility in their decisions (knowing that uncertainties exist) and provides managers with the latitude to change direction and improve understanding of ecological systems to achieve management objectives and is about taking action to improve progress toward desired outcomes.

Adverse weather conditions: Reduced streamflow, lack of rain in the drainage basin, or lower water supply behind a pondage or reservoir dam, resulting in a reduced gross head that limits the production of hydroelectric power or forces restrictions to be placed on multipurpose reservoirs or other water uses.

Alternating current (AC): An electric current that reverses its direction at regularly recurring intervals, usually 50 or 60 times per second.

Alternative fuel: Alternative fuels, for transportation applications, include the following:

- Methanol
- Denatured ethanol and other alcohols
- Fuel mixtures containing 85% or more by volume of methanol, denatured ethanol, and other alcohols with gasoline or other fuels
- Natural gas
- Liquefied petroleum gas (propane)
- Hydrogen
- Coal-derived liquid fuels
- Fuels (other than alcohol) derived from biological materials (biofuels such as soy diesel fuel)
- Electricity (including electricity from solar energy)

The term "alternative fuel" does not include alcohol or other blended portions of primarily petroleum-based fuels used as oxygenates or extenders, i.e., MTBE, ETBE, other ethers, and the 10% ethanol portion of gasohol.

Alternator: A device that turns the rotation of a shaft into alternating current (AC).

Ambient: Natural condition of the environment at any given time.

Ampere (Amp): A unit of electrical current; can be thought of like the rate of water flowing through a pipe (liters per minute).

Ampere-Hour: Amperes times hour, used to measure energy production over time and battery capacity.

Amplitude: Measure the value of signal strength at any given time or frequency level.

Ancillary facilities: Wind turbine generator (WTG) support components (i.e., transformers, electrical collection system, substations, O&M) that allow the electricity produced by the WTG to be connected to the existing electrical grid.

Anemometer: Wind speed measurement device, used to send data to the controller. Also used to conduct wind site surveys. The anemometer is one of the components of a meteorological tower; it is a sensor that measures wind speed and direction.

Angle of attack: In wind turbine operation, the angle of the airflow is relative to the blade.

Angular misalignment: Misalignment of two shafts because of an angle between their centerlines.

Anthropogenic: Made or generated by a human or caused by human activity. The term is used in the context of global climate change to refer to gaseous emissions that are the results of human activities, as well as other potentially climate-altering activities, such as deforestation.

Anti-seize lubricant: Lubricant with polytetrafluoroethylene (PTFE), graphite, or molybdenum disulfide added to prevent galling (a form of wear caused by metal against metal meshing) or cold welding of mating threads.

Apparent power: The product of the voltage (in volts) and the current (in amperes). It comprises both active and reactive power. It is measured in "volt-amperes" and often expressed in "kilovolt-amperes" (kVA) or "megavolt-amperes" (MVA).

Appropriate use: A proposed or existing use on a refuge that meets at least one of the following conditions: (1) the use is a wildlife-dependent one; and (2) the use contributes to fulfilling the refuge purpose(s).

Arc flash: Sudden release of energy from an electrical system because of a short circuit or other fault conditions.

Asynchronous generator: Electrical generator used to produce electricity using a squirrel-cage rotor and a wound stator assembly.

Attendant: Trained, skilled, and authorized individual who provides a safety watch and/or retrieval person for confined space entry.

Availability factor: A percentage representing the number of hours a generating unit is available to produce power (regardless of the amount of power) in a given period, compared to the number of hours in the period.

Axial misalignment: Misalignment of two shafts because of incorrect spacing between the ends.

Axial runout: Movement or displacement of the shaft within the equipment along its centerline.

Axis cabinet: Cabinet that functions to control components of a blade-pitch drive system.

Balance-of-station costs: About 52% of the cost of wind turbine installation includes the cost of assembly, transport and installation, electrical infrastructure, staging, support structure, development, and project management costs.

Bedplate: Structural component or assembly to mount the drive train and other systems to the tower.

Blade: Airfoil used to convert the motion of the oncoming wind into rotational movement of the wind turbine rotor assembly and drive train.

Blade pitch: Controlled movement of wind turbine blades.

Blade root: Large section of the blade used as an attachment assembly to the hub.

Biodiversity conservation: The goal of conservation biology, which is to retain indefinitely as much of the earth's biodiversity as possible, with emphasis on biotic elements most vulnerable to human impacts.

Biota: The plant and animal life of a region.

Body harness: Full-body harness assembly with associated connector and anchorage components used to arrest a fall.

Bonding: Electrical connection method to ensure that components are maintained as the same potential.

Boundary film: Inadequate lubricant film to separate components at rest.

Brake caliper: Clamp assembly used to close brake pads onto a disc to slow or prevent movement of an assembly such as the output of a gearbox.

Breeding habitat: Habitat used by migratory birds or other animals during the breeding season.

British thermal unit (Btu): This is a basic measure of thermal (heat) energy. A Btu is defined as the amount of energy required to increase the temperature of 1 pound of water by 1 degree Fahrenheit, at normal atmospheric pressure. 1 Btu = 1,055 joules.

Btu conversion factor: A factor for converting energy data between one unit of measurement and the British thermal unit (Btu). Btu conversion factors are generally used to convert energy data from physical units of measure (such as barrels, cubic feet, or short tons) into the energy-equivalent measure of Btu (see Appendix A).

Bulk density: Weight per unit of volume, usually specified in pounds per cubic foot.

Bulk modulus: The bulk modulus (K or B) of a substance is a measure of a fluid's compressibility (pounds per square inch, psi, or Pascal, pa).

Candidate species: Plants and animals for which the US Fish and Wildlife Service (FWS) has sufficient information on their biological status and threats to propose them as endangered or threatened under the Endangered Species Act (ESA) but for which development of a proposed listing regulation is precluded by other higher-priority listing activities.

Capacity factor: The ratio of the electrical energy produced by a generating unit for the period of time considered to the electrical energy that could have been produced at continuous full-power operation during the same period.

Capacity, gross: The full-load continuous rating of a generator, prime mover, or other electric equipment under specified conditions as designated by the manufacturer. It is usually indicated on a nameplate attached to the equipment. The **nameplate** indicates not only capacity but also peak power.

Capitol cost: The cost of field development and plant construction and the equipment required for the generation of electricity.

Cathode: The electrode at which reduction (a gain of electrons) occurs. For fuel cells and other galvanic cells, the cathode is the positive terminal; for electrolytic bells (where electrolysis occurs), the cathode is the negative terminal.

Cation: A positively charged ion.

Chained dollars: A measure used to express real prices. Real prices are those that have been adjusted to remove the effect of changes in the purchasing power of the dollar; they usually reflect buying power relative to a reference year. Prior to 1996, real prices were expressed in constant dollars, a measure based on the weights of goods and services in a single year, usually a recent year. In 1996, the US Department of Commerce introduced the chained-dollar measure. The new measure is based on the average weights of goods and services in successive pairs of years. It is "chained" because the second year in each pair, with its weights, becomes the first year of the next pair. The advantage of using the chained-dollar measure is that it is more closely related to any given period covered and is, therefore, subject to less distortion over time.

Characterization: Sampling, monitoring, and analysis activities to determine the extent and nature of contamination at a facility or site. Characterization provides the necessary technical information to develop, screen, analyze, and select appropriate clean-up techniques.

Clearing and grubbing: Process of removing the top layer of soil and vegetation within the areas indicated on the design drawings; cutting and removal of all brush, shrubs, debris, and vegetation to approximately flush with the ground surface or 3 to 6 inches below the surface.

Climate: The average weather (usually taken over a 30-year time period) for a particular region and time period. Climate is different from weather, but rather, it is the average pattern of weather for a particular region. Weather describes the short-term state of the atmosphere. Climatic elements include precipitation, temperature, humidity, sunshine, wind velocity, phenomena such as fog, frost, and hailstorms, and other measures of the weather.

Climate change: The term "climate change" is sometimes used to refer to all forms of climatic inconsistency, but because the earth's climate is never static, the term is more properly used to imply a significant change from one climatic condition to another. In some cases, climate change has been used synonymously with the term global warming; scientists, however, tend to use the term in the wider sense to also include natural changes in the climate.

Climate effects: Impact on residential space heating and cooling (kg CO_2/tree/year) from trees located greater than approximately 15 m (50 ft) from a building (far trees) due to associated reductions in wind speeds and summer air temperatures.

Commercial sector: An energy-consuming sector that consists of service-providing facilities and equipment of businesses; Federal, State, and local governments and other private and public organizations, such as religious, social, or fraternal groups. The commercial sector includes institutional living quarters. It also includes sewage treatment facilities. Common uses of energy associated with this sector include space heating, water heating, air conditioning, lighting, refrigeration, cooking, and running a wide variety of other equipment. Note: This sector includes generators that produce electricity and/or useful thermal output primarily to support the activities of the abovementioned commercial establishments.

Complex grease: Grease with barium and aluminum salts added to soap thickeners to produce desired properties.

Compressible: Fluid property that enables a reduction in volume because of an external force.

Concrete batch plant: A manufacturing plant where cement is mixed before being transported to a construction site, ready to be poured. Equipment and materials including batchers, mixers, sand, aggregate, and cement are required for batching and mixing concrete.

Confined space: A space not designed for human occupancy, capable of partial or full entry by an individual, and with limited ingress and egress.

Conservation: Managing natural resources (includes preservation, restoration, and enhancement) to prevent loss or waste.

Conservation corridor: Connections between suitable habitats that allow passage of plant or animal species.

Conservation easement: A non-possessory interest in real property owned by another, imposing limitations or affirmative obligations with the purpose of retaining or protecting the property's conservation value.

Conservation feature: A feature in the building designed to reduce the usage of energy.

Conservation program: A program in which a utility company furnishes home weatherization services free or at a reduced cost or provides free or low-cost devices for saving energy, such as energy-efficient light bulbs, flow restrictors, weather stripping, and water heater insulation.

Conservation status: Assessment of the status of ecological processes and the viability of species or populations in an ecoregion.

Convection: Transfer of thermal energy to a surrounding fluid of lower temperature.

Critical habitat: According to US Federal law, the ecosystems upon which endangered and threatened species depend; specific geographic areas, whether occupied by a listed species or not, that are essential for its conservation and have been formally designated by rules published in the Federal register.

Cultural resource inventory: A professional study to locate and evaluate evidence of cultural resources within a defined geographic area.

Cultural resources overview: A comprehensive document prepared for a field office that discusses, among other things, project prehistory and cultural history, the nature and extent of known cultural resources, previous research, management objectives, resource management conflicts or issues, and a general statement of how program objectives should be met and conflicts resolved.

Cut-in speed: The speed at which a shaft must turn in order to generate electricity and send it over a wire.

DC: Direct current.

Degradation: The loss of native species and processes due to human activities such that only certain components of the original biodiversity persist, often including significantly altered natural communities.

Demand indicator: A measure of the number of energy-consuming units, or the amount of service or output, for which energy inputs are required.

Demonstrated resources: Same qualifications as identified resources but include measured and indicated degrees of geologic assurance and excludes the inferred.

Dependable capacity: The load-carrying ability of a station or system under adverse conditions for a specified period of time.

Depleted resources: Resources that have been mined: include coal recovered, coal lost in mining, and coal reclassified as subeconomic because of mining.

Diode: A solid-state device that acts as a one-way valve for electricity.

Direct drive: Wind turbine design with rotor assembly connected directly to the generator.

Directional control valve: Component used to alter the path of a fluid within a system.

Distortion: Variation of a signal compared to its true form. Distortion of an AC signal may be determined by the relationship between peak voltage value and RMS (root mean square) voltage value.

Distributed generation (distributed energy resources): Refers to electricity provided by small, modular power generators (typically ranging in capacity from a few kilowatts to 50 MW) located at or near customer demand.

Downwind turbine: A turbine that does not face into the wind and whose direction is controlled directly by the wind.

Drip loop: Loop placed in power and control cables below the yaw deck used to allow slack for the motion of the needle and to enable water to drip from the cables.

Drive train: Components used to connect the rotor assembly to the generator.

Ecological integrity: Native species populations in their historic variety and numbers interacting in naturally structured biotic communities. For communities, integrity is governed by demographics of component species, intactness of landscape-level ecological processes (e.g., natural fire regime), and intactness of internal community processes (e.g., pollination).

Ecological system: Dynamic assemblages of communities that occur together on the landscape at some spatial scale of resolution are tied together by similar ecological processes and form a cohesive, distinguishable unit on the ground. Examples are spruce-fir forest, Great Lakes dune and swale complex, Mojave Desert riparian shrublands.

Economy of scale: The principle that larger production facilities have lower unit costs than smaller facilities.

Ecoregion: A territory defined by a combination of biological, social, and geographic criteria, rather than geopolitical considerations; generally, a system of related, interconnected ecosystems.

Ecosystem: A natural community of organisms interacting with its physical environment, regarded as a unit.

Ecosystem service: A benefit or service provided free by an ecosystem or the environment, such as clean water, flood mitigation, or groundwater recharge.

Efficiency: The ratio of the useful energy output of a machine or other energy-converting plant to the energy input.

EIS (environmental impact statement) corridor: Potential area of impact resulting from the proposed construction activities.

Electric energy: The ability of an electric current to produce work, heat, light, or other forms of energy. It is measured in kilowatt-hours.

Electric power sector: An energy-consuming sector that consists of electricity only and combined heat and power (CHP) plants whose primary business is to sell electricity and heat—i.e., North American Industry Classification System 22 plants.

Electric utility: A corporation, person, agency, authority, or other legal entity or instrumentality aligned with distribution facilities for delivery of electric energy for use primarily by the public. Included are investor-owned electric utilities, municipal and state utilities, Federal electric utilities, and rural electric cooperatives. A few entities that are tariff-based and corporately aligned with companies that own distribution facilities are also included.

Electrical collection system: Consists of underground and overhead cables that carry electricity from and within groups of wind turbines and transmits it to a collection substation and point of interconnection switchyard, which transfers the electricity generated by the project to the regional power grid.

Electromagnet fields (EMF): A combination of invisible electric and magnetic fields of force. They can occur both naturally and due to human constructs.

End user: A firm or individual that purchases products for its own consumption and not for resale (i.e., an ultimate consumer).

Energy: The ability to do work. That is, the capacity for doing work as measured by the capability of doing work (potential energy) or the conversions of this capacity to motion (kinetic energy). Energy has several forms, some of which are easily convertible and can be changed to another form useful for work. Most of the world's convertible energy comes from fossil fuels that are burned to produce heat that is then used as a transfer medium to mechanical or other means in order to accomplish tasks. Electrical energy is usually measured in kilowatt-hours, while heat energy is usually measured in British thermal units (Btu).

Energy efficiency: A ratio of service provided to energy input (e.g., lumens to watts in the case of light bulbs). Services provided can include buildings-sector end uses such as lighting, refrigeration, and heating; industrial processes; or vehicle transportation. Unlike conservation, which involves some reduction of service, energy efficiency provides energy reductions without the sacrifice of service.

Energy source: Any substance or natural phenomenon that can be consumed or transformed to supply heat or power. Examples include petroleum, coal, natural gas, nuclear, biomass, electricity, wind, sunlight, geothermal, water movement, and hydrogen in fuel cells.

Environment: The sum total of all biological, chemical, and physical factors to which organisms are exposed.

Environmental health (abiotic aspects): The composition, structure, and functioning of soil, water, air, and other abiotic features comparable with historic conditions, including the natural abiotic processes that shape the environment.

Environmental impact statement: A document created from a study of the expected environmental effects of a new development or installation.

Environmental restoration: Although usually described as "cleanup," this function encompasses a wide range of activities, such as stabilizing contaminated soil; treating groundwater; decommissioning process buildings, nuclear reactors, chemical separations plants, and many other facilities; and exhuming sludge and buried drums of waste.

Environmental restrictions: In reference to coal accessibility, land-use restrictions that constrain, postpone, or prohibit mining in order to protect environmental resources of an area, for example, surface- or groundwater quality, air quality affected by mining, or plants or animals or their habitats.

Exotic species: A species that is not native to an area and has been introduced intentionally or unintentionally by humans; not all exotics become successfully established.

Fahrenheit: A temperature scale on which the boiling point of water is at 212 degrees above zero on the scale and the freezing point is at 32 degrees above zero at standard atmospheric pressure.

Failure or hazard (electrical power distribution): Any electric power supply equipment or facility failure or other events that, in the judgment of the reporting entity, constitute a hazard to maintaining the continuity of the bulk electric power supply stream such that load reduction action may become necessary and reportable outage may occur. Types of abnormal conditions that should be reported include the imposition of a special operating procedure, the extended purchase of emergency power, other bulk power system actions that may be caused by a natural disaster, a major equipment failure that would impact the bulk power supply, and an environmental and/or regulatory action requiring equipment outages.

Federal land: Public land owned by the Federal Government, including national forests, national parks, and national wildlife refuges.

Federal Power Act: Enacted in 1920, and amended in 1935, the Act consists of three parts. The first part incorporated the Federal Water Power Act administered by the former Federal Power Commission, whose activities were confined almost entirely to licensing non-Federal hydroelectric projects. Parts II and III were added with the passage of the Public Utility Act. These parts extended the Act's authority to include regulating the interstate transmission of electrical energy and rates for its sale as wholesale in interstate commerce. The Federal Energy Regulatory Commission is now charged with the administration of this law.

Fiber optics: Communication system using a controlled light source and photodetector connected by an optical fiber cable.

Flow battery: An electrochemical energy storage device, which utilizes tanks of rechargeable electrolytes to refresh the energy-producing reaction. Since its capacity is limited only by the size of its electrolyte tanks, it is useful for large-scale backup systems to supplement other forms of generation which may be intermittent in nature.

Flow control: Device used to adjust the velocity of the fluid within a fluid system.

Fluid power: Pressurized fluid system used to provide controlled power to remote actuators.

Fluid transfer: Low-pressure system used to move fluid for transportation, processing, cooling, or heating.

Fossil fuels: A general term for combustible geologic deposits of carbon in reduced (organic) form and of biological origin, including coal, oil, natural gas, oil shales, and tar sands. A major concern is that they emit carbon dioxide into the atmosphere when burnt, thus, significantly contributing to the enhanced greenhouse effect.

Frequency: Number of wave oscillations within a second. Units of frequency are in hertz (Hz).

Gallon: A volumetric measure equal to 4 quarts (231 cubic inches) used to measure fuel oil. One gallon equals 3,785 liters; one barrel equals 42 gallons.

Gearbox: A protective casing for a system of gears that converts the wind into mechanical energy. The gears increase the rpm of the low-speed shaft, transferring its energy to the high-speed shaft in order to provide enough speed to generate electricity.

Generation (electricity): The process of producing electric energy from other forms of energy; also, the amount of electric energy produced, expressed in watthours (Wh).

Generator: A device for converting mechanical energy to electrical energy that is located in the nacelle.

Geographic Information Systems (GIS): A computerized system to compile, store, analyze, and display geographically referred information.

Gigawatt (GW): One billion watts or one thousand megawatts.

Gigawatt-electric (GWe): One billion watts of electric capacity.

Gigawatt-hour (GWh): One billion watt-hours.

Global Positioning System (GPS): A navigation system using satellite signals to fix the location of a radio receiver on or above the earth's surface.

Global warming: An increase in the near-surface temperature of the earth. Global warming has occurred in the distant past as a result of natural influence (it is a cyclical event that has occurred throughout the earth's history), but the term is most often used to refer to the warming predicted to occur as a result of increased emissions of greenhouse gases from commercial or industrial resources.

Green pricing/marketing: In the case of renewable electricity, green pricing represents a market solution to the various problems associated with the regulatory valuation of the nonmarket benefits of renewables. Green pricing programs allow electricity customers to express their willingness to pay for renewable energy development through direct payments on their monthly utility bills.

Greenhouse effect: The effect produced as greenhouse gases allow incoming solar radiation to pass through the earth's atmosphere but prevent most of the outgoing infrared radiation from the surface and lower atmosphere from escaping into outer space.

Greenhouse gas: Gases that trap the heat of the sun in the earth's atmosphere, producing the greenhouse effect. The two major greenhouse gases are water vapor and carbon dioxide. Other greenhouse gases include methane, ozone, chlorofluorocarbons, and nitrous oxide.

Grid (also "Power Grid" and "Utility Grid"): A common term referring to an electricity transmission and distribution system.

Gross Domestic Product (GDP): The total value of goods and services produced by labor and property located in the United States. As long as the labor and property are located in the United States, the supplier (that is, the workers and, for property, the owners) may be either US residents or residents of foreign countries.

Gross generation: The total amount of electric energy produced by the generating units at a generation station or stations, measured at the generator terminals.

Habitat: The place or type of site where species and species assemblages are typically found and/or successfully reproduced.

Habitat conservation: Protecting an animal or plant habitat to ensure that the use of that habitat by the animal or plant is not altered or reduced.

Habitat fragmentation: The breaking up of a specific habitat into smaller, unconnected areas.

Horizontal axis wind turbine (HAWT): The most common type of wind turbine where the axis of rotation is oriented horizontally.

Horsepower: A unit for measuring the rate of work (or power) equivalent to 33,000 ft-pounds per minute or 746 W.

Hub: The center part of the rotor assembly, which connects the blades to the low-speed shaft.

Hub adapter: Disc assembly used to mount the rotor assembly to a wind turbine main shaft, gearbox input shaft, or generator shaft.

Hub height: In a horizontal axis wind turbine, the distance from the turbine platform to the rotor shaft.

Hydraulic: System that uses an incompressible fluid as a power transfer medium.

Hydrokinetic energy: The energy possessed by a body of water because of its motion (kinetic energy = ½ mass × velocity^2).

Idle capacity: The component of operable capacity that is not in operation and not under active repair but capable of being placed in operation within 30 days; and capacity not in operation but under active repair that can be completed within 90 days.

Impedance: The opposition to power flow in an AC circuit; that is, any device that introduces such opposition in the form of resistance, reactance, or both. The impedance of a circuit or device is measured as the ratio of voltage to current, where a sinusoidal voltage current of the same frequency is used for the measurement, and it is measured in ohms.

Incandescent lamp: A glass enclosure in which light is produced when a tungsten filament is electrically heated so that it glows. Much of the energy is converted into heat; therefore, this class of lamp is a relatively inefficient source of light. Included in this category are the familiar screw-in light bulbs, as well as somewhat more efficient lamps, such as tungsten halogen lamps, reflector or r-lamps, parabolic aluminized reflector (PAR) lamps, and ellipsoidal reflector (ER) lamps.

Indicator species: A species used as a gauge for the condition of a particular habitat, community, or ecosystem. A characteristic or surrogate species for a community or ecosystem.

Indigenous: Native to an area.

Indigenous species: A species that, other than as a result of an introduction, historically occurred or currently occurs in a particular ecosystem.

Interior roads: Provide lower-volume secondary road access and serves a smaller area than Access Roads and connect to Access Roads. Typically consists of low-volume spur roads that provide point access and connect to the Access Roads. The Interior Roads are along the Turbine Line Corridors used during construction to construct the turbine foundations with concrete, deliver turbine towers and components, and operation personnel to the turbine during operations.

Joule: This is the basic energy unit for the metric system or, in a later, more comprehensive formulation, the International System of Units (SI). It is ultimately defined in terms of meter, kilogram, and second.

kBtu: A unit of work or energy, measured as 1,000 British thermal units. One kBtu is equivalent to 0.293 kWh.

Kaplan turbine: A type of turbine that has two blades whose pitch is adjustable. The turbine may have gates to control the angle of the fluid flow into the blades.

Kilowatt (kW): One thousand watts of electricity.

Kilowatt-hour (kWh): One thousand watt-hours.

Kinetic energy: Energy available as a result of motion that varies directly in proportion to an object's mass and the square of its velocity.

Lamp: A term generally used to describe artificial light. The term is often used when referring to a bulb or tube.

Land use: The ultimate uses to be permitted for currently contaminated lands, waters, and structures at each Department of Energy installation. Land-use decisions will strongly influence the cost of environmental management.

Landform: The physical shape of the land reflecting geologic structure and processes of geomorphology that have sculpted the structure.

Landscape: A heterogeneous land area composed of a cluster of interacting ecosystems that are repeated in similar forms throughout.

Leeward: Away from the direction of the wind. Opposite of windward.

Limiting factor: An environmental limitation that prevents further population growth.

Load: The simultaneous demand of all customers required at any specified point in an electric power system.

Load balancing: Keeping the amount of electricity produced (the supply) equal to the consumption (the demand). This is one of the challenges of wind energy production, which produces energy on a less predictable schedule than other methods.

Low-speed shaft: Connects the rotor to the gearbox.

Machine head: Top level of a HAWT assembly containing power-generation equipment attached to the rotor assembly.

Magneto-Telluri: An electromagnetic method of determining structures below the earth's surface using electrical currents and the magnetic field.

Main lay-down area (also "staging area"): A designated secure area or space(s), adjacent to the construction site, where construction equipment and material/supplies in transit are temporarily stored, assembled, or processed, as part of a construction operation. In addition, temporary construction trailers and vehicles may be parked within the boundary limits of this secure space.

Mean power output (of a wind turbine): The average power output of a wind energy conversion system at a given mean wind speed based on a Raleigh frequency distribution.

Megavolt-amperes (MVA): Millions of volt-amperes, which are a measure of apparent power.

Megawatt (WM): One million watts of electricity.

Megawatt-hour (MWh): million watts per hour

Meteorological mast: One of the components of a meteorological tower, the meteorological mast supports the anemometers and data logger.

Meteorological towers: Wind measurement systems that can be of steel tube or lattice construction and can be free-standing or guyed; they are equipped with sensors to measure wind speed and direction, temperature, and pressure.

Nacelle (or cowling): The cover for the gearbox, drive train, and generator of a wind turbine that converts the energy of the wind into electrical energy. The Nacelle can rotate a full 360 degrees at the top of the tower to capture the prevailing wind and weighs as much as 50 tons.

Natural processes: A complex mix of interactions among animals, plants, and their environment that ensures the maintenance of an ecosystem's full range of biodiversity. Examples include population and predator-prey dynamics, pollination and seed dispersal, nutrient cycling, migration, and dispersal.

Net metering: Arrangement that permits a facility (using a meter that reads inflows and outflows of electricity) to sell any excess power it generates over its load requirement back to the electrical grid to offset consumption.

Net summer capacity: The maximum output, commonly expressed in megawatts (MW), that generating equipment can supply to system load, as demonstrated by a multi-hour test, at the time of summer peak demand. This output reflects a reduction in capacity due to electricity use for station service or auxiliaries.

Nonindustrial private: An ownership class of private lands where the owner does not operate wood-using processing plants.

Nonrenewable fuels: Fuels that cannot be easily made or "renewed," such as oil, natural gas, and coal.

Nonutility generation: Electric generation by nonutility power producers to supply electric power for industrial, commercial, and military operations, or sales to electric utilities.

Nonutility power producer: A corporation, person, agency, authority, or other legal entity or instrumentality that owns electric generating capacity and is not an electrical utility. Nonutility power producers include qualifying congenators, qualifying small power producers, and other nonutility generators without a designated, franchised service area that do not file forms listed in the Code of Federal Regulations, Title 18, Part 141.

Operations and maintenance facilities (O&M): For storing equipment and supplies required during operation. Some maintenance facilities include control functions such as supervisory control and data acquisition (SCADA) to provide two-way communication with each wind turbine.

Overhead transmission line: Potential route for overhead electrical lines connecting from the project substation to a determined point of interconnection at the existing system.

Peak demand: The maximum load during a specified period of time.

Peak watt: A manufacturer's unit indicating the amount of power a photovoltaic cell or module will produce at standard test conditions (normally 1,000 W/m^2 and 25°C).

Peaking plants: Electricity generators that are operated to meet the peak or maximum load on the system. The cost of energy from such plants is usually higher than from baseload plants.

Power factor: The ratio of real power (kW) to apparent power kV-A for any given load and time.

Power grid (also "utility grid"): A common term referring to an electricity transmission and distribution system.

Powerhouse: A structure at a hydroelectric plant site that contains the turbine and generator.

Public land: Land owned by the local, state, or Federal government.

Public Utility Regulatory Policies Act of 1978 (PURPA): National Energy Act, PURPA, contains measures designed to encourage the conservation of energy, more efficient use of resources, and equitable rates. Principal among these was suggested retail rate reforms and new incentives for the production of electricity by congenators and uses of renewable resources.

Quadrillion Btu (Quad): Equivalent to 10^{15} Btu.

Rayleigh frequency distribution: A mathematical representation of the frequency of ratio that specific wind speeds occur within a specified time interval.

Reactance: A phenomenon associated with AC power characterized by the existence of a time difference between volt and current variations.

Reactor: This is the body that provides a reaction to the displacer. It could be a body fixed to the seabed or the seabed itself. It could also be another structure or mass that is not fixed but moves in such a way that reaction forces are created (e.g., moving by a different amount or at different times). A degree of control over the forces acting on each body and/or acting between the bodies (particularly stiffness and damping characteristics) are often required to optimize the amount of energy captured. In some designs, the reactor is actually inside the displacer, while in others it is an external body.

Reclamation: Process of restoring the surface environment to acceptable preexisting conditions. Includes surface contouring, equipment removal, well plugging, revegetation, etc.

Renewable energy: Energy that is produced using resources which regenerate quickly or are inexhaustible. Wind energy is considered inexhaustible, because while it may blow intermittently, it will never stop.

Renewable energy resources: Energy resources that are naturally replenishing but flow limited. They are virtually inexhaustible in duration but limited in the amount of energy that is available per unit of time. Renewable energy resources include biomass, hydro, geothermal, solar, wind, ocean thermal, wave action, and tidal action.

Renewable Portfolio Standard (RPS): A mandate requiring that renewable energy provide a certain percentage of total energy generation or consumption.

Resistivity survey: The measurement of the ability of a material to resist or inhibit the flow of an electrical current, measured in ohmmeters. Resistivity is measured by the voltage between two electrodes, while an electrical current is generated between two other electrodes. Resistivity surveys can be used to delineate the boundaries of geothermal fields.

Rotor: The blades and other rotating components of a wind energy conversion turbine.

SCADA: Supervisory control and data acquisition; collects data throughout the wind farm to monitor and provide control from a remote location.

Shadow flicker: The effect caused by the sun's casting shadows from moving wind turbine blades.

Soil erosion: A natural process in which soil particles are detached and removed by wind or water.

Species: The basic category of biological classification intended to designate a single kind of animal or plant. Any variation among the individuals may be regarded as not affecting the essential sameness which distinguishes them from all other organisms.

Start-up speed: The wind speed at which a rotor begins to rotate.

Staging area (also "main lay-down area"): A designated secure area or space, adjacent to the construction site, where construction equipment and materials/supplies in transit are temporarily stored, assembled, or processed, as a part of a construction operation.

Switching station: A particular type of substation where energy, of the same voltage, is routed either from different sources or to different customers. Switching stations often contain circuit breakers, switches, and other automated mechanisms that switch or divide their output between different distribution lines when system faults occur or shut down transmission altogether in the event of a serious problem.

Tower: Steel structures which support the turbine assembly. Higher towers allow for longer blades and the capture of faster-moving air at higher altitudes.

Transformer: Used to step up or step down AC voltage or AC current.

Transmission/Interconnection facilities: A collection substation terminates collection feeder cables and steps up the voltage to that of the transmission system to which the project ultimately connects.

Transmission line: Structures and conductors that carry bulk supplies of electrical energy from power-generating units.

Transmission system (electric): An interconnected group of electric transmission lines and associated equipment of moving or transferring electric energy in bulk between points of supply and points at which it is transformed for delivery over the distribution system lines to consumers or is delivered to other electrical systems.

Turbine (also see "wind turbine"): A term used for a wind energy conversion device that produces electricity. A machine for the generation of rotary mechanical power from the energy of a stream of fluid (such as water, steam, or hot gas). Turbines convert the kinetic energy of fluids to mechanical energy through the principles of impulse and reaction or a mixture of the two.

Upwind turbine: A turbine that faces into the wind and requires a wind vane and yaw drive in order to maintain proper orientation in relation to the wind.

Utility grid (also see "power grid"): A common term referring to an electricity transmission and distribution system.

Variable-speed wind turbine: Turbines in which the rotor speed increases and decreases with changing wind speed, producing electricity with a variable frequency.

Voltage: The measure of electrical potential difference.

Watt (electric): The electrical unit of power. The rate of energy transfer is equivalent to 1 ampere of eclectic current flowing under a pressure of 1 volt at unity power factor.

Watt (thermal): A unit of power in the metric system, expressed in terms of energy per second, equal to the work done at a rate of 1 joule per second.

Watt-hour (Wh): The electrical energy unit of measure equal to 1 watt of power supplied to, or taken from, an electric circuit steadily for 1 hour.

Wind energy: Energy present in wind motion that can be converted to mechanical energy for driving pumps, mills, and electric power generators. Wind pushes against sails, vanes, or blades radiating from a central rotating shaft.

Wind generator: A wind energy conversion system designed to produce electricity.

Wind load: The lateral pressure on a structure in pounds per square foot, due to wind blowing in any direction.

Wind power (also see "wind energy"): Power generated by converting the mechanical energy of the wind into electrical energy through the use of a wind generator.

Wind power plant: A group of wind turbines interconnected to a common utility system through a system of transformers, distribution lines, and (usually) one substation. Operation, control, and maintenance functions are often centralized through a network of computer-monitoring systems, supplemented by visual inspection. This is a term commonly used in the United States. In Europe, it is called a generating station.

Wind project: Wind projects vary in size, from small projects of one to a few turbines (known as "behind the meter" or "distributed wind systems"), serving individual customers, to large projects ("utility" or "commercial-scale"), designed to provide wholesale electricity to utilities or an electricity market.

Wind rose: A diagram that indicates the average percentage of time that the wind blows from different directions on a monthly or annual basis.

Wind turbine lay-down area: An area adjacent to the wind turbine foundation, where wind turbine components are temporarily stored, assembled, or processed, as part of the wind turbine assembly operation.

Wind turbine or windmill: A device for harnessing the kinetic energy of the wind and using it to do work or generate electricity. Wind turbine generators are smaller plants that may be located in remote locations where wind resources are present and relatively consistent.

Wind vane: Wind direction measurement device, used to send data to the yaw drive.

Windward: Into or facing the direction of the wind. Opposite of leeward.

Wind-wave direction (WWD): This is the direction that the wind-waves are coming from. Wind-waves are produced (or were recently produced) by the local wind. If a swell is present, these waves arrive at a lower period (more frequently) than do the swells. Direction is given on a 16-point compass scale.

Wind-wave height (WWH): This is the average height of the highest one-third of the wind-waves. Again, it is estimated by the process mentioned under "swell height," except that it is calculated from the energies above the separation frequency.

Wind-wave period (WWP): This is the peak period in seconds of the wind-waves.

Yaw: The rotation of a horizontal axis wind turbine around its tower or vertical axis.

Yaw brake: System used to hold the nacelle assembly in place during normal operation.

Yaw deck: Deck below the yaw bearing and nacelle assembly.

Yaw drive: Motor that keeps an upwind turbine facing into the wind.

Yaw system: Wind turbine control and drive system used to position the rotor assembly into the oncoming wind.

Yield strength: Determined by drawing a line to the Young's modulus slope that intersects the stress–strain curve and the 0.002 strain value (0.2%) on the lower strain axis of the graph (see Figure 1.4).

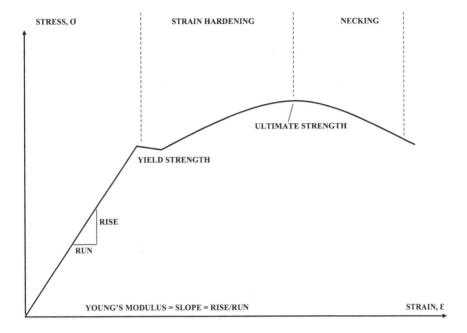

FIGURE 1.4 Wind strain modulus.

Young's modulus (aka modulus of elasticity): Straight-line portion of the stress–strain graph (see Figure 1.4) produced during tensile testing of a ductile material. Within this region of Figure 1.4, material deformation is proportional to the applied load and will return to its original length when the load is removed.

Zerk fitting: Quick-connect device used for adding grease to a component.

Zero-energy condition: State of the condition of a system with all sources of energy isolated, de-energized, and disabled to prevent reenergizing.

WIND TYPES

Winds are defined based on their strength and direction. There are several types of winds, and these are listed and defined as follows:

- **Planetary circulations:**
  - **Jet streams** are fast-flowing, narrow, meandering air currents in the atmosphere of the earth.
  - **Polar jets** ~30,000 to 39,000 ft (9–12 km) are the strongest jet streams.
  - **Trade winds** are permanent east-to-west prevailing winds that blow steadily from the equator from the northeast in Northern Hemisphere or southwest in Southern Hemisphere.
- **Geostrophic wind** is a wind whose speed and direction are determined by a balance of pressure-gradient force and the force due to the earth's rotation.

- **Thermal wind** is the vector difference between winds at higher-altitude geostrophic wind minus that at lower altitudes in the atmosphere.
- **Gradient wind or shear wind** is the rate of increase of wind strength with a unit increase in height above the ground level (Crocker, 2000; Wizelius, 2007).
- **Katabatic wind (aka fall winds)** is a drainage wind, a wind that carries high-density air from a higher elevation down a slope at hurricane speeds under the force of gravity.
- **Anabatic wind (aka upward flow)** is a warm wind during the daytime in calm sunny weather which blows up a steep slope or mountainside, driven by heating the slope through insolation (i.e., the amount of sun rays reaching a given area).
- **Bora** is a north to northeastern katabatic wind in areas near the Adriatic Sea.
- **Foehn wind (aka rain shadow wind)** is a type of warm, dry, downslope wind that occurs in the downward side (lee side) of a mountain range.
- **Chinook wind** is an end-of-winter warm, dry wind which blows down the east side of the Rocky Mountains.
- **Sea breeze (aka onshore breeze)** is more localized than prevailing winds that flow from any large body of water toward or onto a landmass; it develops due to differences in air pressure created by the differing heat capacities of water and dry land.
- **Land breeze (aka offshore breeze)** is the flow of wind from land to sea after sunset.
- **Convective storms (aka thunderstorms)** are storms created by surface heating.
- **Hurricane (aka cyclone, typhoon, tropical depression)** is a tropical cyclone that occurs in the Atlantic Ocean and northeastern Pacific Ocean.
- **Tornado** is a violently rotating column of air that is in contact with both the surface of the earth and a cumulonimbus cloud.
- **Wind gust** is a brief increase in the speed of the wind, usually less than 20 seconds.
- **Dust devil** is a strong, well-formed, and relatively short-lived whirlwind.
- **Microburst** is a downdraft in a thunderstorm that is less than 2.5 miles in scale.
- **Atmospheric wave** is a periodic disturbance in the fields of atmospheric variables.

It is important to note that most of the wind types listed above are not appropriate for optimal wind turbine operation. To unleash the horse at full stride, so to speak, it is important to have a steady flow of wind in the 5 m/s (approximately 11.2 miles/h) range.

SUMMARY

In 1976, energy policy analyst Amor B. Lovins coined the term *soft energy path* to describe an alternative future where energy efficiency and appropriate renewable energy sources steadily replace a centralized energy system based on fossil and nuclear fuels. In 2009, Joshua Green pointed out that in various publications Lovins further argued that the United States had arrived at an important crossroads and could take one of two paths. The first, supported by US policy, promised a future of steadily increasing reliance on dirty fossil fuels and nuclear fission and had serious environmental risks. The alternative, which Lovins called "the soft path," favored "benign" sources of renewable energy like wind power, solar power, biofuels, geo-thermal energy, wave, and tidal power, along with a heightened commitment to energy conservation and energy efficiency.

As a lifelong student, researcher, lecturer, and ardent advocate of and an advocate for the development and use of renewable or alternate energy sources (i.e., eventu-ally excluding all fossil fuel use to the point practicable and possible), I agree with Lovins in many respects, but I take issue with those who state that renewable energy sources are "benign." In my view, the definition of the term benign means something that is hurtless, innocent, innocuous, and/or inoffensive. Thus, the labeling of renew-able energy sources as benign implies that the use of renewable energy sources is without hurt, is innocent, is not innocuous or offensive, and is safe. The truth is the use of renewable energy sources, like the use of wind, has a Dr. Jekyll and Mr. Hyde characteristic(s), and their use is not only not totally fulfilling (i.e., being a complete replacement with the same energy results as fossil fuels) but are not safe or depend-able. Thus, we need to keep in mind the old saying or mantra that sometimes we have to accept the bad with the good, and the good with the bad, meaning that the employment and use of renewable energy has some drawbacks and less than accept-able benefits, but the advantages of clean and sustainable energy sources, such as wind power, outweigh the downsides of renewable energy. It is like the patient who is told that he or she has a cancer growth in the right arm that has spread through-out the entire arm and, therefore, the arm must be amputated to save his or her life. Many would accept the loss of an arm to enable the sustaining and continuance of life—this is an example of the old, accept the bad with the good scenario; that is, nothing is perfect.

Again, I am an advocate for the use of renewable energy. Simply, I think using renewable energy sources instead of fossil fuels is a good thing. However, with any "good thing" usually comes a bad thing or bad side to the thing. Nothing in use by humankind that is human made is absolutely harmless to the environment ... nothing ... absolutely nothing. Only Mother Nature, with her ultimate plan, affects Nature as we know it in beneficial ways; even when she kills millions of us with her designed orchestrations (the earth and life-altering events—earthquakes, hur-ricanes, vulcanism, and so forth) and the changes (adjustments) to life as we know it; these are simply planned and timed mechanizations. Remember, her plan is the ultimate plan. Who are we to argue otherwise?

Anyway, because I do not agree with the connotation that the so-called soft path is the "benign" path, I also cannot say that the impacts of renewable energy sources are bad, baneful, damaging, dangerous, deleterious, detrimental, evil, or harmful to the environment. The question is: What can I say in this book and elsewhere about renewable energy? I can say that I am biased toward the use of renewable energy and am for Lovins' soft path and against the hard path. But I qualify this by also stating that renewable energy sources have impacts on the environment, both good and bad, and although this book projects the benefits of wind power, it also points out the limitations and the Mr. Hyde aspects of impacts, and that is what this book is all about.

So, let's cut to the chase and get down to the bone marrow … my broad thesis is that renewable energy sources are not the panacea for solving many environmental problems that they are popularly perceived to be. In reality, their adverse environmental impacts can be as strongly negative as the impacts of nonrenewable energy sources. On the other hand, or side of the coin, the benefits of using renewable energy sources, including wind power, outweigh the Mr. Hyde aspects of the emerging non-fossil fuel technologies.

REFERENCES

Crocker, D. 2000. *Dictionary of Aeronautical English*. New York: Routledge.
Wizelius, T. 2007. *Developing Wind Power Projects*. London: Earthscan Publications Ltd.

2 From Windmill to Wind Turbine

WINDMILL TO WIND TURBINE

To some, a windmill may be seen as a machine, but for those that live on the open plains (in the United States) a windmill is seen as a life tool. A windmill converts the kinetic energy of the wind to mechanical energy of rotational motion that pumps freshwater from underground.

KINETIC ENERGY → MECHANICAL ENERGY → PUMP ENERGY

On the American prairie, windmills are recognized as a sign of settlement and livestock production. Across the grassland, windmills are a mark of historical and cultural significance. Today, much larger windmills (wind turbines) bring attention to the vastness of the landscape and also provide a source of electric energy (USFS, 2021).

KINETIC ENERGY → MECHANICAL ENERGY → ELECTRIC ENERGY

THE 411 ON WINDMILLS

Several years ago, when I began researching wind power, it became obvious to me that the place to start was at the beginning; that is, to begin at the starting point—the starting point of when humans began using wind for power, for energy, for making life easier. Humans learned early on—at least it says so in recorded history—that making and employing machines can make life easier. So, looking at and studying historical records about wind power became a full-time task for a few months that later turned into several years and is ongoing. However, after grinding out key information about wind power from my previous studies of the topic and from various expert sources, I concluded that two ingredients needed for the mix, so to speak, to put together a comprehensive and comprehensible account of wind power were and are required. First, the need to visit and study various wind power operations was necessary; this was accomplished by visiting several locations where wind power devices of various varieties or types are actively used throughout the United States. The second ingredient I needed was to obtain opinions, several. You might think that I would seek opinions from experts in the field of wind power in general terms and wind power operations from operators. The truth be known I had already accomplished finding and documenting this ingredient(s) for the mix for the previously published works I had fashioned with lots of help. No, experts are fine, and they

DOI: 10.1201/9781003288947-3

certainly know the subject area, and many are more than happy to pass their information on; they only need to be asked.

Again, I already had expert opinions and knowledge gained from experts in the field and my own studies over the years on wind power. What I really needed was an opinion or a fraction of knowledge gained through interviewing just about anyone who allowed me to question them and assented for me to record what they had to say. That's where and when my students came in handy. My students were all upper-level and graduate college students majoring in environmental studies, environmental science, environmental health, and environmental engineering, a very smart and opinionated group, for sure.

So, what question or questions did I ask my students about wind power? I decided to bunch all of what I wanted to know or hear by asking only one question. The question I asked was: What do you think about wind turbines; that is, do you have an opinion about them and/or their use or benefit?

I was surprised (but should not have been) that the 65 students I asked all had opinions about wind power but most notably they talked more about windmills and not wind turbines. The replies on windmills generally ran as:

> "I like windmills."
> "I love them ... always wanted to go to Holland where that guy used his finger or fingers or whatever to plug the leaking dike."
> "The Netherlands are full of windmills, and they are just so beautiful, especially in the wind."
> "I grew up on a farm and we had an old windmill that was used in the old days for pumping water for the farmhouse and the stock."
> "I like photos and paintings that show windmills."
> "I like stories about windmills ... especially the story about that nutcase that mistakes that field of windmills for giants and decides to charge the windmills and do battle with them to gain glory and maybe a lady's hand ... that story really makes me laugh."

Well, there you have it; a small sampling of the dozens of responses that I received from the students. The young woman who said she liked and got a good laugh about the "nutcase" who charged the windmills was speaking, of course, about that gallant want-to-be-knight-errant who desired to revive chivalry and serve his nation under the name Don Quixote. The knight-errant recruited the simple farmer, Sancho Panza, as his squire. Sancho advised Don Quixote that the giants were only windmills and nothing more, and the rest is literary history. When the novel was released in 1605, most readers felt it was a comedy. Today the novel is viewed as a major cornerstone in the evolution of great literary works, a true foundation of great world literature.

Well, that was fiction, but windmills were and are a reality. From windmills to wind turbines is also reality. And it is the reality of wind turbines that this book is all about. One important item, important to me at least, was that out of all the students I queried with the same question posed to all, I received comments from only two students who mentioned wind turbines. Summarizing their comments, they basically stated that they had heard or read about modern-day wind turbines as a source of electrical energy that seemed to be the way of the future and when I joked about Don

Quixote adventures with his tilting at the windmills, one student said he had no idea who Don Quixote was and the other student replied that she had enjoyed reading about his exploits, but he and his exploits were nothing more than fiction, fantasy.

So, at this point, it is time to leave this digression into the world of fiction and to move on to what is real and important in the world of renewable wind power, now and in the future. The discussion begins with a brief history of wind power as it evolved and is still evolving and ends with the highly technical science of wind power generation and its use today and in the future.

IN THE BEGINNING*

From the earliest times of human occupation of the earth, it was well known that in order to survive, certain substances/materials/elements were needed, required, and necessary. Human survival begins with oxygen; without it, there can be no life. Next, humans must have food to maintain life. When air and food are available, humans need some type of shelter for safety. None of these substances/materials/elements can stand alone or in combination to ensure human survival without freshwater. Freshwater for life and survival is an absolute.

Because water is key to maintaining life on the earth, humans from day one of their existence have settled in regions where freshwater is accessible. They settled next to freshwater sources such as springs, rivers, glaciers, and lakes. Later on, around 8,000 BCE, we do not know exactly when, for certain, early humans dug wells and drew freshwater supplied from underground aquifers. They simply dug a hole into the ground to reach the underground aquifers. Then they drew up the water by bucket, raised by hand or mechanically by hand crank drum wound rope, a simple winch assembly, attached to a bucket or bucket-like container, which is lowered down the well shaft to the water level to pick up water and lift to the surface.

Humans are innovative and eventually a number of them decided that a better, easier, and more comfortable means of accessing water was to build stepwells. The most popular location for stepwells was and still is the Indian subcontinent, extending into Pakistan. Stepwells served multiple purposes. First, they made access to drinking water obtainable with relative ease. They also served as storage locations for precious water supplies not only for consumption but also for irrigation needs. The storage function of stepwells was and is also a safeguard during drought when water levels and availability fluctuate. The bottom line on the use of stepwells is that they made it easier for humans to reach the groundwater and to manage and maintain the well.

Humans thrive on making things easy and taking advantage of innovations. One fact that early humans noticed early on is that wind can be used as a source of power. More than 6,000 years ago, our ancestors devised plans for making sails to harness the power of the wind to drive their sailboats. As time passed, wind power became more important and was harnessed more and more. In the 1st century AD, Heron, the

---

* Based on material in F.R. Spellman. 2017. *The Science of Environmental Energy*. Boca Raton, FL: CRC Press.

Greek engineer of Alexandria, devised the first confirmed device that used the wind to power a mechanical machine. He invented a simple windmill wheel to catch the force of the wind and power a musical instrument.

Several centuries passed before windmills of sorts began to be constructed and used in the 7th and 9th centuries in the Middle East. It was not long before windmills started to be used throughout the Muslim world and found their way to India and China, and then the first models arrived in continental Europe. These early designs were not that close to what are used today, but they did include both horizontal and vertical windmills, with varied sets of steps (use-case scenarios) that involved all the so-called traditional jobs of grinding grain and conveying water for irrigation, pumping water from deep wells, and draining water for salt capture and agriculture.

Meanwhile, between the 12th and 16th centuries, Europe started to take a much more serious attitude and approach to exploiting wind power. Eventually, windmills were put to work in various industrial power applications in several European countries, ultimately reaching a total of more than 100,000 windmills performing countless jobs.

After the 16th century, windmills of all shapes, sizes, and types were constructed all around the world. By the time steam and electricity began to be the dominant source of power for industry, there were vertical windmills of three basic shapes: post-mills, tower mills, and smock mills. The post-mill, consisting of a large solid post on which the top-mounted fan that collects wind power powers the machinery inside, is the earliest and was the most popular type of European wooden windmills constructed between the 12th and 19th centuries. Tower mills were wide, tall, and strong because they were built from masonry, bricks, or stone. They had a rotating wooden cap that could rotate and take advantage of the wind that changes direction. After tower mills were put into operation, smock mills, built with refinements based on lessons learned from previous mills, were created with the lower half of the structure made from stone or bricks, and the top half made from light wood covered with a durable tar or metal material and fashioned as an elongated wooden framework called "smock."

After the spread of electricity and fossil fuel-powered internal combustion engines, the need for wind-powered machines decreased dramatically all around the globe, but some wind-powered machines have not become outmoded, antiquated. For instance, wind-powered prairie water pumps (see Figure 2.1) are still commonly used on farms and rangeland because these windmill-type pumps are self-sufficient and require very little maintenance or attention and operate with even a small amount of wind pulling water from deep wells.

AT THE PRESENT TIME

With the advancement of modern science, engineering, and technology, the windmill of the past has morphed into the wind turbine of today. These practical machines, both using wind power in their own way, have served us well in the past, are serving us in the present, and certainly will serve us in the future.

As previously mentioned, the windmill was one of the most effective methods to harness wind power and convert it into wind energy for a variety of uses, including pumping water, grinding grain, and sawing lumber in sawmills—they functioned to accomplish like activities of water wheels used in mills. The enormous blades,

FIGURE 2.1   Water pump windmill, near Zion National Park, Utah. Photo by F. Spellman.

vanes, or sails of a windmill are similar to the wings of an airplane; they both have, to a degree, the same aerodynamic characteristics—when the wind blows over the blades of the windmill, a pressure difference takes place over various sections and causes the blades to turn. The blades move whenever the wind blows against them while turning a camshaft which is connected to a gear train, and it generates wind power for energy for whatever the intended application. Initially, the main application of windmills was to pump water. But later, windmills were used for grinding grains, crushing and shredding rocks, agricultural purposes, and so forth. An advantage of windmills is that they work well in extreme weather conditions.

When windmills morphed into wind turbines, it quickly became obvious that wind turbines (because of their advanced mechanical adaptations using the kinetic energy of wind) were almost perfect generators of electricity for commercial operations. Because they are sources of renewable energy and efficient in operations, they have become an increasing source of electrical energy.

SUMMARY

Although both windmills and wind turbines harness wind power to produce energy, they differ from each other in terms of how they function and their capabilities. Windmills have the advantage of costing little for sectors that require a small amount of power. However, for supplying electricity for large commercial enterprises and housing sectors, wind turbines are the power devices of choice.

SUMMARY

3 The Winds of Change

WINDS OF CHANGE HAVE BEGUN TO BLOW

Climate change, shortage of fossil fuels, pollution generated by fossil fuel usage, and concerns related to environmental conditions have necessitated the need to incorporate renewable energy sources into the overall energy supply. Some say all of this is necessary in order to preserve the globe's environment before the curtain falls on humanity and life in general. The winds of change are blowing and are focused on traditional energy sources and the need to move on with clean, renewable energy sources. This is where wind power enters the scene, the picture, the mood, and the change of attitude toward renewable energy production and use. Simply, it can be said that wind energy is the horse, and the need for a cleaner environment is the saddle—it can be said that we need to saddle up, mount up, harness, and ride the wind.*

THE HORSE OF WIND POWER

The wind turbine (the horse of wind power) harnesses the wind—which is a free, clean, and far and wide available renewable energy source—to generate electric power. When we harness a wind turbine, it turns wind energy into electricity using the aerodynamic force from the rotor blades, which work like an airplane wing or helicopter rotor blade. When wind flows across the blade, the air pressure on one side of the blade decreases. The difference in air pressure across the two sides of the blade creates both lift and drag. Because the force of the lift is stronger than the drag, it causes the rotor to spin. The rotor connects to the generator, either directly (if it's a direct drive turbine) or through a shaft and a series of meshing gears within a gearbox that speed up rotation and allow for a physically smaller generator (physical size is important because the entire wind turbine package needs to be crammed into the nacelle high above the ground). The translation of aerodynamic force to the rotation of a generator creates electricity (EERE, 2021).

WIND TURBINE CONSTRUCTION MATERIALS

With the exception of the blades, metals are the primary class of materials used to construct wind turbines. Most major components are constructed of ferrous alloys (primarily steel). Designers favor ferrous materials because there is extensive design experience with these materials from the mining industry, and they are relatively cheap to purchase and machine; moreover, they are easily fabricated using conventional practices (Sutherland, 1999).

---

* The wind towers addressed in this chapter focus on onshore installations only.

DOI: 10.1201/9781003288947-4

33

Vertical-axis wind turbine (VAWT) technology uses aluminum in blades because the blades in this technology do not require the twisted and tapered sections of horizontal-axis wind turbines (HAWTs) to achieve relatively high aerodynamic efficiencies. Additionally, extrusion technology allows VAWT aluminum blades to be constructed quickly and relatively inexpensively (Kadlec, 1975). Moreover, innovations in the manufacturing process also allow some variations in the aerodynamic cross sections of the blade through step tapering (Berg et al., 1990).

Fiberglass-reinforced composites are currently the blade materials of choice for wind turbine blades. This class of materials is called fiberglass composites or fiber-reinforced plastics (FRP). Usually, E-glass in polyester, vinyl ester, or epoxy matrix is used to construct turbine blades.

WIND FARM

A wind farm or, by another name, wind park or wind power station or wind power plant is an array of wind turbines in the same location (see Figures 3.1–3.2) used to produce electric power. The placement of wind farms is impacted by factors such as wind conditions (see Figure 3.3), the surrounding terrain, access to electric transmission, and other siting considerations. In a utility-scale wind farm, each turbine generates electricity, which flows to a substation where it then transfers to the grid, where it powers communities.

Within the wind farm location (or nearby) are the transmission lines that carry electricity at high voltages over long distances from wind turbines and other energy

FIGURE 3.1 Wind farm in Indiana. Photo by F. Spellman.

FIGURE 3.2 Wind turbines at Tehachapi Pass, California wind farm. The wind farm site consists of 3,200 acres with a nameplate capacity of 705 MW. Photo by F. Spellman.

generators to areas where that energy is needed. To ensure that the electricity generated by the wind turbines and transmitted via transmission lines is kept at a usable voltage and current level for industrial and domestic use, transformers are used. Transformers receive AC (alternating current) electricity at one voltage and increase or decrease the voltage (via step-up and step-down transformers—thus reducing the required current), which decreases the power losses that happen when transmitting large amounts of current over long distances with transmission lines. When the electricity reaches its destination, transformers step-down the voltage to make it safe and useable by buildings and homes in that community.

Linking the transmission system to the distribution system that delivers electricity to the end users is a substation. Within the substation, transformers convert electricity from high voltages to lower voltages, which can then be delivered safely to consumers. Substations are often connected to other substations to either increase or decrease voltage before the electricity is delivered to the end user.

SADDLING THE HORSE

In order to generate wind-powered electricity, we must saddle the horse of wind power, the wind turbine. If we were to lay out the components of a large wind turbine of the type that is commonly used at the present time, the total number of individual components would be counted in the ±8,000 range.

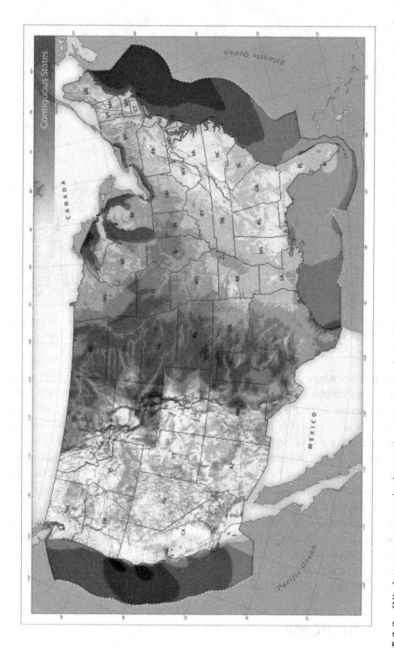

FIGURE 3.3 Wind power resources in the contiguous states and coastlines of the United States. The darkened regions on the map indicate wind speeds of more than 10 m/s at 100 m hub height. Source: NREL 2021. Accessed @ www.NREL.gov/wind.

One thing seems certain; the use of wind power to generate electricity continues to grow, especially given commitments by the United States and other countries throughout the globe to ensure that a significant percentage of energy comes from renewable sources. In order to meet such objectives, increasingly larger turbines with higher capacity are being developed. The engineering aspects of large turbine development tend to focus on design and materials for the blades and towers. However, foundations are also a critical component of large wind turbines and represent a significant cost of wind energy installations.

Note that a wind turbine is typically a high-rise structure with an average tower height on the earth of about 90 m–130 m (~295 ft–427 ft).

WIND TURBINE FOUNDATIONS (ONSHORE)

Wind turbine foundations take on the load transmitted from the wind turbine tower and the turbine on the top, especially the huge overturning moments. The *overturning moment* is taken as the sum of the moments on the column and any shear on the column multiplied by the distance from the base of the column to the base of the footing.

Basically, there are five common types of tower foundations presently used for onshore wind towers, including the shallow mat extension, the ribbed beam basement, the underneath piled foundation, the uplift anchors, and the so-called new type.

Before describing each of these foundation types, it is important to point out that when installing a foundation for a particular wind turbine, there is much more involved in the process than just simply digging a hole in the ground, inserting some type of tower connection device, and pouring several cubic yards of cement to anchor the foundation. First, before construction begins, the requirements of the local building code must be checked on and then abided by. After obtaining a building permit in compliance with applicable codes and before a foundation is poured for an onshore wind turbine tower, a site investigation must be made to determine its suitability for construction. This inspection must include the utilization of the principles of soil mechanics involving geotechnical engineering guidelines along with in situ and laboratory testing. Because the turbine installation transmits the total load to the soil and/or rock foundation, engineering must be employed. The foundation engineering study of the potential site must determine the vertical, horizontal, moment, and torsion loads that will be transmitted by the superstructure to the foundation system. The strength and stress-related deformability of soils and geological conditions of the soil under consideration that will support the foundation system must be determined. Because the design geometry and location of the substructure element often have an effect on how the soil responds, the foundation engineer must be reasonably versed in structural design (Bowles, 2001). In addition, it must be determined that the loading imposed by the foundation elements is sufficiently less than the ultimate capacity of the soil to assure foundation safety (ASCE, 1976). Simply, the wind turbine foundation site must be suitable for construction and the most suitable foundation type to be used must be determined.

In regard to what type of onshore foundation to use, there are five possible choices. The shape can be both round and octagonal, with diameters ranging from 15 m to 22 m (~49 ft to 72 ft). The *shallow mat extension type* turbine foundation is composed of three basic parts: the basement, the mat, and the central pillar. The advantage of the shallow mat extension foundation is the brief construction period. It also is suitable for the plain area and hilly terrain and can be used widely for a large range of wind turbine tower capacity. However, this type of foundation is not environmentally friendly. It also requires more steel and concrete, and extensive digging for a larger basement area that is needed. The *ribbed beam basement type* wind turbine foundation also includes three parts: the basement, the beams, and the central pillar. However, unlike the shallow mat type, the beams are designed to counter overturning moments. The slim foundation is suitable for the plain area and uses less concrete and steel. On the other hand, this type of foundation is complex, and requires a larger basement area and considerably more excavation and digging work. The *underneath pile type* foundation fights against the forces of overturning by using piles to interact between the piles and the soil. This type of foundation has a large carrying capacity, creates a small footprint, experiences minor settlement, and is suitable for soft soil areas such as mudflat and marsh areas. But this complicated foundation type is also costly, and the piles have an impact on the surrounding structures. The *uplift anchors type* turbine foundation's prestressed anchors take advantage of the carrying capacity of a rock. Along with the great carrying capacity, this type of foundation is suitable for mountainous areas, has a small footprint, and uses little concrete and steel. On the downside, these foundations require corrosion resistance coating and suitable rocks to anchor to. The *new type* wind turbine foundation is so named because its innovation in foundation structures is an ongoing process, basically the next design generation that comes about.

WIND TURBINE TOWERS

Wind turbine towers are mostly made from tubular steel, with cement and hybrids of cement and steel beginning to be used in some applications. The tower supports the turbine. Towers come in sections and are assembled on-site. Because wind speed increases with height, taller towers enable turbines to capture more energy and generate more electricity. Winds at elevations of 30 m (roughly 100 ft) or higher are also less turbulent.

Because increasing wind turbine tower height is beneficial in regard to increasing a turbine's ability to generate increased levels of electricity, those involved with producing wind-powered energy are constantly reaching the limits of increasing turbine height. The possibility of increasing a turbine's electrical output with increased height of the turbine to take advantage of stronger winds aloft is not only of interest to industry power providers, investors, and ratepayers (customers) but is a focus of governmental researchers and others. For example, in the United States, the government's National Renewable Energy Laboratory (NREL) has funded studies by various researchers and groups to determine the feasibility of increasing wind turbine tower heights to take advantage of increased wind speed at the increased heights.

One of these NREL-funded studies conducted in 2019 by a group of experts published their findings in a document titled "Increasing Wind Turbine Tower Heights: Opportunities and Challenges." This thoughtful and thorough study presents the opportunities, challenges, and potentials associated with increasing wind turbine tower heights, focusing on land-based wind energy technology. Keep in mind that the following conclusions reached in this analysis include uncertain resource data, which increases at higher aboveground levels, coupled with high sensitivity in terms of the analysis results to the assumed wind shear. Moreover, the capital expenditure and levelized cost of energy estimates are based on cost characterizations that generally reflect modern state-of-the-art technology and do not consider the potential for future innovations to alter the capital expenditures required to achieve a given tower height. The research team's principal conclusions are as follows (Lantz et al., 2019):

- Wind resource quality improves significantly with height above ground.
- Wind speed differences translate to sizeable capacity factor improvements.
- The most wind-rich regions of the country generally show an economic preference for the lowest considered tower height; higher hub heights (e.g., 110 m and 140 m) are often preferred in more moderate wind speed regions.
- Higher nameplate and lower specific power turbines (e.g., 150 W to 175 W/m^2) also show a general economic preference for the lowest considered tower height; however, these larger turbines require tower heights of at least 110 m.
- The highest-capacity turbine considered (4.5 MW) has a relatively greater preference for 140 m (~459 ft) hub heights than smaller 3-MW-class turbines.
- Reducing the cost of realizing taller towers is critical to capturing the value of higher wind speeds at higher aboveground levels.
- Additional factors that could impact tower height include blade tip clearance requirements, balance-of-station costs, turbine nameplate capacity, and specific power.
- When pursuing higher tower heights, a system-level incremental capital cost of less than $500/kW for low specific power turbines and potentially as low as $200/kW, particularly for higher specific power turbine configurations, could support a levelized cost of energy reduction across much of the country and might also push less-energetic wind resource regions further along the path to economic competitiveness.
- To realize taller wind turbine towers, an array of potential concepts remains in play. These concepts rely on various materials spanning rolled tubular steel (currently the most widely used option), concrete, and lattice steel, for space frame designs, as well as hybrid designs that use a combination of concepts.

Along with the benefit of taller wind turbine towers, it is important to point out that there are also a few detriments to consider. For example, the tall towers and blades up to 90 m (295 ft) long are difficult to transport over the roads and in some locations impossible to transport through existing tunnels and under various bridges and over-passes. Tall towers also require very tall and expensive cranes that are difficult to operate, requiring highly skilled operators. Tower construction is massive in order to support the heavy blades, gearbox, and generator. From the perspective of national security, radar and wireless operations may be affected, confused by reflections from tall wind turbine installations.

WIND DYNAMICS

In the study of the science of wind power, the determining factors are directly related to the dynamics of the atmosphere—local weather. These determining factors include the strength of winds, the direction they are blowing, temperature, available sunlight, and the length of time since the last weather event (strong winds) that can do damage to wind turbines and associated ancillaries.

Nondestructive weather events (including milder winds and precipitation) that work to turn wind turbine blades are beneficial, obviously. However, few people and users of wind power will categorize weather events such as tornadoes, hurricanes, and typhoons as beneficial.

Because a wind turbine can't operate without wind, at this point of the discussion, it is important to discuss wind.

On bright clear nights, the earth cools more rapidly than on cloudy nights, because cloud cover reflects a large amount of heat back to the earth, where it is reabsorbed. The earth's air is heated primarily by contact with the warm the earth. When air is warmed, it expands and becomes lighter. Air warmed by contact with the earth rises and is replaced by cold air that flows in and under it. When this cold air is warmed, it too rises and is replaced by cold air. This cycle continues and generates a circulation of warm and cold air called *convection*.

At the earth's equator, the air receives much more heat than the air at the poles. This warm air at the equator is replaced by colder air flowing in from north and south. The warm, light air rises and moves poleward high above the earth. As it cools, it sinks, replacing the cool surface air that has moved toward the equator.

It is the circulating movement of warm and cold air (convection) and the differences in heating that cause local *winds* and *breezes*. Different amounts of heat are absorbed by different land and water surfaces. Soil that is dark and freshly plowed absorbs much more than grassy fields, for example. Land warms faster than water during the day and cools faster at night. Consequently, the air above such surfaces is warmed and cooled, resulting in the production of local winds.

Winds should not be confused with air currents. The wind is primarily oriented toward horizontal flow, while *air currents* are created by air moving upward and downward. Both wind and air currents have a direct impact on the operation of wind turbines. An important factor in determining the locations most affected by wind turbine operation is wind direction. Because wind flow is an important ingredient in the operation of wind turbines, wind direction in the Northern Hemisphere (and elsewhere) is important (see Figure 3.4).

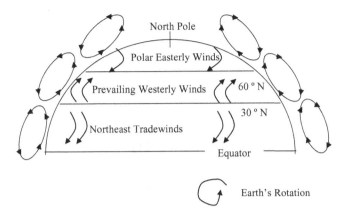

FIGURE 3.4   Wind directions in the Northern Hemisphere.

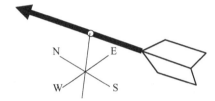

FIGURE 3.5   Weathervane—weighted end faces into the wind.

The wind is a weather element with two measurements that concern meteorologists and wind turbine operators/managers: wind direction and speed. In regard to wind direction, we are all familiar with the weathervane (see Figure 3.5), which is designed to turn itself into an oncoming wind; it tells us the direction the wind is coming *from*. Wind turbines have wind vanes, anemometers, mounted to their structures to measure wind direction and communicate this information to the yaw drive to orient the turbine properly with respect to the wind. Keep in mind that all wind direction information tells us is where the wind is coming from, not where it is going.

Wind speed, or wind velocity, is a fundamental atmospheric rate and is measured by an anemometer (see Figure 3.6); it can also be classified using the older Beaufort scale, which is based on people's observations of specifically defined wind effects.

As shown in Figure 3.6, the rotating cup anemometer usually consists of three hemispherical or cone-shaped cups mounted symmetrically about a vertical axis of rotation. The rate of rotation of the cups is essentially linear over the normal range of measurements, with the linear wind speed being about two to three times the linear speed of a point on the center of a cup, depending on the construction of the cup assembly.

Another type of anemometer commonly used today is the vane-oriented propeller anemometer. It usually consists of a two-, three-, or four-bladed propeller which rotates on a horizontal pivoted shaft that is turned into the wind by a vane. There are several propeller anemometers which employ lightweight molded plastic or

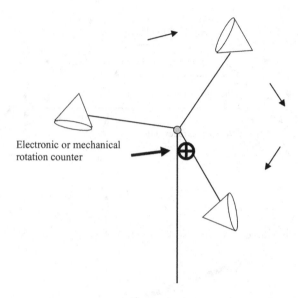

Electronic or mechanical
rotation counter

FIGURE 3.6 Cup anemometer. The mounted electronic or mechanical rotation-counter counts or signals rotations as the wind turns the cups. By counting the signals and measuring the radius of the arm to find the circumference of the circle, one can calculate the distance the wind traveled. Thus, mathematically we can state wind speed as a result of distance/time = miles per hour (mph).

polystyrene foam for the propeller blades to achieve low stating threshold speeds. Some propeller anemometers are not associated with a moving vane. Rather, two orthogonal fixed-mount propellers are used to determine the vector components (i.e., speed and direction) of the horizontal wind. A third propeller with a fixed-mount rotating about a vertical axis may be used to determine the vertical component of the wind if desired.

The bottom line on wind direction is that it determines the design of the turbine. Upwind turbines face the wind while downward turbines face away. Most utility-scale land-based wind turbines are upwind turbines (EERE, 2021).

WIND TURBINE BLADES

Most wind turbines have three blades which are mostly fiberglass. Turbine blades vary in size, but a typical modern land-based wind turbine has blades of over 52 m (170 ft). At the present time, the largest wind turbine is GE's Haliade-X off-shore wind turbine, with blades 107 m (351 ft)—longer than a football field. Much like the wings of an aircraft when wind flows across the blade, the air pressure on one side of the blade decreases. The difference in air pressure across the two sides of the blade creates both lift and drag. The force of the lift is stronger than the drag, and this causes the rotor to spin (EERE, 2021).

INSIDE THE HORSE

LAND-BASED GEARBOX TURBINE

The land-based gearbox is the drivetrain (the stallion or workhorse) of a turbine comprised of the rotor, main bearing, main shaft, gearbox, parking, or secondary brake.

Basically, the gearbox converts the relatively slow speed and high torque of the turbine's blades (about 5–15 rotations per minute for a modern machine) to high speeds (1,000–1,800 rotations per minute) needed to generate electricity using a high-speed induction generator. The relationship between rotational speed (RPM) and torque is expressed as

$$P = \frac{(T \times S)}{9,549} \qquad (3.1)$$

where
   P = power (kW)
   T = torque (N-m)
   S = shaft speed (RPM)
   Constant = 9,549

It is important to note that all the moving parts in the gearbox make it one of the highest-maintenance parts of the wind turbine.

A possible alternative to gearbox-type wind turbines is the direct drive generator that can generate electricity at much lower speeds. Direct drive systems do not require gearboxes and therefore have fewer moving parts. Fewer moving parts equal less maintenance. However, they usually use permanent magnets instead of copper windings. The permanent magnets are expensive, heavy, rare earth materials such as neodymium (chemical symbol Nd and atomic number 60) and dysprosium (chemical symbol Dy and atomic number 66), and they require heavier generators than geared machines for a given turbine capacity (EERE, 2021).

Irrespective of whether it's direct drive or geared, these components are massive (200–320 tons for a 10-MW turbine generator system), and as they are positioned on top of the wind turbine's tower, they also increase the weight and cost of the tower and foundation. They also require large, expensive cranes for installation and have transportation constraints due to their weight.

Adding to these challenges is that wind turbine tower heights have grown from 60 m to 80 m (197 ft to 262 ft) and are expected to exceed 100 m (300 ft) in the coming years. At the same time, average wind turbine capacities have increased from 1 MW to 2–3 MW on land and 5–6 MW offshore, with plans for 10–12 MW offshore wind turbines by the mid-2020s. This increase in capacity means more powerful machines that can generate more electricity, but it also means larger and heavier components.

Nacelle

The nacelle (Fr: small boat) sits atop the tower and contains the gearbox, low- and high-speed shafts, generator, and brake. Some nacelles are larger than a house and for a 1.5-MW-geared turbine and can weigh more than 4.5 tons.

Yaw System

From a change in direction point of view, the yaw system is like the bit and reins in the horse's mouth directing the horse, the turbine in a different direction by rotating the nacelle on upwind turbines to keep them facing the wind. In the turbine, the yaw motors power the yaw drive to make this happen. Downwind turbines don't require a yaw drive because the wind manually blows the rotor away from it.

Pitch System

The advantage of wind turbine pitch control systems is that they are fast and can be used to better regulate power flow, especially when or near the high-speed limit. Figure 3.7 shows a variable-speed wind turbine pitch control system. The generator output can be controlled to follow the commanded power. The pitchable blade controls aerodynamic power. The dashed line shown in Figure 3.7 indicates that the pitch angle can be controlled. Simply stated, the pitch system adjusts the angle of

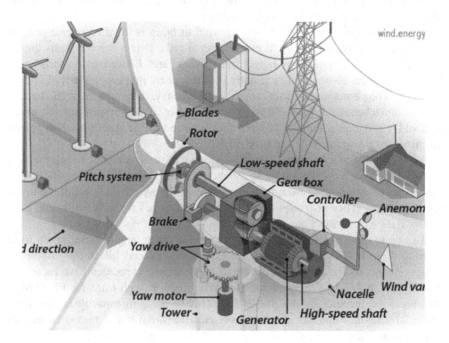

FIGURE 3.7 Inside the land-based wind turbine (the horse). Source: accessed 09/07/2021 @ www.energygov/eere/inside-wind-turbine.

the wind turbine's blades with respect to the wind, controlling the rotor speed. By adjusting the angle of a turbine's blades, the pitch system controls how much energy the blades can extract. The pitch system can also "feather" the blades, adjusting their angle so they do not produce a force that would cause the rotor to spin. Feathering the blades slows the turbine's rotor to prevent damage to the turbine machinery when wind speeds are too high for safe operation.

Turbine Hub

The turbine hub is part of the turbine's drivetrain, and the turbine blades fit into the hub that is connected to the turbine's main shaft. The hub, generally made of welded steel or cast, must withstand all the loads generated by the blades. Hubs vary in design and operation depending on the overall expected performance of the turbine, but in general there are three types: rigid hubs, teetering hubs, and hubs for hinged blades. *Rigid hubs* have all major parts in a fixed position relative to the shaft; however, this design does not exclude variable pitch blades. *Teetering hubs* are complex and used mainly on two-bladed wind turbines. The teetering hub can reduce aerodynamic imbalances or loads due to rotor rotation and yawing of the turbine. *Hubs for hinged blades* are somewhat similar to, or are a cross between a rigid hub and a teetering hub. Teetering hubs are complex but have the advantage that the two blades balance each other.

Turbine Gearbox

The wind turbine gearbox or drivetrain is composed of the rotor, main bearing, main shaft, gearbox, and generator. The gearbox assembly may be designed with multiple gear sets on parallel shafts or in a planetary system. A planetary system (see Figure 3.8) has a central gear or sun gear, outer fixed ring gear, and pinion gears or plant gears located between them mounted to the carrier assembly. The drivetrain converts the low-speed, high-torque rotation of the turbine's rotor (blades and hub assembly) into electrical energy.

Turbine Rotor

The wind turbine rotor assembly consists of several components used to capture energy from the wind and translate it to usable mechanical power to rotate the generator. The wind turbine's rotor is composed of blades and a hub.

Low-Speed Shaft

The wind turbine's low-speed shaft (see Figure 3.8) is part of the turbine's drivetrain; it is connected to the rotor and spins between 8 and 20 rotations per minute.

Main Shaft Bearing

The wind turbine's main shaft bearing is part of the turbine's drivetrain; it supports the rotating low-speed shaft and reduces friction between moving parts so that the forces from the rotor don't damage the shaft.

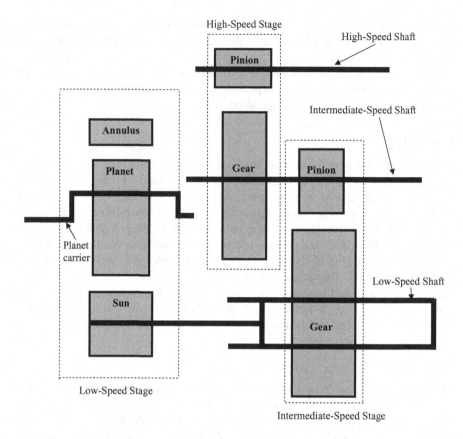

FIGURE 3.8   750-kW gearbox. Adaptation of EERE (2001).

HIGH-SPEED SHAFT

The wind turbine's high-speed shaft (see Figure 3.8) is part of the turbine's drive-train; it connects to the gearbox and is the prime mover of the generator.

WIND TURBINE GENERATORS

The wind turbine generator (WTG; see Figure 3.9) is driven by the high-speed shaft. Copper windings turn through a magnetic field in the generator to produce electricity. Some generators are driven by gearboxes, and others are direct drives where the rotor attaches directly to the generator.

At the present time, the four types of wind turbine generators used in various wind turbine systems are:

• Direct current (DC) generators
• Alternating current (AC) synchronous generators
• AC asynchronous generators
• Switched reluctance generators

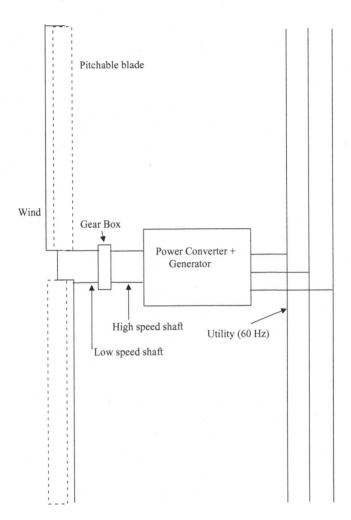

FIGURE 3.9 Wind turbine variable-speed pitch system. Adaptation from E. Muijadi and C.P Butterfield NREL U S. Department of Energy Laboratory at IEEE industry annual meeting October 3–7, 1999, Phoenix, Arizona.

Because of the active nature of wind, it is ideal to operate the WTGs at variable speeds—the physical stress on the turbine drive and blades is reduced which improves aerodynamic efficiency and torque transient behaviors—each of these generators can be run at fixed or variable speeds.

DC Generators

DC generators are not that frequently used in wind power operations except where there are low power demands. Moreover, these generators, because they use commutators and brushes to convert the AC power to DC output, require costly regular maintenance such as brush replacement, cleaning, and undercutting the mica separators between the copper commutator bars.

AC Synchronous Generators (aka Permanent Magnet Synchronous Generators (PMSGs) or Electrically Excited Synchronous Generators (EESGs))

These generators take excitation from either permanent magnets or electromagnets. Whenever the rotor is driven by the wind, three-phase power is produced in this generator via the stator windings, connected to the grid using transformers and power converters. Note that if the generator is of a fixed-speed synchronous type, the rotor speed has to be at exactly the synchronous speed; otherwise, synchronism will be lost—random fluctuations of wind speed and periodic disturbances happen due to tower-shading effects (i.e., alteration in a uniform flow of wind due to the presence of the tower). Another issue with synchronous wind turbine generators is that they tend to have a low damping effect so that they do not allow drivetrain transients (due to sudden changes in the magnitude and/or direction of a torque load) to be absorbed electrically. In addition to being more complicated, expensive, and susceptible to failure compared to induction generators, it is important to note that when synchronous wind turbine generators are incorporated into the power grid, synchronizing their frequency with the grid requires a very skilled operation.

At the present time, permanent magnet (PM) generators are the trend. Also, as stated earlier, a common practice today is to use rare earth magnets in wind turbines, especially in offshore wind turbines, as they allow for high-power density and diminished size (low mass) and weight with peak efficiency at all speeds, offering a high annual production of energy with low lifetime expenditures. Most direct drive turbines are equipped with permanent magnet generators that typically contain the rare earth's neodymium and smaller quantities of dysprosium. Although on a different extent or scale, the same is true for numerous gearbox designs. Using a rather straightforward structure, the PMs are rugged and installed on the rotor to generate a constant magnetic field. The produced electricity is collected from the stator by using the commutator, slip rings, or brushes. To lower cost, the PMs are integrated into a cylindrical cast aluminum rotor. Note that for onshore wind turbine installations it is not necessary to utilize permanent magnet generators because reduced size and weight is not a concern as it is with offshore installations. PM generators are similar in operation to synchronous generators except that PM generators can be operated asynchronously (i.e., induction generators that require the stator to be magnetized from the grid before it works).

PMSGs have some advantages including not needing a commutator, slip rings, and brushes, making them dependable, rugged, and simple. However, PMSGs can't produce electricity with a fixed frequency due to the variability of wind speeds. This means the generators should be connected to the power grid through rectifying AC–DC–AC by power converters; accordingly, the generated AC power containing variable frequency and magnitude is first rectified into fixed DC, and then converted back to AC power. Note that if the wind turbine system includes a battery bank storage system some of the DC power is directed to the bank to charge the batteries. The number one source of wind turbine failure has been the failure of the gearbox. So, when the permanent magnet machines are used for direct drive applications, they can avoid the gearbox problem.

AC Asynchronous Generators

When the object of wind turbine science is to look for well-designed, simple, inexpensive, and dependable wind turbines, planners typically turn to induction generators; this is the trend in modern wind power systems. The induction generators are classified into two types: fixed-speed induction generators (FSIGs) with squirrel-cage rotors and double-fed induction generators (DFIGs) with wound rotors,

The DGIG generator is currently the system of choice for multimegawatt wind turbines. As stated, the DFIG wind turbine is a wound-rotor induction generator that is operated by controlling slip rings or by a power converter interconnected with the grid (see Figure 3.10). The aerodynamic design of this turbine is such that it is capable of operating over a wide wind speed range to achieve optimum aerodynamic efficiency by tracking the ideal tip-speed ratio. As a result, the generator's rotor must operate at a variable rotational speed. As shown in Figure 3.10, the stator is directly connected to the grid and the rotor is interfaced through a crowbar circuit to prevent an overvoltage or surge condition of a power supply in the wind energy system that results in high voltage and current, and also included is an AC–DC–AC power converter that controls the active and reactive power flow.

Switched Reluctance Wind Turbine Generator

Equipped with no permanent magnets of electrical field windings and instead constructed with laminated steel sheets, the switched reluctance wind turbine generator is easy to produce and assemble and basically is quite simple. This type of generator

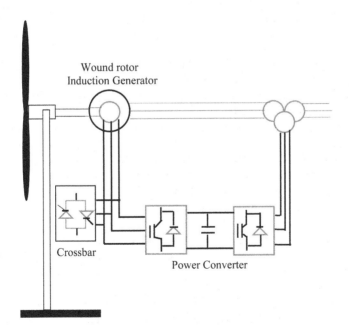

FIGURE 3.10 DFIG wind turbine diagram. Source: Adaptation from Anaya-Lara et al. (2009).

also has very high reliability because it can operate in harsh or high-temperature environments. These generators have strong rotors and stators. When the rotor rotates, the reluctance of the magnetic circuit linking the stator and rotor changes and induces currents in the winding on the stator. Because the reluctance torque is only a fraction of electrical torque, the rotor of a switched reluctance generator is usually larger than the others with electrical excitations for a given rate of voltage. If reluctance generators are connected with direct drive features, the machines are too large and heavy to be practical for use in wind power applications.

WIND TURBINE CONTROLLER

The wind turbine controller is basically the nervous system of the horse; it works to allow the machine to start at wind speeds of about 7–11 miles per hour (mph) and shuts off the machine when wind speeds exceed 55–65 mph. The controller turns off the turbine at higher wind speeds to avoid damage to different parts of the turbine.

WIND TURBINE BRAKE

Unlike the brakes in an automobile, wind turbine brakes keep the rotor from turning after it's been shut down by the pitch system. Once the turbine blades are stopped by the controller, the brake keeps the turbine blades from moving, which is necessary for safety while performing maintenance tasks.

WIND TURBINE LIFE SPAN

The wind turbine horse is designed for a life span of 20–30 years. The actual life span of wind turbines is dependent on not only weather conditions but also the integrity of the manufacturer's turbine parts and preventive and regular maintenance. Maintenance and repair of wind turbine components is expensive. Downtime for repairs and maintenance and the loss of operation due to shutdown have become critical challenges for turbine manufacturers and operators in achieving life span goals and at the same time minimizing costs of operation and repairs. Sounds reasonable and probably not particularly difficult to accomplish, but this is not the case—improving the operational reliability and extending the lifetime of wind turbines—preventing the transition of the horse to a jade, a keffel, or an old nag—is anything but easy for a number of reasons:

- Exposure to extreme, hostile weather conditions such as extreme temperatures, fluctuations of wind speed, solar radiation, dust, humidity, lightning, onslaughts of rain, hail, snow, ice, and sandstorms and for those turbines onshore close to the ocean salt contamination.
- As stated earlier, a modern wind turbine can contain as many as 8,000 individual parts and various systems. A machine containing a large number of components and systems is only as good as the individual performance of its individual components and systems. Each of these components and

systems has its own lifetime. According to the author's modification to and license with the Wooden Bucket Theory, failure must first occur in the component or system with the shortest lifetime.

- Due to wind fluctuations in speed and direction with several starts and stops of the system, a wind turbine is subjected to a variety of dynamic loads. These fluctuations subject primary components to fatigue loads (Sutherland, 1999).

- Superior high-strength, fatigue-resistant materials are imperative to some key components in modern large wind turbines due to the nonstop increase in blade length, hub height, and turbine weight.

- Because wind turbines are complex machines, they should be designed at the system level and not the component level which continues to be a common practice by some turbine manufacturers.

- Improper operation of wind in a wind turbine is another cog in extending the lifetime of the turbine. Improper operation generally occurs when an emergency occurs and a qualified technician is not available. When this occurs, it is not unusual for the manager to call upon someone, anyone at times, to check out the operation of the turbine. This someone or anyone is not always a qualified operator or technician but is assumed to be a Jack or Jill of all trades and in a crunch will do just fine at that moment. However, often the so-called Jack or Jill of all trades turns out to be a Joe and Floe who knows little about turbine operations and/or troubleshooting equipment malfunctions. In some cases, instead of keeping a wind turbine operational or troubleshooting a technical problem, Joe and Floe-types (because of their lack of experience and training) might make the operation worse instead of better.

- Another problem that is more common than not is the failure to perform scheduled preventive maintenance. Wind turbines require regular maintenance for optimal performance—to ensure they produce energy over expected life. Preventive maintenance actions and procedures using industry best working practices are required. Performing any wind turbine maintenance activity willy-nilly without an approved checklist and work procedure or by an inexperienced operator/technician should not be allowed or practiced. Preventive maintenance is an ongoing function that should never be allowed to fall by the wayside. Preventive maintenance actions reduce energy and maintenance costs, ensure fewer breakdowns, and reduce downtime. Also, written safety procedures must be in place and followed, but unfortunately, they often are ignored, not practiced, or nonexistent.

- In regard to performing preventive maintenance activities, there is another side of the coin, so to speak. That is, while preventive maintenance must be performed as required, there are also instances where "too much" preventive maintenance is performed. What is really needed and required is a balance between performing preventive maintenance but not overdoing it—we must seek a happy medium, a so-called Goldilocks situation where things are just right; not underdone and not overdone, instead just right. Performance of too much preventive maintenance leads to overspending, wear and tear increases, technician time wasted, unnecessary equipment and parts used, and inaccurate information collected. In maintenance activities, a golden rule that is often followed states in plain English: "when it is working fine do not mess with it."

• In order to keep that Goldilocks balance in wind turbines between performing preventive maintenance and not performing preventive maintenance, it is best to abide by the simple concept of *condition-based maintenance*. Condition-based maintenance (CBM) involves monitoring wind turbine equipment performance with visual inspections, scheduled tests (e.g., vibration analysis—real-time data collection), and sensor devices (e.g., utilizing SCADA system—real-time data collection) to determine the most cost-efficient time to perform maintenance.

THE BOTTOM LINE

Wind turbines are complex machines that (like the horse) when properly operated and tended to are capable of performing as per design or as desired. However, one thing the experienced operator of any machine (or horse) learns quickly (hopefully this is the case) is that equipment can be unpredictable, and failures and stumbles occur. Ensuring adequate operator training and running preventive or condition-based maintenance at the right time can work toward a better overall operational scheme—keeping the wind turbine running in tip-top shape. Note: The focus of this chapter has been on land-based wind turbines. Note that offshore wind turbines, while similar to land-based turbines, are different in their use of direct drive machinery.

REFERENCES

ASCE. 1976. *Subsurface Investigation for Design and Construction of Foundations of Buildings*, Manuals and reports on engineering practice—No 56. New York, NY: American Society of Civil Engineers, 61 pp.

Berg, D.E., Kilmas, P.C., and Stephenson, W.A. 1990. Aerodynamic design and initial performance measurements for the Sandia 34-m diameter vertical-axis wind turbine, Ninth ASME Wind Energy Symposium, D.E., Berg, ed., SED-Vl. 9, ASME, p. 85.

Bowles, J.E. 2001. *Foundation Analysis and Design*, 5th Edition. New York: McGraw-Hill.

EERE. 2021. How a Wind Turbine Works. Accessed @ https//wwwenergy.gov/EERE/wind/how/-wind-turbine-works.

Kadlec, E.G. 1975. *Characteristics of Future Vertical Axis Wind Turbines*, SAND79–1068. Albuquerque, New Mexico: Sandia National Laboratories.

Lantz, E., Roberts, O., Nunemaker, J., DeMeo, E., Dykes, K., and Scott, G. 2019. *Increasing Wind Turbine Tower Heights: Opportunities and Challenges*. Golden, CO: National Renewable Energy laboratory NREL/TP-5000-73629.

Sutherland, H.J. 1999. *A Summary of the Fatigue Properties of Wind Turbine Materials*. Albuquerque, N.M.: Sandia Nation Laboratories.

4 Wind Turbines, Cybersecurity, and SCADA

What will you do when you are hacked?
What will you do when the power goes out?
What will you do when the wind turbine gets a mind of its own?
What will happen when they hack the wind?
Has anyone ever been killed by a cyberterrorist?
Is SCADA the answer?
Is there an answer?
Is there?
?

—F. Spellman (2011)

PARALYZING THE HORSE

On April 23, 2000, police in Queensland, Australia, stopped a car on the road and found a stolen computer and radio inside. Using commercially available technology, a disgruntled former employee had turned his vehicle into a pirate command center of sewage treatment along Australia's Sunshine Coast. The former employee's arrest solved a mystery that had troubled the Maroochy Shire wastewater system for two months. Somehow the system was leaking hundreds of thousands of gallons of putrid sewage into parks, rivers, and the manicured grounds of a Hyatt Regency hotel—marine life died, the creek water turned black, and the stench was unbearable for residents. Until the former employee's capture—during his 46th successful intrusion—the utility's managers did not know why.

> Utility managers and specialists in engineering, environmental science, and environmental health, along with law enforcement agencies, studied this case of cyber-terrorism because, at the time, it was the only one known in which someone used a digital control system to deliberately cause harm. The former employee's intrusion shows how easy it is to break in—and how unrestrained he was with his power.

DOI: 10.1201/9781003288947-5

To sabotage the system, the former employee set up the software on his laptop to identify itself as a pumping station and then suppressed all alarms. The former employee was the "central control station" during his intrusions, with unlimited command of 300 SCADA nodes governing sewage and drinking water alike.

The bottom line: as serious as the former employee's intrusions were, they pale in comparison with what he could have done to the fresh-water system—he could have done anything he liked.

—Barton Gellman (2002)

Other reports of cyber exploits illustrate the debilitating effects such attacks can have on the nation's security, economy, and public health and safety.

- In May 2015, media sources reported that data belonging to 1.1 million health insurance customers in the Washington, DC, area were stolen in a cyber-attack on a private insurance company. Attackers accessed a database containing names, birth dates, e-mail addresses, and subscriber ID numbers of customers.
- In December 2014, the Industrial Control Systems Cyber Emergency Response Team (ICS-CERT; works to reduce risks within and across all critical infrastructure sectors by partnering with law enforcement agencies) issued an updated alert on a sophisticated malware campaign compromising numerous industrial control system environments. Their analysis indicated that this campaign had been ongoing since at least 2011.
- In the January 2014 to April 2014 release of its Monitor report, ICS-CERT reported that a public utility had been compromised when a sophisticated threat actor gained unauthorized access to its control system network through a vulnerable remote access capability configured on the system. The incident highlighted the need to evaluate security controls employed at the perimeter and ensure that potential intrusion vectors are configured with appropriate security controls, monitoring, and detection capabilities.

IN THE WORDS OF MASTER SUN TZU FROM "THE ART OF WAR":

Those who are first on the battlefield and await the opponents are at ease; those who are last and head into battle are worn out.

In 2000, the FBI identified and listed threats to critical infrastructure. These threats are listed and described in Table 4.1. In 2015, the GAO described the sources of cyber-based threats. These threats are listed and described in detail in Table 4.2.

TABLE 4.1
Threats to Critical Infrastructure Observed by the FBI

Threat	Description
Criminal groups	There is an increased use of cyber intrusions by criminal groups who attack systems for purposes of monetary gain
Foreign intelligence services	Foreign intelligence services use cyber tools as part of their information-gathering and espionage activities
Hackers	Hackers sometimes hack into networks for the thrill of the challenge or for bragging rights in the hacker community. While remote hacking once required a fair amount of skill or computer knowledge, hackers can now download attack scripts and protocols from the Internet and launch them against victim sites. Thus, while attack tools have become more sophisticated, they have also become easier to use
Hacktivists	Hacktivism refers to politically motivated attacks on publicly accessible Web pages or e-mail servers. These groups and individuals overload e-mail servers and hack into Web sites to send a political message
Information warfare	Several nations are aggressively working to develop information warfare doctrine, programs, and capabilities. Such capabilities enable a single entity to have a significant and serious impact by disrupting the supply, communications, and economic infrastructures that support military power—impacts that, according to the Director of Central Intelligence, can affect the daily lives of Americans across the country
Inside threat	The disgruntled organization insider is a principal source of computer crimes. Insiders may not need a great deal of knowledge about computer intrusions because their knowledge of a victim system often allows them to gain unrestricted access to cause damage to the system or to steal system data. The insider threat also includes outsourcing vendors
Virus writers	Virus writers are posing an increasingly serious threat. Several destructive computer viruses and "worms" have harmed files and hard drives, including the Melissa Macro Virus, the Explore.Zip worm, the CIH (Chernobyl) Virus, Nimda, and Code Red

Source: FBI, 2007.

DID YOU KNOW?

Presidential Policy Directive 21 defined "All hazards" as a threat to an incident natural or manufactured that warrants action to protect life, property, the environment, and public health or safety, and to minimize disruptions of government, social, or economic activities.

Threats to systems supporting critical infrastructure are evolving and growing. As shown in Table 4.2, cyber threats can be unintentional or intentional. Unintentional

TABLE 4.2
Common Cyber Threat Sources

Source	Description
Non-adversarial-malicious	
Failure in information technology equipment	Failures in displays, sensors, controllers, and information Information technology hardware responsible for data storage, processing, and communications
Failure in environmental controls	Failures in temperature/humidity controllers or power supplies
Software coding errors	Failures in operating systems, networking, and general-purpose and mission-specific applications
Natural or manufactured disaster	Events beyond an entity's control such as fires, floods/tsunamis, tornadoes, hurricanes, and earthquakes
Unusual or natural event	Natural events beyond the entity's control that are not considered to be disasters (e.g., sunspots)
Infrastructure failure or outage	Failure or outage of telecommunications or electrical power
Unintentional user errors	Failures resulting from erroneous, accidental actions taken by individuals (both system users and administrators) in the course of executing their everyday responsibilities
Adversarial	
Hackers or hacktivists	Hackers break networks for the challenge, revenge, stalking, or monetary gain, among other reasons. Hacktivists are ideologically motivated actors who use cyber exploits to further political goals
Malicious insiders	Insiders (e.g., disgruntled organization employees, including contractors) may not need a great deal of knowledge about computer intrusions because their position with the organization often allows them to gain unrestricted access and cause damage to the target system or to steal system data. These individuals engage in purely malicious activities and should not be confused with non-malicious insider accidents
Nations	Nations, including nation-state, state-sponsored, and state-sanctioned programs, use cyber tools as part of their information-gathering and espionage activities. In addition, several nations are aggressively working to develop information warfare doctrine, programs, and capabilities
Criminal groups and organized crime	Criminal groups seek to attack systems for monetary gain. Specifically, organized criminal groups use cyber exploits to commit identity theft, online fraud, and computer extortion
Terrorist	Terrorists seek to destroy, incapacitate, or exploit critical infrastructures in order to threaten national security, cause mass casualties, weaken the economy, and damage public morale and confidence
Unknown malicious outsiders	Unknown malicious outsiders are threat sources or agents that, due to a lack of information, agencies are unable to classify as being one of the five types of threat sources or agents listed above

Source: GAO analysis of unclassified government and nongovernmental data. GAO 16-79.

or non-adversarial threats include equipment failures, software coding errors, and the actions of poorly trained employees. They also include natural disasters and failures of critical infrastructure on which the organization depends but are outside of its control. Intentional threats include both targeted and untargeted attacks from a variety of sources, including criminal groups, hackers, disgruntled employees, foreign nations engaged in espionage and information warfare, and terrorists. These threat adversaries vary in terms of the capabilities of the actors, their willingness to act, and their motives, which can include seeking monetary gain or seeking an economic, political, or military advantage (GAO, 2015).

CYBERSPACE

Today's developing "information age" technology has intensified the importance of critical infrastructure protection, in which cybersecurity has become as critical as physical security to protecting virtually all critical infrastructure sectors. The Department of Defense (DoD) determined that cyber threats to contractors' unclassified information systems represented an unacceptable risk of compromise to DoD information and posed a significant risk to US national security and economic security interests.

So what do cyber threats have to do with wind power? More specifically, what do cyber threats have to do with wind turbines and wind farms? Well, quite a lot actually. The skilled hacker can bring an online wind turbine to a screeching halt and paralyze it from operating as designed. Moreover, it has been demonstrated that this same hacker can paralyze an entire wind farm without too much trouble. An example of how this can occur and did occur will be given in short order. For now, however, it is important to get on board and in tune with the language and techniques of the hackers (aka the exploiters) before discussing the potential impacts on wind power operations.

In the past few years, especially since 9/11, it has been somewhat routine for us to pick up a newspaper or magazine or view a television news program where a major topic of discussion is cybersecurity or the lack thereof. For example, recently there has been a discussion about Russian hackers trying to influence the US 2016 elections. Many of the cyber intrusion incidents we read or hear about have added new terms or new uses for old terms to our vocabulary. For example, old terms such as Botnets (short for robot networks, also balled bots, zombies, botnet fleets, and many others) are groups of computers that have been compromised with malware such as Trojan Horses, worms, backdoors, remote control software, and viruses. These have taken on new connotations with regard to cybersecurity issues. Relatively new terms such as scanners, Windows NT hacking tools, ICQ hacking tools, mail bombs, sniffer, logic bomb, nukers, dots, backdoor Trojan, key loggers, hackers' Swiss knife, password crackers, blended threats, Warhol Worms, Flash Threats, Targeted Attacks, and BIOS crackers are now commonly read or heard. New terms have evolved along with various control mechanisms. For example, because many control systems are vulnerable to attacks of varying degrees, these attack attempts range from telephone line sweeps (wardialing) to wireless network sniffing (wardriving), to physical network port scanning, and to physical monitoring and intrusion. When wireless network sniffing is performed at (or near) the target point by a pedestrian (warwalking), meaning that instead of a person being in an automotive vehicle, the potential intruder may be sniffing the network for weaknesses or vulnerabilities on foot, posing as a person walking, but they may have a handheld PDA device or laptop computer (Warwalking, 2003). Further, adversaries can leverage common computer software programs, such as Adobe Acrobat and Microsoft Office, to deliver a threat by embedding exploits within software files that can be activated when a user opens a file within its corresponding program. Finally, the communications infrastructure and the utilities are extremely dependent on the Information Technology Sector. This dependency is due to the reliance of the communications systems on the software that runs the control mechanism of the operations systems, the management software, the billing software, and any number of other software packages used by the industry. Table 4.3 provides descriptions of common exploits or techniques, tactics, and practices used by cyber adversaries.

HACKING A WIND FARM

Greenberg (2017) reports that two researchers from the University of Tulsa, with permission of wind turbine and wind farm owners/administrators, used a proof-of-concept (PoC) malware (that is, malware used to demonstrate that their idea and method works) they built to test if they could paralyze, or damage individual wind turbines and an entire wind farm with their hacking activities. At the first turbine they chose to hack and test, they had no trouble picking the simple lock on the turbine's metal door, and upon entering, opened the unsecured server closet inside.

While inside the server closet, the two researchers quickly unplugged a network cable and inserted it into a Raspberry Pi minicomputer (a low-cost, modular, and open designed series of small single-board computers the size of a deck of cards) that had been fitted with a Wi-Fi antenna. Then the Pi was switched on and attached to another Ethernet cable from the microcomputer into an open port on a microwave-sized programmable automation controller (computer) that controlled the turbine. The two researchers then closed the door behind them and walked back to their van.

Sitting in the van, one of the researchers opened a MacBook Pro and typed into the laptop's command line and soon saw a list of IP addresses representing every networked turbine in the field. A few minutes later he typed another command, and the researchers (hackers) watched as the single turbine emitted a muted screech like the brakes of an 18-wheeler, slowed, and came to a complete stop.

Mission accomplished—not that difficult to hack a wind turbine and later the entire wind farm.

MONITORING AND CONTROLLING THE HORSE

Not all relatively new and universally recognizable cyber terms like the ones mentioned earlier have sinister connotation or meaning, of course. Consider, for example, the following digital terms: backup, binary, bit byte, CD-ROM, CPU, database, e-mail, HTML, icon, memory, cyberspace, modem, monitor, network, RAM, Wi-Fi

TABLE 4.3
Common Methods of Cyber Exploits

Exploit	Description
Watering hole	A method by which threat actors exploit the vulnerabilities of carefully selected websites frequented by users of the targeted system. Malware is then injected into the targeted system via the compromised websites
Phishing and spear phishing	A digital form of social engineering that uses authentic-looking e-mails, websites, or instant messages to get users to download malware, open malicious attachments, or open links that direct them to a website that requires information or executes malicious code
Credentials based	An exploit that takes advantage of a system's insufficient user authentication and/or any elements of cybersecurity supporting it to include not limiting the number of failed login attempts, the use of hard-coded credentials, and the use of a broken or risky cryptographic algorithm
Trusted third parties	An exploit that takes advantage of the security vulnerabilities of trusted third parties to gain access to an otherwise secure system
Classic buffer overflow	An exploit that involves the intentional transmission of more data than a program's input buffer can hold, leading to the deletion of critical data and subsequent execution of malicious code
Cryptographic weakness	An exploit that takes advantage of a network employing insufficient encryption when either storing or transmitting data, enabling adversaries to read and/or modify the data stream
Structured Query Language (SQL) Injection	An exploit that involves the alteration of a database search in a web-based application, which can be used to obtain unauthorized across to sensitive information in a database, resulting in data loss at corruption, denial or service, or complete host takeover
Operating system command injection	An exploit that takes advantage of a system's inability to properly neutralize special elements used in operating system commands, allowing the adversaries to execute unexpected commands on the system by either modifying already evoked commands or evoking their own
Cross-site scripting	An exploit that uses third-party web resources to run lines of programming code (referred to as scripts) within the victim's web browser or scriptable application. This occurs when a user, using a browser, visits a malicious website or clicks a malicious link. The most dangerous consequences can occur when this method is used to exploit additional vulnerabilities that may permit an adversary to steal cookers (data exchanged between a web server and a browser), log keystrokes, capture screen shots, discover, and collect network information, or remotely access and control the victim's machine
Cross-site request forgery	An exploit takes advantage of an application that cannot, or does not, sufficiently verify whether a well-formed, valid, consistent request was intentionally provided by the user who submitted the request, tricking the victim into executing a falsified requires that results in the system or data being compromised
Path traversal	An exploit that seeks to gain access to files outside of a restricted directory by modifying the directory pathname in an application that does not properly neutralize special elements (e.g., "...," "/," ".../," etc.)

(Continued)

TABLE 4.3 (CONTINUED)
Common Methods of Cyber Exploits

Exploit	Description
Integer overflow	An exploit where malicious code is inserted that leads to unexpected integer overflow, or wraparound, which can be used by adversaries to control looping or make security decisions in order to cause program crashes, memory corruption, or the execution of arbitrary code via buffer overflow
Uncontrolled format string	Adversaries manipulate externally controlled format strings in print-style functions to gain access to information and/or execute unauthorized code or commands
Open redirect	An exploit where the victim is tricked into selecting a URL (website location) that has been modified to direct them to an external, malicious site which may contain malware that can compromises the victim's machine
Heap-based buffer overflow	Similar to a classic buffer overflow, but the buffer that is overwritten is allocated in the heap portion of memory, generally meaning that the buffer was allocated using a memory allocation routine, such as "malloc ( )"
Unrestricted upload of files	An exploit that takes advantage of insufficient upload restrictions, enabling adversaries to upload malware (e.g., .php) in place of the intended file type (e.g., .jpg)
Inclusion of functionality from untrusted sphere	An exploit that uses trusted, third-party executable functionality (e.g., web widget or library) as a means of executing malicious code in software whose protection mechanisms are unable to determine whether functionality is from a trusted source, modified in transit, or if they are being spoofed
Certificate and certificate authority compromise	Exploits facilitated via the issuance of fraudulent digital certificates (e.g., transport layer security and Secure Socket Layer). Adversaries use these certificates to establish secure connections with the target organization or individual by mimicking a trusted third party
Hybrid of others	An exploit which combines elements of two or more of the aforementioned techniques

Source: GAO (2015).

(wireless fidelity), record, software, and World Wide Web—none of these terms normally generate thoughts of terrorism in most of us.

There is, however, one digital term, SCADA, that most people have not heard of. This is not the case, however, for those who work with the nation's critical infrastructure, including communications and energy sectors infrastructure. SCADA, or **S**upervisory **C**ontrol **A**nd **D**ata **A**cquisition System (also sometimes referred to as Digital Control Systems or Process Control Systems), provides the real-time control mechanisms for most power and utility facilities and plays an important role in computer-based control systems; it is used to monitor and control wind farms. From coordinating music and lights in the proper sequence with controlling vaulting of spray from water fountains, to controlling systems used in the drilling and refining of oil and natural gas, control systems perform many functions. Many energy distribution networks use computer-based systems to remotely control sensitive feeds, processes, and system equipment previously controlled manually. These systems (commonly known as SCADA) allow an energy utility (or operation) to monitor hydraulic oil tank levels, to ensure that contents are stored at correct levels or to monitor tank levels, and, as mentioned, collect data from sensors and control equipment located at remote sites. Common communication sector and customer system sensors measure many parameters, depending on the nature of the industrial manufacturing entity. Common SCADA customer industry system equipment includes valves, pumps, and switching devices for the distribution of electricity. The critical infrastructure of many wind farms is increasingly dependent on SCADA systems.

If we were to ask the specialist to define SCADA, the technical response could be outlined as follows:

- SCADA is a multi-tier system or interfaces with multi-tier systems.
- It is used for physical measurement and control endpoints via a Remote Terminal Unit (RTU) and Program Logic Controller PLC to measure voltage, adjust a value, or flip a switch.
- It is an intermediate processor normally based on commercial third-party OSes—VMS, Unix, Windows, and Linux.
- It human interfaces, for example, with graphical user interface (Windows GUIs).
- Its communication infrastructure consists of a variety of transport mediums such as analog, serial, Internet, radio, and Wi-Fi.

How about the non-specialist response?—for the rest of us who are non-specialists. Well, for those of us in this category we could simply say that SCADA is a computer-based control system that remotely controls and monitors processes previously controlled and monitored manually. The philosophy behind SCADA control systems can be summed up by the phrase, "If you can measure it, you can control it." SCADA allows an operator to use a central computer to supervise (control and monitor) multiple networked computers at remote locations. Each remote computer can control mechanical processes (mixers, pumps, valves, etc.) and collect data from sensors at its remote location. Thus, the phrase: Supervisory Control and Data Acquisition, or SCADA.

The central computer is called the Master Terminal Unit, or MTU. The MTU has two main functions: periodically obtain data from RTUs/PLCs and control remote devices through the operator station. The operator interfaces with the MTU using software called Human Machine Interface, or HMI. The remote computer is called Program Logic Controller (PLC) or Remote Terminal Unit (RTU). The RTU activates a relay (or switch) that turns mechanical equipment "on" and "off." The RTU also collects data from sensors. Sensors perform measurement, and actuators perform control.

In the initial stages, utilities ran wires, also known as hardwire or land lines, from the central computer (MTU) to the remote computers (RTUs). Because remote locations can be located hundreds of miles from the central location, utilities began to use public phone lines and modems, leased telephone company lines, and radio and microwave communication. More recently, they have also begun to use satellite links, the Internet, and newly developed wireless technologies.

DID YOU KNOW?

Modern RTUs typically use a ladder-logic approach to programming due to their similarity to standard electrical circuits. An RTU that employs this ladder-logic programming is called a Programmable Logic Controller (PLC).

Because SCADA systems' sensors provided valuable information, many critical infrastructure entities, utilities, and other industries established "connections" between their SCADA systems and their business system. This allowed Utility/Industrial management and other staff access to valuable statistics, such as chemical usage. When utilities/industries later connected their systems to the Internet, they were able to provide stakeholders/stockholders with usage statistics on the communications segment and utility/industrial web pages. Figure 4.1 provides a basic illustration and example of a representative SCADA network. Note that security system

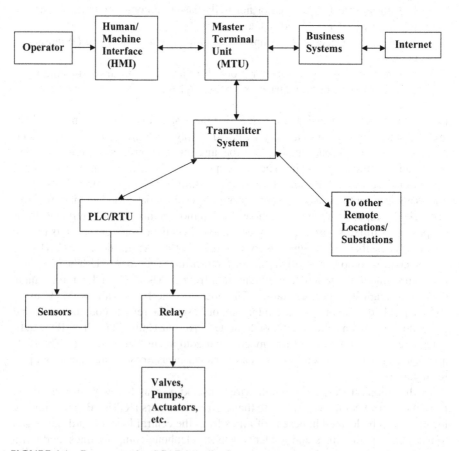

FIGURE 4.1   Representative SCADA network.

protection would normally be placed between the Internet and the business system and between the business system and the MTU.

SCADA APPLICATIONS

Generally speaking and as mentioned, SCADA systems can be designed to measure a variety of equipment operating conditions and parameters or volumes and flow rates of electricity, natural gas and oil, or oil and petrochemical mixture quality parameters, and to respond to change in those parameters either by alerting operators or by modifying system operation through a feedback loop system without having personnel to physically visit each valve, process, or piece of other equipment on a daily basis to check it and/or ensure that it is functioning properly. Automation and integration of diverse large-scale assets required SCADA systems to provide the utmost flexibility, scalability, openness, and reliability. SCADA systems are used to automate certain wind energy production functions; these can be performed without initiation by an operator. In addition to process equipment, in many applications, SCADA systems can also integrate specific security alarms and equipment, such as cameras, motion sensors, lights, and data from card reading systems, thereby providing a clear picture of what is happening at areas throughout a facility. Finally, SCADA systems also provide constant, real-time data on processes, equipment, location access, etc., and the necessary response to be made quickly. This can be extremely useful during emergency conditions, such as when energy distribution lines or piping breaks or when potentially disruptive chemical reaction spikes appear in chemical processing operations. Continuous monitoring of wind turbine health using automated failure detection algorithms can improve turbine reliability and reduce maintenance costs by detecting failures before they reach a catastrophic stage, and by eliminating unnecessary scheduled maintenance. A SCADA data–based condition monitoring system uses data already collected at the wind turbine controller. Currently, it can be said that SCADA has evolved from a simple indicating light and pushbutton control system into a comprehensive operation and handling system for very complex process control and safety shutdown systems. It is a cost-effective way to monitor wind turbines for early warning of failure and performance issues (Kim et al., 2011). In a nutshell, SCADA results in an oversight system that requires fewer operators.

Today, many common wind turbines and other general industrial applications for SCADA systems include but are not limited to those shown below:

- Main bearing
- Generator
- Gearbox
- Hydraulics system
- Boiler controls
- Bearing temperature monitor (electric generators and motors)
- Gas processing
- Plant monitoring
- Plant energy management

- Power distribution monitoring
- Electric power monitoring
- Fuel oil handling system
- Hydroelectric load management
- Petroleum pilot plants
- Plant monitoring
- Process controls
- Process stimulators
- Tank controls
- Utility monitoring
- Safety parameter display systems and shutdown systems
- Tank level control and monitoring
- Turbine controls
- Turbine monitoring
- Virtual annunciator panels
- Alarm systems
- Security equipment
- Event logging

Because these systems can monitor multiple processes, equipment, and wind turbines and then provide quick notification of, or response to, problems or upsets, SCADA systems typically provide the first line of detection for atypical or abnormal conditions. For example, a SCADA system connected to sensors that measure specific machining quality parameters is measured outside of a specific range. A real-time customized operator interface screen could display and control critical systems monitoring parameters.

The system could transmit warning signals back to the operators, such as by initiating a call to a personal pager. This might allow the operators to initiate actions to prevent power outages or contamination and disruption of the energy supply. Further automation of the system could ensure that the system-initiated measures rectify the problem. Preprogrammed control functions (e.g., shutting a valve, controlling flow, throwing a switch, or adding chemicals) can be triggered and operated based on SCADA utility.

SCADA VULNERABILITIES

US ELECTRIC GRID GETS HACKED INTO

The Associated Press (AP) reported on April 9, 2009, that spies hacked into the US energy grid and left behind computer programs (Trojan Horses) that would enable them to disrupt service, exposing potentially catastrophic vulnerabilities in key pieces of national infrastructure.

Even though terrorists, domestic and/or foreign, tend to aim their main focus around the critical devices that control actual energy/power system delivery activities, according to USEPA (2005), SCADA networks were developed with little attention paid to security, making the security of these systems often weak. Studies have found that while technological advancements introduced vulnerabilities, many energy production sector facilities and utilities have spent little time securing their

SCADA networks. As a result, many SCADA networks may be susceptible to attacks and misuse. SCADA systems languished in obscurity, and this was the essence of their security; that is, until technological developments transformed SCADA from a backroom operation to a front-and-center visible control system.

Remote monitoring and supervisory control of processes began to develop in the early 1960s and adopted many technological advancements. The advent of mini-computers made it possible to automate a vast number of once manually operated switches. Advancements in radio technology reduced the communication costs associated with installing and maintaining buried cables in remote areas. SCADA systems continued to adopt new communication methods, including satellite and cel-lular. As the price of computers and communications dropped, it became economi-cally feasible to distribute operations and to expand SCADA networks to include even smaller facilities.

Advances in information technology and the necessity of improved efficiency have resulted in increasingly automated and interlinked infrastructures and created new vulnerabilities due to equipment failure, human error, weather and other natu-ral causes, and physical and cyber-attacks. Some areas and examples of possible SCADA vulnerabilities include (Wiles et al., 2007):

- Human—People can be tricked or corrupted and may commit errors.
- Communications—Messages can be fabricated, intercepted, changed, deleted, or blocked.
- Hardware—Security features are not easily adapted to small self-contained units with limited power supplies.
- Physical—Intruders can break into a facility to steal or damage SCADA equipment.
- Natural—Tornadoes, floods, earthquakes, and other natural disasters can damage equipment and connections.
- Software—Programs can be poorly written.

Specific SCADA weaknesses and potential attack vectors include:

- Does not require any authorization.
- Does not use encryption;
- Does not properly recognize and properly handle errors and exceptions
- No authentication is required
- Data could be intercepted
- Manipulation of data
- Service denial
- IP address spoofing—Internet protocol packets with a false source IP address
- Session hijacking
- Unsolicited responses
- Packet fuzzing—Involves imputing false information, etc.
- Unauthorized control
- Log data medication

SCADA system computers and their connections are susceptible to different types of information system attacks and misuse, such as those mentioned above. The Computer Security Institute and Federal Bureau of Investigation conduct an annual Computer Crime and Security Survey (FBI, 2007). The survey reported on ten types of attacks or misuse and reported that virus and denial of service had the greatest negative economic impact. The same study also found that 15% of the respondents reported abuse of wireless networks, which can be a SCADA component. On average, respondents from all sectors did not believe that their organization invested enough in security awareness. For example, utilities as a group reported a lower average computer security expenditure/investment per employee than many other sectors such as transportation, telecommunications, and financial.

Sandia National Laboratories' *Common Vulnerabilities in Critical Infrastructure Control Systems* described some of the common problems it has identified in the following five categories (Stamp et al., 2003):

1. **System Data**: Important data attributes for security include availability, authenticity, integrity, and confidentiality. Data should be categorized according to its sensitivity, and ownership and responsibility must be assigned. However, SCADA data are often not classified at all, making it difficult to identify where security precautions are appropriate (for example, which communication links to secure, databases requiring protection, etc.).

2. **Security Administration**: Vulnerabilities emerge because many systems lack a properly structured security policy (security administration is notoriously lax in the case of control systems), equipment and system implementation guides, configuration management, training, and enforcement and compliance auditing.

3. **Architecture**: Many common practices negatively affect SCADA security. For example, while it is convenient to use SCADA capabilities for other purposes such as fire and security systems, these practices create single points of failure. Also, the connection of SCADA networks to other automation systems and business networks introduces multiple entry points for potential adversaries.

4. **Network** (including communication links): Legacy systems' hardware and software have very limited security capabilities, and the vulnerabilities of contemporary systems (based on modern information technology) are publicized. Wireless and shared links are susceptible to eavesdropping and data manipulation.

5. **Platforms**: Many platform vulnerabilities exist, including default configurations retained, poor password practices, shared accounts, inadequate protection for hardware, and nonexistent security monitoring controls. In most cases, important security patches are not installed, often due to concern about negatively impacting system operation; in some cases, technicians are contractually forbidden from updating systems by their vendor agreements.

The following incident helps to illustrate some of the risks associated with SCADA vulnerabilities:

- During the course of conduction, a vulnerability assessment, a contractor stated that personnel from his company penetrated the information system of a utility within minutes. Contractor personnel drove to a remote substation and noticed a wireless network antenna. Without leaving their vehicle, they plugged in their wireless radios and connected to the network within 5 minutes. Within 20 minutes they had mapped the network, including SCADA equipment, and accessed the business network and data. This illustrates what a cybersecurity advisor from Sandia National Laboratories specializing in SCADA stated that utilities are moving to wireless communication without understanding the added risks.

THE INCREASING RISK

According to GAO (2015), historically, security concerns about control systems (SCADA included) were related primarily to protecting against physical attack and misuse of refining and processing sites or distribution and holding facilities. However, more recently there has been a growing recognition that control systems are now vulnerable to cyber-attacks from numerous sources, including hostile governments, terrorist groups, disgruntled employees, and other malicious intruders.

In addition to control system vulnerabilities mentioned earlier, several factors have contributed to the escalation of risk to control systems, including (1) the adoption of standardized technologies with known vulnerabilities, (2) the connectivity of control systems to other networks, (3) constraints on the implementation of existing security technologies and practices, (4) insecure remote connections, and (5) the widespread availability of technical information about control systems.

ADOPTION OF TECHNOLOGIES WITH KNOWN VULNERABILITIES

When a technology is not well known, not widely used, not understood or publicized, it is difficult to penetrate it and thus disable it. Historically, proprietary hardware, software, and network protocols made it difficult to understand how control systems operated—and therefore how to hack into them. Today, however, to reduce costs and improve performance, organizations have been transitioning from proprietary systems to less expensive, standardized technologies such as Microsoft's Windows and Unix-like operating systems and the common networking protocols used by the Internet. These widely used standardized technologies have commonly known vulnerabilities, and sophisticated and effective exploitation tools are widely available and relatively easy to use. As a consequence, both the number of people with the knowledge to wage attacks and the number of systems subject to attack have increased. Also, common communication protocols and the emerging use of Extensible Markup Language (commonly referred to as XML) can make it easier for a hacker to interpret the content of communications among the components of a control system.

Control systems are often connected to other networks—enterprises often integrate their control system with their enterprise networks. This increased connectivity has significant advantages, including providing decision makers with access to real-time information and allowing engineers to monitor and control the process control system from different points on the enterprise network. In addition, enterprise networks are often connected to the networks of strategic partners and to the Internet. Further, control systems are increasingly using wide area networks and the Internet to transmit data to their remote or local stations and individual devices. This convergence of control networks with public and enterprise networks potentially exposes the control systems to additional security vulnerabilities. Unless appropriate security controls are deployed in the enterprise network and the control system network, breaches in enterprise security can affect the operation of control systems.

According to industry experts, the use of existing security technologies as well as strong user authentication and patch management practices are generally not implemented in control systems because control systems operate in real time, typically are not designed with cybersecurity in mind, and usually have limited processing capabilities.

Existing security technologies such as authorization, authentication, encryption, intrusion detection, and filtering of network traffic and communications require more bandwidth, processing power, and memory than control system components typically have. Because controller stations are generally designed to do specific tasks, they use low-cost, resource-constrained microprocessors. In fact, some devices in the electrical industry still use the Intel 8088 processor, introduced in 1978. Consequently, it is difficult to install existing security technologies without seriously degrading the performance of the control system.

Further, complex passwords and other strong password practices are not always used to prevent unauthorized access to control systems, in part because this could hinder a rapid response to safety procedures during an emergency. As a result, according to experts, weak passwords that are easy to guess, shared, and infrequently changed are reportedly common in control systems, including the use of default passwords or even no password at all.

In addition, although modern control systems are based on standard operating systems, they are typically customized to support control system applications. Consequently, vendor-provided software patches are generally either incompatible or cannot be implemented without compromising service, shutting down "always-on" systems, or affecting interdependent operations.

Potential vulnerabilities in control systems are exacerbated by insecure connections. Organizations often leave access links—such as dial-up modems to equipment and control information—open for remote diagnostics, maintenance, and examination of system status. Such links may not be protected with authentication of encryption, which increases the risk that hackers could use these insecure connections to break into remotely controlled systems. Also, control systems often use wireless communications systems, which are especially vulnerable to attack, or leased lines that pass through commercial telecommunications facilities. Without encryption to protect data as it flows through these insecure connections or authentication mechanisms to limit access, there is limited protection for the integrity of the information being transmitted.

Public information about infrastructures and control systems is available to potential hackers and intruders. The availability of this infrastructure and vulnerability data was demonstrated by a university graduate student, whose dissertation reportedly mapped every business and industrial sector in the American economy to the fiber-optic network that connects them—using material that was available publicly on the Internet, none of which was classified.

CYBER THREATS TO CONTROL SYSTEMS

There is a general consensus—and increasing concern—among government officials and experts on control systems about potential cyber threats to the control systems that govern our critical infrastructures. As components of control systems increasingly make critical decisions that were once made by humans, the potential effect of a cyber threat becomes more devastating. Such cyber threats could come from numerous sources, ranging from hostile governments and terrorist groups to disgruntled employees and other malicious intruders.

In July 2002, National Infrastructure Protection Center (NIPC) reported that the potential for compound cyber and physical attacks, referred to as "swarming attacks," is an emerging threat to the US critical infrastructure. As NIPC reports, the effects of a swarming attack include slowing or complicating the response to a physical attack. For instance, a cyber-attack that disabled the water supply or the electrical system in conjunction with a physical attack could deny emergency services the necessary resources to manage the consequences—such as controlling fires, coordinating actions, and generating light.

Control systems, such as SCADA, can be vulnerable to cyber-attacks. Entities or individuals with malicious intent might take one or more of the following actions to successfully attack control systems:

- Disrupt the operation of control systems by delaying or blocking the flow of information through control networks, thereby denying the availability of the networks to control system operations.
- Make unauthorized changes to programmed instructions in PLCs, RTUs, or DCS controllers, change alarm thresholds, or issue unauthorized commends to control equipment, which could potentially result in damage to equipment (if tolerances are exceeded), premature shutdown of processes (such as prematurely shutting down transmission lines), or even disabling of control equipment.
- Send false information to control system operators either to disguise unauthorized changes or to initiate inappropriate actions by system operators.
- Modify the control system software, producing unpredictable results.
- Interfere with the operation of safety systems.

In addition, in control systems that cover a wide geographic area, the remote sites are often unstaffed and may not be physically monitored. If such remote systems are physically breached, the attackers could establish a cyber connection to the control network.

SECURING CONTROL SYSTEMS

Several challenges must be addressed to effectively secure control systems against cyber threats. These challenges include: (1) the limitations of current security technologies in securing control systems; (2) the perception that securing control systems may not be economically justifiable; and (3) the conflicting priorities within organizations regarding the security of control systems.

A significant challenge in effectively securing control systems is the lack of specialized security technologies for these systems. The computing resources in control systems that are needed to perform security functions tend to be quite limited, making it very difficult to use security technologies within control system networks without severely hindering performance.

Securing control systems may not be perceived as economically justifiable. Experts and industry representatives have indicated that organizations may be reluctant to spend more money to secure control systems. Hardening the security of control systems would require industries to expend more resources, including acquiring more personnel, providing training for personnel, and potentially prematurely replacing current systems that typically have a lifespan of about 20 years.

Finally, several experts and industry representatives indicated that the responsibility for securing control systems typically includes two separate groups: IT security personnel and control system engineers and operators. IT security personnel tend to focus on securing enterprise systems, while control system engineers and operators tend to be more concerned with the reliable performance of their control systems. Further, they indicate that, as a result, those two groups do not always fully understand each other's requirements and collaborate to implement secure control systems.

STEPS TO IMPROVE SCADA SECURITY

The President's Critical Infrastructure Protection Board and the Department of Energy (DOE) have developed the steps outlined below to help organizations improve the security of their SCADA networks. DOE (2001) points out that these steps are not meant to be prescriptive or all-inclusive. However, they do address essential actions to be taken to improve the protection of SCADA networks. The steps are divided into two categories: specific actions to improve implementation and actions to establish essential underlying management processes and policies.

21 STEPS TO INCREASE SCADA SECURITY (DOE, 2001)

The following steps focus on specific actions to be taken to increase the security of SCADA networks:

1. **Identify all connections to SCADA networks.**
   Conduct a thorough risk analysis to assess the risk and necessity of each connection to the SCADA network. Develop a comprehensive understanding of

all connections to the SCADA network, and how well those connections are protected. Identify and evaluate the following types of connections:

- Internal local area and wide area networks, including business networks
- The Internet
- Wireless network devices, including satellite uplinks
- Modem or dial-up connections
- Connections to business partners, vendors, or regulatory agencies

2. **Disconnect unnecessary connections to the SCADA network.**

   To ensure the highest degree of security of SCADA systems, isolate the SCADA network from other network connections to as great a degree as possible. Any connection to another network introduces security risks, particularly if the connection creates a pathway from or to the Internet. Although direct connections with other networks may allow important information to be passed efficiently and conveniently, insecure connections are simply not worth the risk; isolation of the SCADA network must be a primary goal to provide needed protection. Strategies such as utilization of "demilitarized zones" (DMZs) and data warehousing can facilitate the secure transfer of data from the SCADA network to business networks. However, they must be designed and implemented properly to avoid introduction of additional risk through improper configuration.

3. **Evaluate and strengthen the security of any remaining connections to the SCADA networks.**

   Conduct penetration testing or vulnerability analysis of any remaining connections to the SCADA network to evaluate the protection posture associated with these pathways. Use this information in conjunction with risk management processes to develop a robust protection strategy for any pathways to the SCADA network. Since the SCADA network is only as secure as its weakest connecting point, it is essential to implement firewalls, intrusion detection systems (IDSs), and other appropriate security measures at each point of entry. Configure security system rules to prohibit access from and to the SCADA network and be as specific as possible when permitting approved connections. For example, an Independent System Operator (ISO) should not be granted "blanket" network access simply because there is a need for a connection to certain components of the SCADA system. Strategically place IDSs at each entry point to alert security personnel of potential breaches of network security. Organization management must understand and accept responsibility or risks associated with any connection to the SCADA network.

4. **Harden SCADA networks by removing or disabling unnecessary services.**

   SCADA control servers built on commercial or open-source operating systems can be exposed to attack default network services. To the greatest degree possible, remove or disable unused services and network demons to reduce the risk of direct attack. This is particularly important when SCADA networks are interconnected with other networks. Do not permit a

service or feature on a SCAA network unless a thorough risk assessment of
the consequences of allowing the service/feature shows that the benefits of
the service/feature far outweigh the potential for vulnerability exploitation.
Examples of services to remove from SCADA networks include automated
meter reading/remote billing systems, e-mail services, and Internet access.
An example of a feature to disable is remote maintenance. Another example is the numerous secure configurations such as the National Security
Agency's series of security guides. Additionally, work closely with SCADA
vendors to identify secure configurations and coordinate any and all changes
to operational systems to ensure that removing or disabling services does
not cause downtime, interruption of service, or loss of support.

5. **Do not rely on proprietary protocols to protect your system.**
   Some SCADA systems are unique, proprietary protocols for communications between field devices and servers. Often the security of SCADA
   systems is based solely on the secrecy of these protocols. Unfortunately,
   obscure protocols provide very little "real" security. Do not rely on proprietary protocols or factor default configuration setting to protect your system. Additionally, demand that vendors disclose any backdoors or vendor
   interfaces to your SCADA systems and expect them to provide systems that
   are capable of being secured.

6. **Implement the security features provided by device and system vendors.**
   Older SCADA systems (most systems in use) have no security features
   whatsoever; SCADA system owners must insist that their system vendor
   implement security features in the form of product patches or upgrades.
   Some newer SCADA devices are shipped with basic security features, but
   these are usually disabled to ensure ease of installation.

   Analyze each SCADA device to determine whether security features are
   present. Additionally, factory default security settings (such as in computer
   network firewalls) are often set to provide maximum usability, but minimal
   security. Set all security features to provide the maximum security only
   after a thorough risk assessment of the consequences of reducing the security level.

7. **Establish strong controls over any medium that is used as a backdoor
   into the SCADA network.**
   Where backdoors or vendor connections do exist in SCADA systems,
   strong authentication must be implemented to ensure secure communications. Modems, wireless, and wired networks used for communications
   and maintenance represent a significant vulnerability to the SCADA network and remote sites. Successful "war dialing" or "war driving" attacks
   could allow an attacker to bypass all of the other controls and have direct
   access to the SCADA network or resources. To minimize the risk of such
   attacks, disable inbound access and replace it with some type of callback
   system.

8. **Implement internal and external intrusion detection systems and
   establish 24-hour-a-day incident monitoring.**

To be able to effectively respond to cyber-attacks, establish an intrusion detection strategy that includes alerting network administrators of malicious network activity originating from internal or external sources. Intrusion detection system monitoring is essential 24 hours a day; this capability can be easily set up through a pager. Additionally, incident response procedures must be in place to allow an effective response to any attack. To complement network monitoring, enable logging on all systems and audit system logs daily to detect suspicious activity as soon as possible.

9. **Perform technical audits of SCADA devices and networks, and any other connected networks, to identify security concerns.**
   Technical audits of SCADA devices and networks are critical to ongoing security effectiveness. Many commercial and open-sourced security tools are available that allow system administrators to conduct audits of their systems/networks to identify active services, patch level, and common vulnerabilities. The use of these tools will not solve systemic problems but will eliminate the "paths of least resistance" that an attacker could exploit. Analyze identified vulnerabilities to determine their significance and take corrective actions as appropriate. Track corrective actions and analyze this information to identify trends. Additionally, retest systems after corrective actions have been taken to ensure that vulnerabilities were actually eliminated. Scan non-production environments actively to identify and address potential problems.

10. **Conduct physical security surveys and assess all remote sites connected to the SCADA network to evaluate their security.**
    Any location that has a connection to the SCADA network is a target, especially unmanned or unguarded remote sites. Conduct a physical security survey, and inventory access points at each facility that has a connection to the SCADA system. Identify and assess any source of information including remote telephone/computer network/fiber-optic cables that could be tapped; radio and microwave links that are exploitable; computer terminals that could be accessed; and wireless local area network access points. Identify and eliminate single points of failure. The security of the site must be adequate to detect or prevent unauthorized access. Do not allow "live" network access points at remote, unguarded sites simply for convenience.

11. **Establish SCADA "Red Teams" to identify and evaluate possible attack scenarios.**
    Establish a "Red Team" to identify potential attack scenarios and evaluate potential system vulnerabilities. Use a variety of people who can provide insight into weaknesses of the overall network, SCADA system, physical systems, and security controls. People who work on the system every day have great insight into the vulnerabilities of your SCADA network and should be consulted when identifying potential attack scenarios and possible consequences. Also, ensure that the risk from a malicious insider is fully evaluated, given that this represents one of the greatest threats to an organization. Feed information resulting from the "Red Team" evaluation

into risk management processes to assess the information and establish appropriate protection strategies.

The following steps focus on management actions to establish an effective cybersecurity program:

12. **Clearly define cybersecurity roles, responsibilities, and authorities for managers, system administrators, and users.**

Organization personnel need to understand the specific expectations associated with protecting information technology resources through the definition of clear and logical roles and responsibilities. In addition, key personnel need to be given sufficient authority to carry out their assigned responsibilities. Too often, good cybersecurity is left up to the initiative of the individual, which usually leads to inconsistent implementation and ineffective security. Establish a cybersecurity organizational structure that defines roles and responsibilities and clearly identifies how cybersecurity issues are escalated and who is notified in an emergency.

13. **Document network architecture and identify systems that serve critical functions or contain sensitive information that requires additional levels of protection.**

Develop and document robust information security architecture as part of a process to establish an effective protection strategy. It is essential that organizations design their network with security in mind and continue to have a strong understanding of their network architecture throughout its lifecycle. Of particular importance, an in-depth understanding of the functions that the systems perform and the sensitivity of the stored information is required. Without this understanding, risk cannot be properly assessed, and protection strategies may not be sufficient. Documenting the information security architecture and its components is critical to understanding the overall protection strategy and identifying single points of failure.

14. **Establish a rigorous, ongoing risk management process.**

A thorough understanding of the risks to network computing resources from denial-of-service attacks, and the vulnerability of sensitive information to compromise, is essential to an effective cybersecurity program. Risk assessments form the technical basis of this understanding and are critical to formulating effective strategies to mitigate vulnerabilities and preserve the integrity of computing resources. Initially, perform a baseline risk analysis based on the current threat assessment to use for developing a network protection strategy. Due to rapidly changing technology and the emergence of new threats on a daily basis, an ongoing risk assessment process is also needed so that routine changes can be made to the protection strategy to ensure it remains effective. Fundamental to risk management is the identification of residual risk, with a network protection strategy in place and acceptance of that risk by management.

15. **Establish a network protection strategy based on the principle of defense-in-depth.**

A fundamental principle that must be part of any network protection strategy is defense-in-depth. Defense-in-depth must be considered early in the design phase of the development process and must be an integral consideration in all technical decision-making associated with the network. Utilize technical and administrative controls to mitigate threats from identified risks to as great a degree as possible at all levels of the network. Single points of failure must be avoided, and cybersecurity defense must be layered to limit and contain the impact of any security incidents. Additionally, each layer must be protected against other systems at the same layer. For example, to protect against the inside threat, restrict users to access only those resources necessary to perform their job functions.

16. **Clearly identity cybersecurity requirements.**
   Organizations and companies need structured security programs with mandated requirements to establish expectations and allow personnel to be held accountable. Formalized policies and procedures are typically used to establish and institutionalize a cybersecurity program. A formal program is essential to establishing a consistent, standards-based approach to cybersecurity through an organization and eliminates sole dependence on individual initiatives. Policies and procedures also inform employees of their specific cybersecurity responsibilities and the consequences of failing to meet those responsibilities. They also provide guidance regarding actions to be taken during a cybersecurity incident and promote efficient and effective actions during a time of crisis. As part of identifying cybersecurity requirements, include user agreements and notification and warning banners. Establish requirements to minimize the threat from malicious insiders, including the need for conducting background checks and limiting network privileges to those absolutely necessary.

17. **Establish effective configuration management processes.**
   A fundamental management process needed to maintain a secure network is configuration management. Configuration management needs to cover both hardware configurations and software configurations. Changes to hardware or software can easily introduce vulnerabilities that undermine network security. Processes are required to evaluate and control any change to ensure that the network remains secure. Configuration management begins with well-tested and documented security baselines for your various systems.

18. **Conduct routine self-assessments.**
   Robust performance evaluation processes are needed to provide organizations with feedback on the effectiveness of cybersecurity policy and technical implementation. A sign of a mature organization is one that is able to self-identify issues, conduct root cause analyses, and implement effective corrective actions that address individual and systemic problems. Self-assessment processes that are normally part of an effective cybersecurity program include routine scanning for vulnerabilities, automated auditing of the network, and self-assessments of organizational and individual performance.

19. **Establish system backups and disaster recovery plans.**
   Establish a disaster recovery plan that allows for rapid recovery from any emergency (including a cyber-attack). System backups are an essential part

of any plan and allow rapid reconstruction of the network. Routinely exercise disaster recovery plans to ensure that they work and that personnel are familiar with them. Make appropriate changes to disaster recovery plans based on lessons learned from exercises.

20. **Senior organizational leadership should establish expectations for cybersecurity performance and hold individuals accountable for their performance.**
Effective cybersecurity performance requires commitment and leadership from senior managers in the organization. It is essential that senior management establish an expectation for strong cybersecurity and communicate this to their subordinate managers throughout the organization. It is also essential that senior organizational leadership establish a structure for the implementation of a cybersecurity program. This structure will promote consistent implementation and the ability to sustain a strong cybersecurity program. It is then important for individuals to be held accountable for their performance as it relates to cybersecurity. This includes managers, system administrators, technicians, and users/operators.

21. **Establish policies and conduct training to minimize the likelihood that organizational personnel will inadvertently disclose sensitive information regarding SCADA system design, operations, or security controls.**
Release data related to the SCADA network only on a strict, need-to-know basis and only to persons explicitly authorized to receive such information. "Social engineering," the gathering of information about a computer or computer network via questions to naïve users, is often the first step in a malicious attack on computer networks. The more information revealed about a computer or computer network, the more vulnerable the computer/network is. Never divulge data revealed to a SCADA network, including the names and contact information about the system operators/administrators, computer operating systems, and/or physical and logical locations of computers and network systems over telephones or to personnel unless they are explicitly authorized to receive such information. Any requests for information by unknown persons need to be sent to a central network security location for verification and fulfillment. People can be a weak link in an otherwise secure network. Conduct training and information awareness campaigns to ensure that personnel remain diligent in guarding sensitive network information, particularly their passwords.

THE BOTTOM LINE

This chapter pointed out that every SCADA system has hardware, firmware, software, and personnel involved in its operation. Attacks are targeted at the weak points in these systems. Because all weak points cannot be eliminated 100%, it is important to implement the ways to protect SCADA systems from cyber threats. Obviously, the most effective solution to cyber threats is to physically disconnect from the Internet and external devices. A final word: Based on my study and inspection of wind turbine SCADA systems throughout the country, I found that some installations that had inadequate separation between IT (information technology) and OP (operation technology) were exposed to significant vulnerability; they were leaving the turbine door wide open, so to speak.

REFERENCES

Associated Press (AP). 2009. *Goal: Disrupt.* From the 04/04/09 The Virginian-Pilot, Norfolk, Va.

DOE. 2001. *21 Steps to Improve Cyber Security of SCADA Networks.* Washington, DC: Department of Energy.

FBI. 2007. *Ninth Annual Computer Crime and Security Survey.* FBI: Computer Crime Institute and Federal Bureau of Investigations. Washington, DC.

GAO. 2015. *Critical Infrastructure Protection: Sector-Specific Agencies Need to Better Measure Cybersecurity Progress.* Washington, DC: United States Government Accountability Office.

Gellman, B. 2002. Cyber-Attacks by Al Qaeda Feared: Terrorists at Threshold of Using Internet as Tool of Bloodshed, Experts Say. *Washington Post*, June 27, p. A01.

Greenberg, A. 2017. Researchers Found They Could Hack an Entire Wind Farm. Accessed 10/31/2021 @ www.wired.com.

Kim, K. et al. 2011. Performance assessment of a wind turbine using SCADA. Accessed 11/10/2021 @ https://papers.phonsicity.or/index/php.

Stamp, J. et al., 2003. *Common Vulnerabilities in Critical Infrastructure Control Systems*, 2nd Edition. White Sands, New Mexico: Sandia National Laboratories.

USEPA. 2005. Securing SCADA. Accessed 11/5/2021 @ https://www.epa.gov/office-inspector-general/2005.org-reports.

Warwalking. 2003. Unwired. Accessed 11/2/2021 @ paloaltoonline.com.weekly/2003/.

Wiles, J. et al., 2007. *Techno Security's™ Guide to Securing SCADA.* Burlington, MA: Elsevier, Inc.

Part II

The Nuts and Bolts of Wind Power

$$\text{Wind Power} = k \, C_p \, \tfrac{1}{2} \rho \, AV^3$$

5 Wind Energy

The wind goeth toward the south, and turneth about unto the north; it whirleth about continually, and the wind returneth again.

—Ecclesiastes 1:6

The Good, Bad, and Ugly of Wind Energy:

> *Good*: As long as Earth exists, the wind will always exist. The energy in the winds that blow across the U.S. each year could produce more than 16 billion GJ of electricity—more than one and one-half times the electricity consumed in the U.S. in 2000.
> *Bad*: Turbines are expensive. Wind doesn't blow all the time, so they have to be part of a larger plan. Turbines make noise. Turbine blades kill birds.
> *Ugly*: Some look upon giant wind turbine blades cutting through the air as grotesque scars on the landscape, visible polluters.
> *The bottom line*: Do not expect Don Quixote, mounted in armor on his old nag, Rocinante, with or without Sancho Panza, to charge those windmills. Instead, expect—you can count on it, bet on it, and rely on it—the charge will be not to tilt at the windmills, but instead to build those windmills. It will be done by the rest of us, to satisfy our growing, inexorable need for renewable energy. What other choice do we have?

F.R. Spellman (2015)

INTRODUCTION*

Obviously, wind energy or power is all about wind. In simple terms, wind is the response of the atmosphere to uneven heating conditions. In regard to the earth's atmosphere, and, again, to state the obvious, it is constantly in motion. Anyone observing the constant weather changes and cloud movement around them is well aware of this phenomenon. Although its physical manifestations are obvious, the importance of the dynamic state of our atmosphere is much less obvious.

The constant motion of the earth's atmosphere (air movement) consists of both horizontal (*wind*) and vertical (*air currents*) dimensions. The atmosphere's motion is the result of thermal energy produced from the heating of the earth's surface and the air molecules above. Because of differential heating of the earth's surface, energy flows from the equator poleward.

Hanson (2004) points out that the energy resources contained in the wind in the United States are well known and mapped in detail. Even though this is the case,

---

* Parts of this section are adapted from F.R. Spellman (2015) *Environmental Impacts of Renewable Energy*. Boca Raton, FL: CRC Press.

DOI: 10.1201/9781003288947-7

and it is clear that air movement plays a critical role in transporting the energy of the lower atmosphere, bringing the warming influences of spring and summer and the cold chill of winter, and generally that wind and air currents are fundamental to how nature functions, the effects of air movements on our environment are often overlooked. All life on the earth has evolved or has been sustained with mechanisms dependent on-air movement: pollen is carried by winds for plant reproduction; animals sniff the wind for essential information; wind power was the motive force that began the earliest stages of the industrial revolution. Now we see the effects of winds in other ways, too: Wind causes weathering (erosion) of the earth's surface; wind influences ocean currents; air pollutants and contaminants such as radioactive particles transported by the wind impact our environment.

DID YOU KNOW?

Wind speed is generally measured in m/s, but Americans usually think in mph. A good rule of thumb to keep at hand is that to convert m/s to mph, double the value in m/s and add 10%.

AIR IN MOTION

In all dynamic situations, forces are necessary to produce motion and changes in motion—winds and air currents. The air (made up of various gases) of the atmosphere is subject to two primary forces: gravity and pressure differences from temperature variations.

Gravity (gravitational forces) holds the atmosphere close to the earth's surface. Newton's law of universal gravitation states that everybody in the universe attracts another body with a force equal to:

$$F = G\frac{m_1 m_2}{r^2} \tag{5.1}$$

where
   F = force
   m_1 and m_2 = the masses of the two bodies
   G = universal constant of 6.67×10^{-11} N × m²/kg²
   r = distance between the two bodies

The force of gravity decreases as an inverse square of the distance between the two bodies. Thermal conditions affect density, which, in turn, cause gravity to affect vertical air motion and planetary air circulation. This affects how air pollution is naturally removed from the atmosphere.

Although forces in other directions often overrule gravitational force, the ever-present force of gravity is vertically downward and acts on each gas molecule, accounting for the greater density of air near the earth.

Atmospheric air is a mixture of gases, so the gas laws and other physical principles govern its behavior. The pressure of a gas is directly proportional to its temperature. Pressure is force per unit area (P = F/A), so a temperature variation in the air generally gives rise to a difference in pressure of force. This difference in pressure resulting from temperature differences in the atmosphere creates air movement—on both large and local scales. This pressure difference corresponds to an unbalanced force, and when a pressure difference occurs, the air moves from a high- to a low-pressure region.

In other words, horizontal air movements (called *advective winds*) result from temperature gradients, which give rise to density gradients and, subsequently, pressure gradients. The force associated with these pressure variations (*pressure gradient force*) is directed at right angles to (perpendicular to) lines of equal pressure (called *isobars*) and is directed from high to low pressure.

In Figure 5.1A, the pressures over a region are mapped by taking barometric readings at different locations. Lines drawn through the points (locations) of equal pressure are called isobars. All points on an isobar are of equal pressure, which means no air movement along the isobar. The wind direction is at right angles to the isobar in the direction of the lower pressure. In Figure 5.1B, notice that air moves down a pressure gradient toward a lower isobar like a ball rolls down a hill. If the isobars are close together, the pressure gradient force is large, and such areas are characterized by high wind speeds. If isobars are widely spaced, the winds are light because the pressure gradient is small.

Localized air circulation gives rise to *thermal circulation* (a result of the relationship based on a law of physics whereby the pressure and volume of a gas are directly related to its temperature). A change in temperature causes a change in the pressure and/or volume of a gas. With a change in volume comes a change in density, since P = m/V, so regions of the atmosphere with different temperatures may have different air pressures and densities. As a result, localized heating sets up air motion and gives rise to *thermal circulation.*

Once the air has been set into motion, secondary forces (velocity-dependent forces) act. These secondary forces are caused by the earth's rotation (Coriolis force) and contact with the rotating earth (friction). The *Coriolis force,* named after its discoverer, French mathematician Gaspard Coriolis (1772–1843), is the effect of rotation on the atmosphere and all objects on the earth's surface. In the Northern Hemisphere, it causes moving objects and currents to be deflected to the right; in the Southern Hemisphere, it causes deflection to the left, because of the earth's rotation. Air, in large-scale north or south movements, appears to be deflected from its expected path. That is, air moving poleward in the Northern Hemisphere appears to be deflected toward the east; air moving southward appears to be deflected toward the west.

Friction (drag) can also cause the deflection of air movements. This friction (resistance) is both internal and external. The friction of its molecules generates internal friction. Friction is also generated when air molecules run into each other. External friction is caused by contact with terrestrial surfaces. The magnitude of the frictional force along a surface is dependent on the air's magnitude and speed, and the opposing frictional force is in the opposite direction of the air motion.

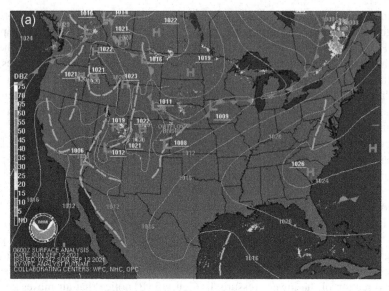

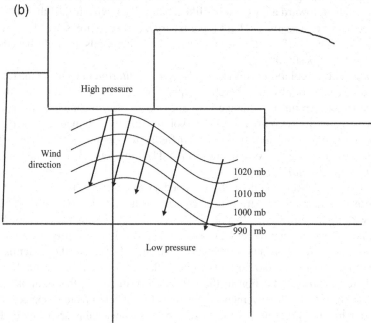

FIGURE 5.1   (a) Isobars drawn through locations having equal atmospheric pressures. The air motion, or wind direction, is at right angles to the isobars and moves from a region of high pressure to a region of low pressure. (b) Air moves down a pressure gradient toward a lower isobar. Source: NOAA (2021) weather maps accessed 09/12/2021 @ https://www.gov/ffc/maplast.

WIND ENERGY*

Wind energy is the movement of the wind to create power. Since early recorded history, people have been harnessing the energy of the wind for milling grain, pumping water, and other mechanical power applications. Wind energy propelled boats along the Nile River as early as 5000 BC. By 200 BC, simple windmills in China were pumping water, while vertical-axis windmills with woven reed sails were grinding grain in Persia and the Middle East.

The use of wind energy spread around the world, and by the 11th century, people in the Middle East were using windmills extensively for food production; returning merchants and crusaders carried this idea back to Europe. The Dutch refined the windmill and adapted it for draining lakes and marshes in the Rhine River Delta. When settlers took this technology to the North American continent in the later 19th century, they began using windmills to pump water for farms and ranches and, later, to generate electricity for homes and industry. The first known wind turbine for producing electricity was by Charles F. Brush, in Cleveland, Ohio, in 1888; it was a 12-kW unit used to charge batteries in the cellar of a mansion. The first wind turbine outside of the United States to generate electricity was built by Poul la Cour in 1891 in Denmark. He used electricity from his wind turbines to electrolyze water to make hydrogen for the gas lights at the schoolhouse. By the 1930s and 1940s, in the United States, hundreds of thousands were in use in rural areas not yet served by the grid. The oil crisis of the 1970s created a renewed interest in the wind until the US government stopped giving tax credits. Today, there are several hundred thousand windmills in operation around the world, many of which are used for water pumping. But it is the use of wind energy as a pollution-free means of generating electricity on a significant scale that is attracting the most current interest in the subject. As a matter of fact, with the present and pending shortage and high cost of fossil fuels to generate electricity and the green movement toward the use of cleaner fuels, wind energy is the world's fastest-growing energy source and wind power industry, businesses, and homes with clean, renewable electricity for many years to come. In the United States since 1970, wind-based electricity generating capacity has increased markedly, although (at present) it remains a small fraction of total electric capacity and consumption. But this trend is beginning to change—with the advent of $4/gal gasoline, high heating and cooling costs and subsequent increases in the cost of electricity, worldwide political unrest, or uncertainty in oil-supplying countries, one only needs to travel the "wind corridors" of the United States encompassing parts of Arizona, New Mexico, Texas, Missouri, and north through the Great Plains to the Pembina Escarpment and Turtle Mountains of North Dakota and elsewhere to witness the considerable activity and the seemingly exponential increase in wind energy development and wind turbine installations; these machines are being installed to produce and provide electricity to the grid.

Table 5.1 highlights wind energy's quadrillion Btu ranking in current renewable energy source use. As pointed out in Table 5.1, the energy consumption by energy

---

* Much of the information in this section is from USDOE-EERE 2005. *History of wind energy*. Accessed 06/14/21@http://www1/eere/emergu/gpv/womdamdjudrp/printable_versions/wind_hisotry.htm.

TABLE 5.1

US Energy Consumption by Energy Source, 2019 (Quadrillion Btu)

Energy Source	2019
Renewable	11.4
Biomass (biofuels, waste, wood, and wood-derived)	5.0
Biofuels	2.3
Waste	2.7
Geothermal	0.2
Hydroelectric conventional	2.5
Solar/PV	1/0
Wind	**2.7**

Source: EIA 2021. *U.S. Energy consumption by Energy Source.* Accessed 06/12/21 @http://www.eia.doe.gov/enaf/ alternate/page/renew_energy_conump/table 1.html.

source computations is from the year 2007. Thus, the 0.319 wind quadrillion Btu figure is expected to steadily rise to an increasingly higher level, and this should be reflected in the 2008–2010 figures when they are released.

DID YOU KNOW?

We can classify wind energy as a form of solar energy. As mentioned, winds are caused by uneven heating of the atmosphere by the sun, irregularities of the earth's surface, and the rotation of the earth. As a result, winds are strongly influenced and modified by local terrain, bodies of water, weather patterns, vegetative cover, and other factors. The wind flow, or motion of energy when harvested by wind turbines, can be used to generate electricity.

WIND POWER BASICS

The term "wind energy" or "wind power" describes the process by which the wind is used to generate mechanical power or electricity. Wind turbines convert the kinetic energy in the wind into mechanical power. This mechanical power can be used for specific tasks (such as grinding grain or pumping water), or a generator can convert this mechanical power into electricity (EERE, 2006a).

DID YOU KNOW?

During the rotation of the nacelle, there is a possibility of twisting the cables inside the tower; the cables all become more and more twisted if the turbine

keeps turning in the same direction, which can happen if the wind keeps changing in the same direction. The wind turbine is therefore equipped with a cable
twist counter, which notifies the controller that it is time to straight the cables.

—Khaligh and Onar (2010)

WIND ENERGY/POWER CALCULATIONS

A wind turbine is a machine that converts the kinetic energy in the wind into the
mechanical energy of a shaft. Calculating the energy and power available in the wind
relies on knowledge of basic physics and geometry. The kinetic energy of an object
is the extra energy which it possessed because of its motion. It is defined as the work
needed to accelerate a body of a given mass from rest to its current velocity. Once
in motion, a body maintains its kinetic energy unless its speed changes. The kinetic
energy of a body is given by Equation (5.2).

$$KE = 1/2 \, mv^2 \tag{5.2}$$

where
  KE = kinetic energy
  m = mass
  v = velocity

Example 5.1: Determining Power in the Wind

Step 1
For the purpose of finding the kinetic energy of moving air (wind), let's say we
have a large packet of wind (i.e., a geometrical package of air passing through the
plain of a wind turbine's blades (which sweep out a cross-sectional area A), with a
thickness (D); see Figure 5.2) passing through the plane over a given time.

Step 2
To determine power in wind we begin by considering the kinetic energy of the
packet of air along with mass m moving at velocity v. Next, we need to divide by
time to get power:

$$\text{Power though area A} = \frac{1}{2} \left(\frac{m \, \text{passing through A}}{t} \right) v^2 \tag{5.3}$$

Step 3
The mass flow rate is (ρ is air density—the mass per unit volume of the earth's
atmosphere).

$$m = \frac{1}{2} \left(\frac{m \, \text{passing through A}}{t} \right) \rho A v \tag{5.4}$$

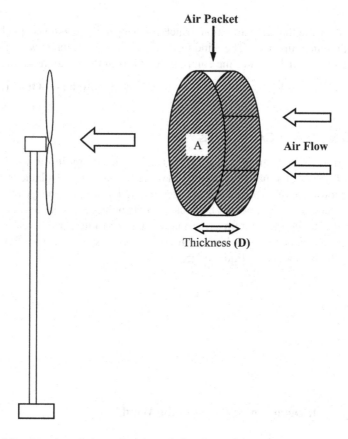

FIGURE 5.2 A packet of air passing through the plane of a wind turbine's blades, with the thickness (D) passing through the plane over a given time.

Step 4
To obtain the power in the wind equation, we must combine

$$\text{Power through area } A = \tfrac{1}{2}\left(\rho A v\right)v^2$$

$$\text{Power in the wind} = P_W = \tfrac{1}{2}\,\rho A v^3 \qquad\qquad (5.5)$$

where
  P_W (W) = power in the wind
  ρ (kg/m³) = air density (2.225 kg/m³ at 15°C and 1 atm)
  A (m²) = cross-sectional area that wind passes through perpendicular to the wind
  v (m/s) = wind speed normal to A (1 m/s = 2.237 mph)

DID YOU KNOW?

Air density decreases with increasing altitude, as does air pressure. It also changes with variance in temperature or humidity.

From Equation (5.5), it is evident that power increases as the cube of wind speed. That is, doubling the wind speed increases the power by eight. Moreover, the energy in 1 hour of 20 mph winds is the same as the energy in 8 hours of 10 mph winds. When we speak of power (W/m²) and wind speed (mph), we cannot use average wind speed because the relationship is nonlinear. Power in the wind is also proportional to A. For a conventional horizontal-axis wind turbine, $A = (\pi/4)D^2$, so wind power is proportional to the blade diameter squared. Because cost is approximately proportional to blade diameter, larger wind turbines are more cost-effective.

AIR DENSITY CORRECTION FACTORS

Earlier we pointed out that air density is affected by different temperatures and pressures.

Air density correction factors can correct air density for temperature and altitude. Correction factors for both temperature and altitude correction can be found in standardized tables.

Equation (5.6) is used to determine air density for different temperatures and pressures.

$$\rho = \frac{P \times MW \times 10^{-3}}{RT} \tag{5.6}$$

where

  P = absolute pressure (atm) starts at zero gas pressure (no molecules—a vacuum) and has no practical maximum limit

  MW = molecular weight of air (g/mol) = 28.97 g/mol

  T = absolute temperature (K) [the temperature in Kelvin (273 + °C) or Rankine (absolute zero = −460°F or 0°R)]

  R = ideal gas constant = $8.2056 \times 10^{-5} \times m^3 \times atm \times K^{-1} \times mol^{-1}$

DID YOU KNOW?

Air density is greater at lower temperatures.

ELEVATION AND THE EARTH'S ROUGHNESS

Wind speed is affected by its elevation above the earth and the roughness of the earth. Because power increases like the cube of wind speed, we can expect a significant economic impact from even a moderate increase in wind speed; thus, in the operation of wind turbines, wind speed is a very important parameter. The earth's surface features can't be ignored when deciding where to place wind turbines and in the calculation of their output productivity. Natural obstructions such as mountains and forests and human-made obstructions such as buildings provide friction as winds flow over them. The point is, there is a lot of friction in the first few hundred meters above ground—smooth surfaces (like water) are better. Thus, the greater the elevation above the earth, the greater the wind speed; tall wind turbine towers are better.

When actual measurements are not possible or available, it is possible to characterize or approximate the impact of rough surfaces and height on wind speed. This is accomplished using Equation (5.7).

$$\frac{v}{v_0} = \left(\frac{H}{H_0} \right)^{\alpha}$$
                                                                                (5.7)

where

α = friction coefficient—obtained from standardized tables; typical value of α in open terrain is 1/7; for a large city $\alpha=0.4$; for calm water, $\alpha=0.1$

v = windspeed at height H

v_0 = windspeed at height H_0 (H_0 is usually 10 m)

DID YOU KNOW?

As pointed out, the energy in the wind is a cubed function of wind speed, which means that if the wind speed doubles there is eight times as much available energy, not twice as much as one might expect.

WIND TURBINE ROTOR EFFICIENCY

Generally, when we think or talk about efficiency we think about input vs. output and know that if we put 100% into something and get 100% output, we have a very efficient machine, operation, or process. In engineering, we can approximate efficiency input vs output by performing mass balance calculations. In using these calculations, we know that according to the laws of conservation, materials cannot disappear or be destroyed; thus, they must exist and be somewhere. However, in this input vs. output concept, there are a couple of views on maximum rotor efficiency that the ill-informed or misinformed have taken; neither makes sense but are stated here to point out what really does make sense. The two wrong assumptions: First, it can be assumed that downwind velocity is zero; the turbine extracted all of the power from the wind; second, downwind velocity is the same as the upwind velocity—turbine extracted no power. Again, in a word: Wrong! It was Albert Betz in 1919 who set the record straight when he theorized that there must be some ideal slowing of the wind so that the turbine extracts the maximum power. Betz theorized in his wind turbine efficiency theory (i.e., to their power input vs power output and overall efficiency in general) that the efficiency rule or laws of conservation, as stated above, do not apply. Betz's law states the maximum possible energy that can be derived from a wind turbine. To understand Betz's law and its derivation, we have provided the following explanation.

Derivation of Betz's Law

To understand Betz's law, we must first understand the constraint on the ability of a wind to convert the kinetic energy in the wind into mechanical power. Visualize

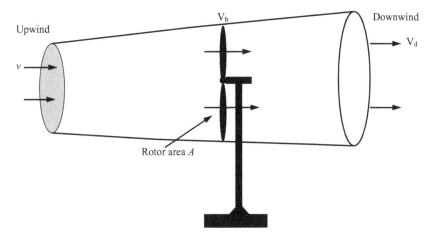

FIGURE 5.3   Wind passing through a turbine.

wind passing through a turbine (see Figure 5.3)—it slows down, and the pressure is reduced, so it expands.

Equation (5.8) is used to determine power extracted by the blades; another step in the derivation of Betz's law.

$$P_b = \tfrac{1}{2}m\left(v^2 - v_d^{\,2}\right) \tag{5.8}$$

where
   m = mass flow rate of air within stream tube
   v = upwind undisturbed windspeed
   v_d = downwind wind speed

Using Equation (5.9), determining mass flow rate is the next step in Betz's law derivation. In making the determination, it is easiest to use the cross-sectional area A at the plane of the rotor because we know what this value is. The mass flow rate is

$$m = \rho\, A v_b \tag{5.9}$$

Assume the velocity through the rotor v_b is the average of upwind velocity v and downwind velocity v_d:

$$v_b = \frac{v + v_d}{2} \quad\rightarrow\quad m = \rho\, A\left(\frac{v + v_d}{2}\right)$$

Then use Equation (5.10)

$$P_b = \tfrac{1}{2}\rho\, A\left(\frac{V + v_d}{2}\right)(v + v_d^2) \tag{5.10}$$

Before moving on to the derivation process, it is important that we define the speed/wind speed ratio, λ.

$$\lambda = \frac{v_d}{v} \qquad (5.11)$$

Then use Equation (5.12).

$$P_b = \tfrac{1}{2}\rho A \left(\frac{v + \lambda v}{2} \right) \left(v^2 - \lambda^2 v^2 \right)$$

$$\left(\frac{v + \lambda v}{2} \right) \left(v^2 - \lambda^2 v^2 \right) = \frac{v^3}{2} - \frac{\lambda^2 v^3}{2} + \frac{\lambda v^3}{2} - \frac{\lambda^3 v^3}{2} \qquad (5.12)$$

$$= \frac{v^3}{2} \left[\left(1 + \lambda \right) - \lambda^2 \left(1 + \lambda \right) \right]$$

$$= \frac{v^3}{2} \left[\left(1 + \lambda \right) \left(1 - \lambda^2 \right) \right]$$

$$\longrightarrow P_b = \underbrace{\left(\tfrac{1}{2}\rho A v^3 \right)}_{\substack{\downarrow \\ P_W = \text{Power in the Wind}}} \quad \underbrace{\tfrac{1}{2} \left[\left(1+\lambda \right) \left(1-\lambda^2 \right) \right]}_{\substack{\downarrow \\ C_P = \text{Rotor Efficiency}}}$$

The next step is to find the speed wind speed ratio λ, which maximizes the rotor efficiency, C_P.

$$C_P = \tfrac{1}{2} \left[\left(1+\lambda \right) \left(1-\lambda^2 \right) \right] = \tfrac{1}{2} - \frac{\lambda^2}{2} + \frac{\lambda}{2} - \frac{\lambda^3}{2}$$

Set the derivative of rotor efficiency to zero and solve for λ.

$$\frac{\partial C_P}{\partial \lambda} = -2\lambda + 1 - 3\lambda^2 = 0$$

$$\frac{\partial C_P}{\partial \lambda} = 3\lambda + 2\lambda - 1 = 0$$

$$\frac{\partial C_P}{\partial \lambda} = \left(3\lambda + 2\lambda \right) \left(\lambda + 1 \right) = 0 \longrightarrow \lambda = 1/3$$

Maximizes rotor efficiency

When we plug the optimal value for λ back into C_P to find the maximum rotor efficiency:

$$C_P = \tfrac{1}{2}\left[(1+1/3)(1-1/3)\right] = 16/27 = 59.3\% \qquad (5.13)$$

The maximum efficiency of 59.3% occurs when air is slowed to one-third of its upstream rate. Again, this factor and value is called the "Betz efficiency" or "Betz's law" (Betz, 1966). In plain English, Betz's law states that all wind power cannot be captured by the rotor; otherwise, air would be completely still behind the rotor and not allow more wind to pass through. For illustrative purposes, in Table 5.2, we list wind speed, power of the wind, and power of the wind based on the Betz limit (59.3%).

TIP-SPEED RATIO

Efficiency is a function of how fast the rotor turns. The tip-speed ratio (TSR) is an extremely important factor in wind turbine design. Tip-speed ratio is the ratio of the speed of the rotating blade tip to the speed of the free-stream wind (see Figure 5.4). Stated differently, TSR is the speed of the outer tip of the blade divided by wind speed. There is an optimum angle of attack which creates the highest lift to drag ratio. If the rotor of the wind turbine spins too slowly, most of the wind will pass straight through the gap between the blades, therefore, giving it no power! But if

TABLE 5.2
Betz Limit for 80 M Rotor Turbine

Wind Speed (mph/m/s)	Power (kW) of Wind	Power (kW) Betz Limit
5/2.2	36	21
10/4.5	285	169
15/6.7	962	570
20/8.9	2,280	1,352
25/11.2	4,453	2,641
28/12.5	6,257	3,710
30/13.4	7,695	4,563
35/15.6	12,220	7,246
40/17.9	18,241	10,817
45/20.1	25,972	15,401
50/22.4	35,626	21,126
55/24.6	47,419	28,119
56/25.0	50,053	29,681
60/26.8	61,563	36,507

Source: Adapted from Devlin, L, 2007. *Wind Turbine Efficiency*. Accessed @ http://k0lee.com/2007/11/wind-turbine-efficiency/

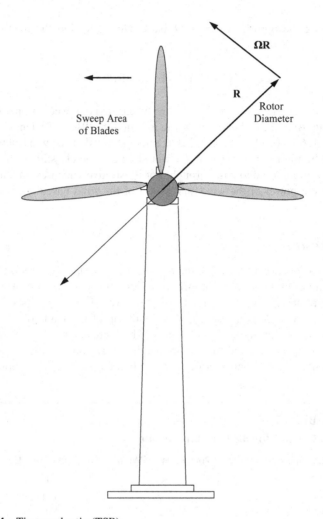

FIGURE 5.4 Tip-speed ratio (TSR).

the rotor spins too fast, the blades will blur and act like a solid wall to the wind. Moreover, rotor blades create turbulence as they spin through the air. If the next blade arrives too quickly, it will hit that turbulent air. Thus, it is actually better to slow down the blades. Because the angle of attack is dependent on wind speed, there is an optimum tip-speed ratio.

$$TSR = \Omega R/V \qquad (5.14)$$

where
    Ω = rotational speed in radians/s
    R = rotor radius
    V = wind "free-stream" velocity

SMALL-SCALE WIND POWER

In meeting your needs for electricity, a small home-size wind machine may be the answer. A small home-size wind turbine has rotors between 8 ft and 25 ft in diameter, stands upward of 30 ft, and can supply the power needs of an all-electric home or small business. Utility-scale turbines range in size from 50 kW to 750 kW. Single small turbines below 50 kW are used for homes, telecommunication dishes, or water pumping.

According to EERE (2006b), there are many questions potential users of small home-size wind turbines should ask and answer before installing a small wind system operation. In this section, we provide some of the answers and explanations EERE provides in addressing these questions.

We begin by asking the obvious question: What are the benefits to homeowners from using wind turbines? Wind energy systems provide a cushion against electricity price increases. Wind energy systems reduce US dependence on fossil fuels, and they don't emit greenhouse gases. If you are building a home in a remote location, a small wind energy system can help you avoid the high cost of extending utility power lines to your site.

Although wind energy systems involve a significant initial investment, they can be competitive with conventional energy sources when accounted for a lifetime of reduced or altogether avoided utility costs. The length of payback time—the time before the savings resulting from your system equal the system cost—depends on the system you choose, the wind resource in your site, electric utility rates in your area, and how you use your wind system.

DID YOU KNOW?

The difference between power and energy is that power [kilowatts (kW)] is the rate at which electricity is consumed, while energy [kilowatt-hours (kWh)] is the quantity consumed.

Another frequently asked question: Is wind power practical for me? Small wind energy systems can be used in connection with an electricity transmission and distribution system (called grid-connected systems) or in stand-alone applications that are not connected to the utility grid. A grid-connected wind turbine can reduce your consumption of utility-supplied electricity for lighting, appliances, and electric heat. If the turbine cannot deliver the amount of energy you need, the utility makes up the difference. When the wind system produces more electricity than the household requires, the excess can be sold to the utility. With the interconnections available today, switching takes place automatically. Stand-alone wind energy systems can be appropriate for homes, farms, or even entire communities (a cohousing project, for example) that are far from the nearest utility lines.

Stand-alone systems (systems not connected to the utility grid) require batteries to store excess power generated for use when the wind is calm. They also need a charge controller to keep the batteries from overcharging. As mentioned earlier in the text, deep-cycle batteries, such as those used for golf carts, can discharge and recharge 80% of their capacity hundreds of times, which makes them a good option for remote renewable energy systems. Automotive batteries are shallow-cycle batteries and should not be used in renewable energy systems because of their short life in deep-cycling operations.

Small wind turbines generate direct current (DC) electricity. In very small systems, DC appliances operate directly off the batteries. If you want to use standard appliances that use conventional household alternating current (AC) you must install an inverter to convert DC electricity from the batteries to AC (see Figure 5.5). Although the inverter slightly lowers the overall efficiency of the system, it allows the home to be wired for AC, a definite plus with lenders, electrical code officials, and future homebuyers.

SAFETY NOTE

For safety, batteries should be isolated from living areas and electronics because they contain corrosive and explosive substances. Lead-acid batteries also require protection from temperature extremes.

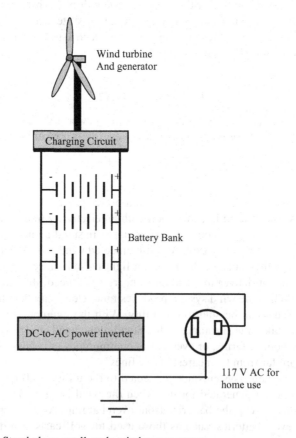

FIGURE 5.5   Stand-alone small-scale wind power system.

In grid-connected systems (or interactive systems), the only additional equipment required is a power condition unit (inverter) that makes the turbine output electrically compatible with the utility grid. Usually, batteries are not needed.

This type of system can be practical if the following conditions exist.

CONDITIONS FOR STAND-ALONE SYSTEMS

- You live in an area with average annual wind speeds of at least 4.0 m/s (9 mph).
- A grid connection is not available or can only be made through an expensive extension. The cost of running a power line to a remote site to connect with the utility grid can be prohibitive, ranging from $15,000 to more than $50,000 per mile, depending on terrain.
- You have an interest in gaining energy independence from the utility.
- You would like to reduce the environmental impact of electricity production.
- You acknowledge the intermittent nature of wind power and have a strategy for using intermittent resources to meet your power needs.

CONDITIONS FOR GRID-CONNECTED SYSTEMS

- You live in an area with average annual wind speeds of at least 4.5 m/s (10 mph).
- Utility-supplied electricity is expensive in your area (about 10 to 15 cents per kWh).
- The utility's requirements for connecting your system to its grid are not prohibitively expensive.
- Local building codes or covenants allow you to legally erect a wind turbine on your property.
- You are comfortable with long-term investments.

After comparing stand-alone systems and grid-connected systems and determining which is best suited for your particular circumstance, the next question to consider is whether your location is the right site to install a small-scale wind turbine system? That is, is there enough wind on your site, is it legal to install the system on your property, and are there environmental and/or economic issues?

In determining if your site is right for wind turbine installation, it must be determined if the wind blows hard enough at your site to make a small wind turbine system economically worthwhile? That is a key question and not always easily answered. The wind resource can vary significantly over an area of just a few miles because of local terrain influences on the wind flow. Yet there are steps you can take that will go a long way toward answering the above question.

As a first step, wind resource maps like the ones included in DOE's *Wind Energy Resource Atlas of the United States* (RREDC, 2010) can be used to estimate the wind resource in your region. The highest average wind speeds in the United States are generally found along seacoasts, on ridgelines, and on the Great Plains; however, many areas have wind resources strong enough to power a small wind turbine economically. The wind resource estimates on the maps in the Wind Energy Resource Atlas generally apply to terrain features that are well exposed to the wind, such as plains, hilltops, and ridge crests. Local terrain features may cause the wind resource at a specific site to differ considerably from these estimates.

Average wind speed information can be obtained from a nearby airport. However, caution should be used because local terrain influences and other factors may cause the wind speed recorded at an airport to be different from your particular location. Airport wind data are generally measured at heights about 20–33 ft (6–10 m) above ground. Average wind speeds increase with height and may be 15–25% greater at a typical wind turbine hub height of 80 ft (24 m) than those measured at airport anemometer heights. The Wind Energy Resource Atlas contains data from airports in the United States and makes wind data summaries available.

Again, it is important to have site-specific data to determine the wind resource at your exact location. If you do not have on-site data and want to obtain a clearer, more predictable picture of your wind resource, you may wish to measure wind speeds at your location for a year. You can do this with a recording anemometer, which generally costs $500 to $1500. The most accurate readings are taken at "hub height" (i.e., the elevation at the top of the wind turbine tower). This requires placing the anemometer high enough to avoid turbulence created by trees, buildings, and other obstructions. The standard wind sensor height used to obtain data for the DOE maps is 10 m (33 ft).

Within the same property, it is not unusual to have varied wind resources. If you live in complex terrain, take care in the selection of the installation site. If you site your wind turbine on the top or on the windy side of a hill, for example, you will have more access to prevailing winds than in a gully or on the leeward (sheltered) side of a hill on the same property. Consider existing obstacles and plan for future obstructions, including trees and buildings, which could block the wind. Also, recall that the power in the wind is proportional to its speed (velocity) cubed (v^3). This means that the amount of power you get from your generator goes up exponentially as the wind speed increases. For example, if your site has an annual average wind speed of about 5.6 m/s (12.6 mph), it has twice the energy available as a site with a 4.5 m/s (10 mph) average ($12.6/10^3$).

Another useful indirect measurement of the wind resource is the observation of an area's vegetation. Trees, especially conifers or evergreens, can be permanently deformed by strong winds. The deformity, known as "flagging," has been used to estimate the average wind speed for an area.

In addition to ensuring the proper siting of your small wind turbine system, there are also legal, environmental, and economic issues that must be addressed. For example, you should also

- Research potential legal and environmental obstacles
- Obtain cost and performance information from manufacturers
- Perform a complete economic analysis that accounts for a multitude of factors

- Understand the basics of a small wind system
- Review possibilities for combining your system with other energy sources, backups, and energy efficiency improvements

In regard to economic issues, because energy efficiency is usually less expensive than energy production, making your house more energy-efficient first will likely result in being able to spend less money, since you may need a smaller wind turbine to meet your needs. But, a word of caution, before you spend any money, research potential legal and environmental obstacles to installing a wind system. Some jurisdictions, for example, restrict the height of the structures permitted in residentially zoned areas, although variances are often obtainable. Your neighbors might object to a wind machine that blocks their view, or they might be concerned about noise. Consider obstacles that might block the wind in the future (large, planned developments or saplings, for example). Saplings that will grow into large trees can be a problem in the future. As mentioned, trees can affect wind speed (see Figure 5.6). If you plan to connect the wind generator to your local utility company's grid, find out its requirements for interconnections and buying electricity from small independent power producers.

When you are convinced that a small wind turbine is what you want and there are no obstructions restricting its installation, approach buying a wind system as you would any major purchase.

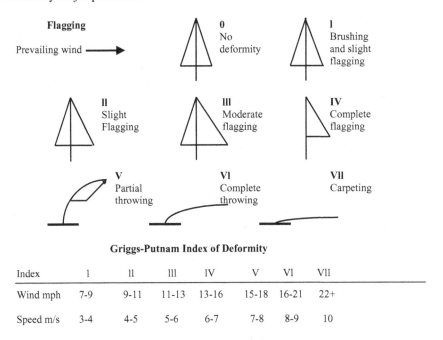

Index	1	ll	lll	IV	V	VI	VII
Wind mph	7-9	9-11	11-13	13-16	15-18	16-21	22+
Speed m/s	3-4	4-5	5-6	6-7	7-8	8-9	10

FIGURE 5.6 A crude method of approximating average annual wind speed from the deformation of trees (and other foliage).

THE BOTTOM LINE ON WIND POWER

According to Archer and Jacobson (2004), technology is much more advanced today in utilizing our wind resource, and the United States is home to one of the best wind resource areas in the world, the Midwest states of North and South Dakota, Nebraska, Kansas, Montana, Iowa, and Oklahoma. However, as with any other source of energy, nonrenewable or renewable, there are advantages and disadvantages associated with their use. On the positive side, it should be noted that wind energy is a free, renewable resource, so no matter how much is used today, there will still be the same supply in the future. Wind energy is also a source of clean, nonpolluting, electricity. Wind turbines can be installed on farms or ranches, thus benefiting the economy in rural areas, where most of the best wind sites are found. Moreover, farmers and ranchers can continue to work the land because the wind turbines use only a fraction of the land—the height and distance between turbines mean that land used for wind turbines can also be used for agriculture and grazing; i.e., only about 5% of the land in a wind farm is actually occupied by the turbines themselves. One huge advantage of wind energy is that it is a domestic source of energy produced in the United States or countries where installed and where the wind is abundant. Again, in the United States, the wind supply is abundant.

On the other side of the coin, wind energy does have a few negatives—wind projects face opposition. Wind power must compete with conventional generation sources on a cost basis. Even though the cost of wind power has decreased dramatically in the past 10 years, the technology requires a higher initial investment than fossil-fueled generators. The challenge to using wind as a source of power is that the wind is intermittent, and it does not always blow when electricity is needed. Wind energy cannot be stored (unless batteries are being used), and not all winds can be harnessed to meet the timing of electricity demands. Another problem is that good sites are often located in remote locations, far from cities where electricity is needed. Moreover, wind resource development may compete with other uses for the land and those alternative uses may be more highly valued than electricity generation. Finally, in regard to the environment, wind power plants have relatively little impact on the environment compared to other conventional power plants. There is some concern over the noise produced by the rotor blades and aesthetic (visual) impacts, and sometimes birds have been killed by flying into the rotors. Most of these problems have been resolved or greatly reduced through technological development or by properly siting wind plants. But keep in mind that the "not in my backyard" (NIMBY) point of view is still alive and strong in the United States and has succeeded in killing many projects, even renewable, "clean" energy projects.

We understand some of the issues related to the NIMBY point of view, although we don't agree with them. While it is questionable and definitely arguable that the United States has reached its apogee in attaining the so-called good life for most of its citizenry, and many feel that our best days are behind us, it is not questionable about our future if we do not innovate and free ourselves from the bondage of fossil fuels and the accompanying economics, politics, and potential turmoil. With fossil fuel supplies dwindling and world political pressures for accessibility ramping up to dangerous levels, we simply need to find reliable alternative energy sources for the future. We may never be able to improve on the so-called good life, but it would be beneficial for most of us if we are able to maintain the level of living that we have obtained to this point, while at the same time reducing oil imports and carbon emissions.

One thing is certain; wind energy production continues to be one of the fastest-growing energy technologies; it looks set to become a major generator of electricity throughout the world.

REFERENCES

Archer, C. and Jacobson, M.Z. 2004. *Evaluation of Global Wind Power.* Stanford, CA: Department of Civil and Environmental Engineering, Stanford University.

Betz, A. 1966. *Introduction to the Theory of Flow Machines.* (D.G. Randall, Trans.) Oxford: Pergamon Press.

EERE. 2006a. *Wind & Hydropower Technologies Program.* Accessed 02/20/10 @ http://www1.3343.energy.gov/windandhydro/wind_how.html.

EERE. 2006b. *How Wind Turbines Work.* Accessed 02/20/10 @ http://www1.eere.energy.gov/Windandhydro/wind_how.html.

Hanson, B.J. 2004. *Energy Power Shift.* Maple, Wisconsin. Lakota Scientific Press.

Khaaligh, A. and Onar, O.C. 2010. *Energy Harvesting.* Boca Raton, FL: Taylor and Francis.

RREDC. 2010. *Wind Energy Resource Atlas of the U.S.* Accessed 02/25/10 @ rredc.nrel.gov/Wind/pubs/atlas.

6 Blueprint Reading

GROPING IN THE DARK

During an evening in December, Rachel's Creek Wind Farm's SCADA alarm system sent a system fault signal to the on-site control station expert control panel where Jodi Price-Aspirin, the duty wind farm on-watch monitor/technician/repair responder, was working. The alarm alert indicated that turbine #66 (out of the total farm population of 150 turbines) had malfunctioned; she also noticed that the brake-set indicator was lighted (indicating that the turbine brake was set) along with several other turbine system component indicators. In addition, a normally green indicator light was now amber, meaning an overload had tripped the electrical panel main circuit breaker, cutting off power to the entire tower/turbine complex.

The lack of electricity, lighting, and the failure of the battery-operated emergency lanterns left the on-duty wind farm monitor/operator unable to perform her normal duties; after arriving at turbine #66, she was left literally groping in the dark.

The sudden, unexpected interruption of electrical power and simultaneous shutdown of the farm's wind turbine #66, the tower's lighting and ventilation, and other systems brought a halt to all things electrical but did not stop the flow of site wind into the farm. Fortunately, she did not have to worry about the blades and attached gearbox and ancillaries rotating or operating because of the wind—the turbine was equipped with an automatic pitch to 90 degrees, meaning the wind had no effect on turning the blades. Meanwhile, the turbine monitor/operator, well trained for such contingencies, immediately contacted the on-call senior maintenance operator and then quickly grabbed her tool kit and high-power flashlight. After the senior maintenance operator arrived, Jodi, guided by the beam of her flashlight, followed behind the operator. They located the bottom tower platform level electrical power panel and noticed that the main circuit breaker had tripped. Jodi watched as the senior operator reset the breaker; however, electrical power was not restored. She correctly discerned that power to the entire panel had been tripped—an unusual situation.

The operators realized that they would have to make the 280-ft ladder system climb up to the top, to the turbine, to make a quick inspection of the equipment within the turbine nacelle. Before climbing, they made sure they donned their safety harnesses, lightened their tool load, and made the long climb to the top and to the inside of the turbine nacelle. Using their flashlights, they made a quick survey of the interior and enclosed machinery. They also used their sense of smell in order to determine if there was any unusual odor like ozone (given off by burning electrical wire or components). Their initial visual search found nothing out of order. So, after not observing any obvious problem, the senior operator asked Jodi for the manufacturer's technical so the senior maintenance operator could view the electrical/hydraulic/pneumatic drawings to trace the systems and check out each component in

DOI: 10.1201/9781003288947-8

succession to find the problem. Unfortunately, Jodi realized that she had forgotten to bring the technical manual along with her, and so being the junior person, she knew she had to go back down the 280-ft ladder, go over to the control station building, pick up the manual, come back, and reclimb the 280-ft back up to the nacelle—not a lot of fun, but necessary.

Jodi and the senior maintenance technician, with the aid of the blueprints, drawings, and schematics within the technical manual, were able to trace the systems and eventually pinpoint the failed component, which turned out to be an electrical device. After about another hour of work and another trip down the ladder to pick up the space component and then back up and Jodi handing the device to the technical, they were able to make the necessary repair and then put the turbine back on line; it worked as per design.

The lesson learned in this particular incident points to the fact that even in a well-lighted compartment (which in this incident was not the case) trying to pinpoint any problem without the blueprint, wiring diagram, or schematic of the system or systems is analogous to groping in the dark.

Obviously, maintenance operators who can provide such good service are valuable to wind turbine and wind farm operations. Upon arrival at the wind farm, the on-call senior maintenance operator, who had been cross-trained as an electrician—a typical Jack or Jill of all trades—went to the electrical panel first to check for damage to the circuit breaker. After determining that the main circuit breaker appeared to be undamaged, she knew that whatever the problem was, it had to be far above her head in the turbine nacelle. The maintenance operator cut the individual power switches to all equipment before resetting the main circuit breaker at the substation. Then the maintenance operator put the lighting, ventilation, and other equipment on line one by one, avoiding resetting the bar screen breaker.

After restoring electrical power to lighting, ventilation, and other circuits, the maintenance operator checked out the moving parts associated with the turbine to determine if anything was jammed. The maintenance operator then examined the hydraulic lines and components. Then she moved on to the electrical components and traced the system using the manufacturer's drawings and her test equipment and finally identified the electrical component that had failed.

Now, this discovery of the component failure in the electrical system and the component's replacement in kind with a new device might appear to be problem solved.

Not really.

Remember, all the lights and ventilation in the tower were also not functioning. So, more troubleshooting was involved, and eventually by the next day, all systems were functional again. And the emergency lighting battery-pack system that had failed was also repaired.

The initial response to the problem(s) with wind turbine #66 (i.e., turbine failure, lighting, ventilation, etc.) was to put it back online again. Even though the maintenance operator and senior maintenance technical both were highly skilled and had considerable local knowledge (experience), they needed a lot of information to get this job done—and they needed it quickly.

The information came from the manufacturer's technical manual with its blueprints, drawings, schematics, and troubleshooting hints/guidance.

Blueprints, drawings, and schematics are used almost everywhere in industrial systems to troubleshoot a problem. The wind turbine malfunction just described points out that blueprints are among the most important forms of communication among people involved in plant maintenance operations.

ILLUSTRATING THE SYSTEM

You have heard the old saying, "A picture is worth a thousand words." This is certainly true when referring to industrial machinery, plant process machinery, electrical motor controllers, and wind turbine systems.

It would be next to impossible for a maintenance supervisor (or any other knowledgeable person) to describe in words the shape, size, configuration, relations of the various components of a machine, and/or its operation in sufficient detail for wind turbine technician/maintenance operator to troubleshoot the process or machine properly. Blueprints are the universal language used to communicate quickly and accurately the necessary information to understand process operations or to disassemble, service, and reassemble process equipment.

The original drawing is seldom used in the plant or field but copies, commonly called "blueprints," are made and distributed to maintenance operators who need them. These blueprints are used extensively in industrial operations to convey the ideas relating to the design, manufacture, and operation of equipment and installations. Simply, blueprints are reproductions or copies of original drawings. Blueprints are made by a special process that produces a white image on a blue background from drawings having dark lines on a light background. With the advancements in the digital world with computers and handheld electronic devices, many of the blueprints and drawings used in troubleshooting are stored in digital memory or can be accessed online via the Internet.

In addition to understanding applicable blueprints, schematic diagrams are also important "pictorial" representations with which the maintenance operator should be familiar. A schematic is a line drawing made for a technical purpose that uses symbols and connecting lines to show how a particular system operates.

Blueprints and schematics are particularly important to a maintenance operator because they provide detailed information (or views) for troubleshooting; that is, they help familiarize the troubleshooter with the overall characteristics of systems and equipment.

In this chapter, the focus is on blueprints and schematics representative of major wind turbine and wind turbine farm support equipment, systems, and processes. Major support equipment/systems included are machine parts, machines, hydraulic and pneumatic systems, piping and plumbing systems, electrical systems, welding, and air conditioning (i.e., if an AC system is part of the turbine tower or maintenance building).

BLUEPRINTS: THE UNIVERSAL LANGUAGE

Technical information about the shape and construction of a simple part, mechanism, or system may be conveyed from one person to another by the spoken or

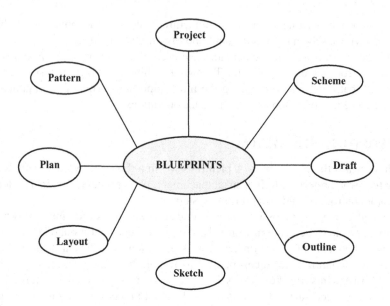

FIGURE 6.1   Blueprints by other names.

written word. As the addition of details makes the part, mechanism, or system more complex, the wind turbine technician and maintenance operator must have a precise method available that describes the object adequately.

The methodology? Blueprints.

Blueprints provide a universal language by which all information about a part, mechanism, or system is furnished to the operator and others.

Keep in mind that when the term "blueprints" is used in this book, we could also be referring to a hard copy or digital copy of a blueprint by another name, as shown in Figure 6.1. Whatever the detailed drawing or blueprint is called, it is used not only to perform troubleshooting functions but also to assemble key parts and components making up the systems involved and the entire unit itself. Understanding blueprints is important to understanding wind turbine systems.

BLUEPRINT STANDARDS

In order to provide a universal language, blueprints must communicate ideas to many different people. It logically follows that to facilitate this communication, all industrialized nations need to develop technical drawings according to universally adopted standards. Moreover, such drawing standards must also include symbols, technical data, and principles of graphic representation.

Universal standardization practices allow blueprints to be uniformly interpreted throughout the globe. The standardization implication should be obvious: Parts, structures, machines, and all other products (designed according to the same system of measurement) may be actually manufactured and interchangeable.

Important Point: Universal standardization of blueprints and drawings is sometimes called *Drawing Conventions*, that is, standard ways of drawing things so that everyone understands the information being conveyed.

In this modern era, with its global economy, the interchangeability of manufactured parts is of increasing importance. Consider, for example, a machine manufactured in Europe that subsequently ends up being used in a factory in North America. Though such a machine is manufactured on one continent and used on another, getting replacement repair parts normally is not a significant problem. However, if the user company has its own machine shop or access to one, it may decide, for one reason or another, that it wants to manufacture its own replacement parts. Without standardized blueprints, such an operation might be difficult to accomplish.

STANDARDS-SETTING ORGANIZATIONS

Two standards-setting organizations have developed drafting standards that are accepted and widely used throughout the globe: The *American National Standards Institute* (ANSI) and the *International Organization Standardization* (metric; ISO). Incorporated into these systems are other engineering standards generated and accepted by professional organizations dealing with specific branches of engineering, science, and technology. These organizations include the *American Society of Mechanical Engineers* (ASME), the *American Welders Society* (AWS), the *American Institute of Architects* (AIA), the *U.S. Military* (MIL), and others. Moreover, to suit their own standards, some large corporations have adopted their own standards.

Important Point: All references to blueprints in this text closely follow the ANSI and ISO standards and the current industrial practices (see Table 6.1).

ANSI STANDARDS FOR BLUEPRINT SHEETS

ANSI has established standards for the sheets onto which blueprints are made (see Table 6.1).

FINDING INFORMATION

In the previous section, the importance of using universal standards or drawing conventions in correctly interpreting blueprints was pointed out. Along with the need

TABLE 6.1
Blueprint Sheet Size (ANSI Y14.1—1980)

Standard US Size (inch)	Nearest International Size (millimeter)
A 8.5 × 11.0	A4 210 × 297
B 11.0 × 17.0	A3 297 × 420
C 17.0 × 22.0	A2 420 × 594
D 22.0 × 34	A1 594 × 841
E 34.0 × 44.0	A0×841 × 1189

to know the conventions, we must also know where to look for information. In this section, we explain how to find the information needed in blueprints.

Typically, designers use technical shorthand in their drawings. However, there generally is too much information to be included in a single drawing sheet. For this reason, several blueprints are often assembled to make a set of *working drawings*.

Working drawings—drawings that furnish all the information required to construct an object—consist of two basic types: *Detail Drawings* for the parts produced and an *Assembly Drawing* for each unit or subunit to be put together.

DETAIL DRAWINGS

A *detail drawing* is a working drawing that includes a great deal of data including the size and shape of the project, what kinds of materials should be used, how the finishing should be done, and what degree of accuracy is needed for a single part. Each detail must be given.

Important Point: Usually, a detail drawing contains only dimensions and information needed by the department for which it is made. The only part that may not need to be drawn is a *standard part*, one that can probably be purchased from an outside supplier more economically than it can be manufactured.

ASSEMBLY DRAWINGS

Many machines and systems contain more than one part. Simply, *assembly drawings* show how the parts fit together. In addition to showing how parts fit together, the overall look of the construction, dimensions needed for installation, and assembly drawings also include a *parts list*, which identifies all the pieces needed to build the item. A parts list is also called a *bill of material*.

Important Point: Because assembly drawings show the working relationship of the various parts of a machine or structure as they fit together, usually, each part in the assembly is numbered and listed in a table on the drawing.

TITLE BLOCK

The first place to look for information on a blueprint is in the *title block*, an outlined rectangular space located in the lower right corner of the sheet. (**Note:** The title block is placed in the lower right corner so that, when the print is correctly folded, it may be seen for easy reference and for filing). The purpose of the title block is to provide supplementary information on the part or assembly to be made and to include in one section of the print, information that aids in the identification and filing of the print.

Although there is some variation among title blocks used by different organizations, certain information is basic. The following paragraphs describe the information we usually find in a title block. Figure 6.2 shows a blank blueprint

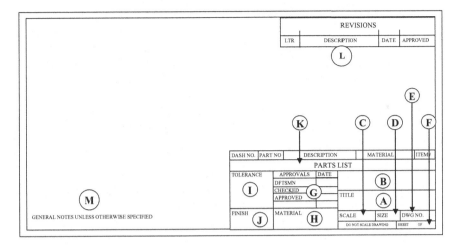

FIGURE 6.2 Blank blueprint sheet.

sheet, with its title block and other features. The letters after the descriptions refer to Figure 6.2.

1) **Title of Drawing**. This box identifies the part or assembly illustrated (A).
2) **The Name of Company and Its Location**. The space above the title is reserved for the name and location (complete address) of the designing or manufacturing firm (B).
3) **Scale**. The drawing *scale* indicates the relationship between the size of the image and the size of the actual object. Some parts are shown at actual size; others are either too big or too small to show conveniently at full size. For example, we could not show a large machine full size on an ordinary sheet of paper. The designer has the choice of drawing a machine, mechanical part, or other objects, or larger or smaller than the actual size.

   Typical scale notations are: $1/2'' = 1''$ (one-half actual size), FULL (actual size), 1:1 (actual size), 2:1 (twice size), 2, 3, 4, etc. (2, 3, 4, etc. times true size). When the scale is shown as NOTED, it means that several scales have been used in making the drawing and each is indicated below the particular view to which it pertains (C).

Important Point: Measurements on a blueprint should never be used because the print may have been reduced in size or stretched. Work only from the dimensions given on the print.

4) **Drawing Size**. Drawings are prepared on standard size sheets, Table 6.1, in multiples of 8.5×11 and 9×12 inches and are designated by a letter to indicate size (D).
5) **Drawing Number**. The *drawing number* is used to identify and control the blueprint. It is also used to designate the part or assembly shown on the

blueprint (i.e., it becomes the number of the part itself). The number is usually coded (to the particular industry and not universally applicable to all industries) to indicate department, model, group, serial number, and dash numbers. This number is also used to file the drawings, making it easier to locate them later on (E).

Important Point: A *dash number* is a number preceded by a dash after the drawing number; it indicates right- or left-hand parts as well as neutral parts and/or detail and assembly drawings.

6) **Sheet Number**. The *sheet numbering* is used on multisheet blueprints to indicate the consecutive order, the total number of prints, and which one of the series this particular drawing happens to be (F).

7) **Approvals Block**. This block is for the signatures and date of release by those who have responsibility for making or approving all or certain facets of the drawing or the manufacture of the part. The block may include signature and date blocks for the following:
   - **Draftsperson**
   - **Checked** (the engineer who checked the drawing for completeness, accuracy, and clarity)
   - **Design** (person responsible for the design of the part)
   - **Stress** (is the place for the engineer who ran the stress calculations for the part)
   - **Materials** (is where the person signs whose responsibility it is to see that the materials needed to make the part available)
   - **Production** (is where the engineer who approved the producibility of the part approves the drawing)
   - **Supervisor** (this is where the person in charge of drafting indicates approval by signing and dating this block)
   - **Approved** (these lines are to record any other required approvals). Each person signs the document and fills in the date on the appropriate line when his or her portion of the work is finished or approved (G).

8) **Materials Block**. This block specifies exactly what the part is made of (e.g., the type of steel to be used) and often includes the size of raw stock to be used (H).

9) **Tolerance Block**. This block indicates the general tolerance limits for one, two, and three-place decimal and angular dimensions. The tolerance limits are often necessary because nothing can be made to the exact size specified on a drawing. Normal machining and manufacturing processes allow for slight deviations. These limits are applicable unless the tolerance is given along with dimension callout (I).

Important Point: *Tolerance* is defined as the total amount of variation permitted from the design size of a part. Parts may have a tolerance given in fractions or decimal inches or decimal millimeters.

10) **Finish Block**. This block gives information on how the part is to be finished (buffed, painted, plated, anodized, or other). Specific finish requirements would be a callout on the drawing with the word NOTED in the finish block (J).

11) **Parts List (Bill of Materials)**. A *parts list* is a tabular form usually appearing right above the title block on the blueprint; it is used only on assembly and installation drawings. The purpose of the parts list is to provide specific information on the quantity and types of materials used in the manufacturing and assembling of parts of a machine or structure.

  The parts list enables a purchasing department to requisition the quantity of materials needed to produce a given number of the assemblies. Individual component parts, their part numbers, and the quantity required for each unit are listed (K).

Important Point: The list is built from the bottom-up. The columns are labeled at the bottom (just above the words "Parts List"), and parts are listed in reverse numerical order above these categories. This allows the list to grow into the blank space as additional parts are added. If a drawing is complicated, containing many parts, the parts list may be on a separate piece of paper that is attached to the drawing.

12) **Revision (Change Block)**. On occasion, after a blueprint has been released, it is necessary to make design revisions/changes. The *revision* or *change* block is a separate block positioned in the upper-right-hand corner of the drawing. It is used to note any changes that have been made to the drawing after its final approval. It is placed in a prominent position because we need to know which revision we are using and what features have been revised/changed. We also need to know whether the changes have been approved; the initials of the draftsperson making the change and those approving it are required.

  When a Drawing Revision/Change Notice has been prepared, the drawing is revised, and the pertinent information is recorded in the Revision/Change Block (L).

The following items are usually included in the Revision/Change Block. The letters after the descriptions refer to Figures 6.3–6.4.

(1) **Sequence Letter**. The *Sequence Letter* is assigned to the change/revision and recorded in the Change/Revision Block (see Figure 6.3). This index letter is also referenced to the field of the drawing next to the changed effected, for example—A 1. BREAK ALL SHARP EDGES

(2) **Zone**. Used on larger-sized prints, this column aids in locating changes (B).

(3) **Description**. This column provides a concise description of change; for example, when a note is removed from the drawing, the type of note is

NOTICE OF CHANGE						
LTR	ZONE	DESCRIPTION	SERIAL	DATE	DR	APP
(A)	(B)	(C)	(D)	(E)	(F)	(G)

FIGURE 6.3 Revision/change block.

REVISIONS				
ZONE	LTR	DESCRIPTION	DATE	APPROVED
E-3	B	INCORP. FEI I REVISED MARKING B18 WAS A4 N/A WAS 12303		

FIGURE 6.4 Change block/other items.

referred to in the description block—PLATING NOTE REMOVED, or when a dimension is changed WAS .975-1.002 €.

(4) **Serial Number**. This column lists the serial number of the assembly or machine on which the change becomes effective (D).

(5) **Date**. The date column is for the date the change was written €.

(6) **Drafter (DR)**. This column is to be initialed by the drafter making the change (F).

(7) **Approved (APP)**. This column carries the initials or name of the engineer approving the change (G).

Other Items: A few industries will include other items in their Revision/Change Block to further document the changes made in the original blueprint. Some of these are:

(8) **Checked**. This column is for the initials or signature of the person who checks and approves the revision/change.

(9) **Authority**. This column usually consists of recording approved engineering change request numbers.

(10) **Change Number**. The *change number* is a listing of the Drawing Change Notice (DCN) number.

(11) **Disposition**. This column carries the coded number indicating the *disposition* of the change request.

(12) **Microfilm**. This column is used to indicate the date the revised drawing was placed on *microfilm*.

(13) **Effective On**. This column (sometimes a separate block) gives the serial number or ship number of the machine, assembly, or part on which the change becomes effective. The change may also be indicated as effective on a certain date.

Important Point: A drawing that has been extensively revised/changed may be redrawn and carry an entry to that effect in the Revised/Change Block or the drawing number may carry a dash letter (-A) indicating a revised/changed drawing.

Important Point: Considerable variation exists among industries in the form of processing and recording changes in prints. The information presented in this section will enable maintenance operators to develop an understanding of the change system in general.

DRAWING NOTES

Notes on drawings provide information and instructions that supplement the graphic presentation as well as the information in the title block and list of materials. Notes on drawings convey many kinds of information (e.g., the size of holes to be drilled, type fasteners to be used, removal of machining burrs, etc.). Specific notes like these are tied by *leaders* directly to specific features.

GENERAL NOTES

General notes refer to the whole drawing. They are located at the bottom of the drawing, to the left of the title block. General notes are neither referenced in the list of materials nor from specific areas of the drawing. Some examples of general notes are given in Figure 6.5.

Important Point: When there are exceptions to general notes on the field of the drawing, the general note will usually be followed by the phrase EXCEPT AS SHOWN or UNLESS OTHERWISE SPECIFIED. These exceptions will be shown by local notes or data on the field of the drawing.

1. BREAK SHARP EDGES .030 R UNLESS OTHERWISE SPECIFIED.

2. THIS PART SHALL BE PURCHASED ONLY FROM SOURCES
   APPROVED BY THE TREATMENT DEPARTMENT.

3. FINISH ALL OVER.

4. REMOVE BURRS.

5. METALLURGICAL INSPECTION REQUIRED BEFORE MACHINING.

FIGURE 6.5   Examples of general notes on drawings.

LOCAL NOTES

Specific notes or *local notes* apply only to certain features or areas and are located near, and directed to, the feature or area by a leader (see Figure 6.6). Local notes may also be referenced from the field of the drawing or the list of materials by the note number enclosed in a *flag* (equilateral triangle; see Figure 6.7). Some examples of local notes are given in Figure 6.8.

CASE STUDY 6.1: THE MAINTENANCE OPERATOR'S "TOOLBOX"

When we place a trouble service call for a heating, air conditioning, television, washing machine/dryer, or other household appliance or system malfunction, usually the repair person responds in short order. The usual practice is for the repair person to check out or look over the appliance or system to get a "feel" for what the problem is. If the repair person determines that the problem is more than just placing the "ON" switch in the correct position and that the system is properly aligned for operation (e.g., valves opened or closed as per design), he/she usually opens their toolbox and begins troubleshooting to determine the problem. After determining the cause of the malfunction, the repair person makes the necessary repair or adjustment and then tests the unit to ensure proper operation.

There is nothing that unusual about the scenario just described; it is nothing more than a routine practice that most of us are familiar with—unfortunately, some of us much more than others. In fact, this practice is so common and familiar that we don't give it much thought. There is an important point we are trying to make here, however. Let's take another look at the routine service call procedure described above.

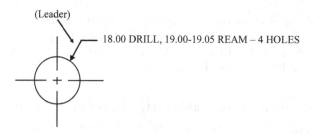

FIGURE 6.6   Local notes.

FIGURE 6.7   Local notes.

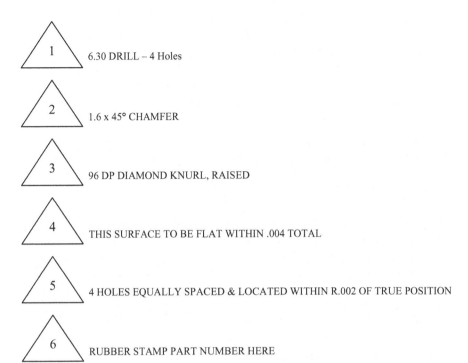

FIGURE 6.8   Local notes.

- We have a problem with a home appliance or system.
- We place a service call.
- A repair person responds.
- The repair person checks out the appliance/system.
- The repair person corrects the problem immediately
- Or the repair person determines the malfunction is a bit more complex.
- The repair person opens his/her toolbox, and the troubleshooting process begins.
- Eventually, the fault is found and corrected (hopefully).

The repair person opens his/her toolbox. This is the part of the routine repair procedure we want to focus on. Why? Good question.

The best way to answer this question is to provide an illustration. In our experience, whenever we have hired a repair person to repair a home appliance, a carpenter to repair a wooden deck, a plumber to unclog our pipes, or an electrician to install a new lighting fixture, these skilled technicians always responded with a toolbox in hand. Have you noticed this too?

We once asked a carpenter we hired to replace several windows with new ones. During this replacement process, we noticed that the carpenter lugged a huge, heavy, clumsy toolbox from window to window. We also noticed that he was able to remove the old windows and install the new ones using only a few basic hand tools—screwdrivers, chisel, and hammer. Why then the big, clumsy toolbox? Eventually, we got around to asking him this question: "Why the big toolbox?" After looking at us like we were somewhat out to lunch, he replied as follows:

> I lug this big toolbox around with me everywhere I go because I never know for sure what tool will be needed to do the job at hand. It is easier to have my complete set of tools in easy reach ... so I can get the job done without running all over the place to get this or that tool. The way I see it ... just makes good common sense.

The statement "I lug this big toolbox around with me everywhere I go because I never know for sure what tool will be needed to do the job at hand" not only makes good common sense but is also highly practicable. For those of us who have worked in wind turbine operation and maintenance operations, I have seen this same routine practiced over and over again by the wind turbine maintenance operator who responds to a plant trouble call. "... just makes good common sense."

At this point the reader might be thinking: So, beyond the obvious, what is the author's point? Another good question.

Simply, the maintenance operator must have an extensive assortment of tools in his/her toolbox to perform many of the turbine maintenance actions they are called upon to respond to. "... just makes good common sense."

But there is more. There are many types of toolboxes and a large variety of tools. One thing is certain: if we fill a standard-sized portable toolbox with tools (the best tools that money can buy), they do us little good unless we know how to use them properly.

Tools are designed to assist us in performing certain tasks. Like electricity, the internal combustion engine, the computer, and tools, when properly used, are extremely helpful to us. They not only make tasks easier but also save much time. On the other hand, few would argue against the adage that any tool is only as good as the person's ability or skill in properly using it.

"Ability and skill." We feel these (and other) qualities are also important tools. They are important tools in the sense that any wind turbine maintenance operator without a certain amount of ability and/or skill is just, no matter the sophistication of the tool in hand, another unskilled user of the tool.

If we agree that ability and skill are tools, we must also agree that they are tools kept in a different kind of "toolbox." The toolbox we are referring to is obvious, of course. Moreover, this kind of toolbox is one we don't have to worry about forgetting to bring to any job we are assigned to perform; we automatically carry it with us (hopefully) everywhere we go.

Ability and skill are not normally innate qualities; instead, they are characteristics that have to be learned. They are also general terms, their connotations wide and various.

Simply, ability and skill entail more than just knowing how to properly use a handsaw or other portable tool. For example, with a little practice, just about anyone can use a handsaw to cut a piece of lumber. But what if we need to cut the lumber to a particular size or dimension? Obviously, to make such a cut to proper size or specification, we need to know not only how to use the cutting tool but also how to measure the stock properly. Moreover, in order to properly measure and determine the proper size of the cut to be made, we also need to know how to use basic math operations to determine how much needs to be actually cut. "… just makes good common sense."

MATH REVIEW

At this point, it is important to provide a review of the basic math needed and often found most useful in reading blueprints. We are likely to find that we already know most of it. Some of it may be new, but only because we haven't worked with blueprints before.

Although blueprints ordinarily give us sizes, we occasionally have to do some calculating to get the exact dimension of what we are particularly concerned with. Usually, we find it by adding and subtracting. At other times, we may have to calculate the number of pieces of a given length we can get from a particular piece of wood, pipe, or bar of steel or aluminum or other material. That is usually a matter of multiplying and dividing. We may want to know how many square feet there are in a particular room, doorway, or roof area.

In addition to basic math operations of adding, subtracting, multiplying, and dividing, the maintenance operator must also know how to work with fractions and decimals. A basic understanding of angles, areas of rectangles, and the radius of circles is also important.

Units of Measurement

Basic knowledge of units of measurement and how to use them is essential. Wind turbine operators and maintenance operators should be familiar both with the US Customary System (USCS) or English System and the International System of Units (SI). Table 6.2 gives conversion factors between SI and USCS systems for three of the basic units that are encountered in blueprint reading.

The basic units used in blueprint reading are for straight line (linear) measurements. That is, most of the calculating we do is with numbers of yards, feet, and inches, or parts of them. What we actually do is find the distance between two or more points and then use numbers to express the answer in terms of yards, feet, and inches (or parts of them). Twelve inches make 1 foot (ft); 3 ft make 1 yard (yd). The symbol indicating an inch is " and that for a ft is '. Yard is simply abbreviated "yd."

TABLE 6.2

Commonly Used Units and Conversions

Quantity	SI Units	SI Symbol	× Conversion Factor	USCS Units
Length	Meter	M	3.2808	ft
Area	Square meter	m^2	10.7639	ft^2
Volume	Cubic meter	m^3	35.3147	ft^3

As technology has improved, so has the need for closer measurement. As we develop new improved measuring tools, it becomes possible to make parts to greater accuracy. Moreover, now that we are in the age of interchangeable parts, we have developed standards of various kinds. This, in turn, allows us to know what is needed and meant by a given specification. Various thread specifications are a good example—as are the conventions and symbols used on blueprints themselves.

Wind turbine maintenance and operations familiar to us today would not be possible without our ability to make close measurements and to do so accurately. That is why basic math operations are used and why it is important to be familiar with them. The basic unit of linear measurement in the United States is the yard, which we break down into feet and inches, and parts of them. In our maintenance work, we are concerned with all of them. We may work more with yards and feet in using plant building drawings. At other times, when we work with plant or pumping station machinery, we find them dimensioned in inches or parts of inches.

FRACTIONS AND DECIMAL FRACTIONS

The number 8 divided by 4 gives an exact quotient of 2. This may be written 8/4 = 2. However, if we attempt to divide 5 by 6, we are unable to calculate an exact quotient. This division may be written 5/6 (read "five-sixths"). This is called a *fraction*. The fraction 5/6 represents a number, but it is not a whole number. Therefore, our idea of numbers must be enlarged to include fractions.

In blueprint reading, we are specifically concerned with fractions of some units involved with measurements. For example, a half-inch is one of two parts needed to make up one inch. That could be written down as 1/2 inch, the bottom number (2) meaning that it will take two parts to make up the whole unit and the top number (1) meaning that we have one of the two parts needed. One-quarter of an inch, which would be shown on a blueprint as 1/4″, means that we need four of them to make up one inch. Since one is all we need, one is all that is shown. It could just as easily have 3/4, 5/8, or 11/16. The basic meaning of the numbers would still be that we need *three*-quarters of an inch, or FIVE-eighths of an inch, or ELEVEN-sixteenths of an inch. In maintenance practice, the inch is further broken down into 32nds (thirty-seconds) and 64ths (sixty-fourths).

Important Point: A 64th is the smallest fraction we'll use; it's the smallest fraction shown on a machinist's ruler, sometimes incorrectly called a "scale."

A *decimal fraction* is a fraction that may be written with 10; 100; 1,000; 10,000 or some other multiple of 10 as its denominator. Thus 47/100, 4,256/10,000, 77/1,000, 3,437/1,000, for example, are decimal fractions.

Important Point: The word "decimal" comes from the Latin word for tenth or tenth part.

In writing a decimal fraction, it is standard procedure to omit the denominator and instead merely indicate what that denominator is by placing a *decimal point* in the numerator so that there are as many figures to the right of this point as there are zeros in the denominator. Thus 47/100 is written 0.47; 4,256/10,000 = 0.4256; 77/1,000 = 0.077; 3,437/1,000 = 3.437.

Important Point: **Decimal fractions** are useful in shortening calculations. They are called decimal fractions because they are small parts of a whole unit. Most technologies now use decimal fractions as a matter of course.

Table 6.3 shows a list of all the fractions we are likely to see on our plant machine prints. The figures at the right-hand side of each column mean exactly the same thing, except that the numbers are expressed as **decimal** parts of an inch.

With the passage of time and corresponding improvements in technology, greater accuracy became possible (measurements in fractions of an inch were no longer exact enough). Smaller parts of an inch were needed and provided by dividing the inch into 1,000 parts, the parts being referred to as "thousandths of an inch." One-thousandth of an inch is written as .001″. Common measurements of an inch, for example, are expressed as follows:

One inch	1.000″
One-thousandth of an inch	.001″
One ten-thousandth	.0001″
One millionth	.000001″

ALPHABET OF LINES

CASE STUDY 6.2: JUST LINES

Over the years we've heard seasoned maintenance operators state that there wasn't any type of blueprint or technical diagram that they couldn't use and/or understand. We often thought this was more braggadocio than truth. We continued to think this way until we cornered one of these seasoned maintenance types and finally asked: "What makes you so confident that you can read and understand any technical blueprint or drawing?" At first, somewhat peeved that we would ask such an obviously dumb question, the maintenance operator answered in her condescending way:

> I know that I can read and understand any print or drawing … because any print or drawing is nothing more than a bunch of drawn lines. Even the components the lines are hooked to are nothing more than lines shown in a different fashion. It all comes down to a bunch of lines … and nothing more.

TABLE 6.3

Common Fractions and Their Decimal Equivalents

Fractions	Decimals	Fractions	Decimals
1/64	.015625	33/64	.515625
1/32	.03125	17/32	.53215
3/64	.046875	35/64	.546875
1/16	.0625	9/16	.5625
5/64	.078125	37/64	.578125
3/32	.09375	19/32	.59375
7/64	.109375	39/64	.609375
1/8	.125	5/8	.625
9/64	.140625	41/64	.640625
5/32	.15625	21/32	.65625
11/64	.171875	43/64	.671875
3/16	.1875	11/16	.6875
13/64	.203125	45/64	.703125
7/32	.21875	23/32	.71875
15/64	.234375	47/64	.734375
¼	.25	¾	.75
17/64	.265625	49/64	.765625
9/32	.28125	25/32	.78125
19/64	.296875	51/64	.796875
5/16	.3125	13/16	.8125
21/64	.328125	53/64	.828125
11/32	.34375	27/32	.84375
23/64	.359375	55/64	.859375
3/8	.375	7/8	.875
25/64	.390625	57/64	.890625
13/32	.40625	29/32	.90625
27/64	.421875	59/64	.921875
7/16	.4375	15/16	.9375
29/64	.453125	61/64	.953125
15/32	.46875	31/32	.96875
31/64	.484375	63/64	.984375
½	.5	1/1	1.0

JUST A BUNCH OF DRAWN LINES?

Notwithstanding the summation provided by the seasoned maintenance operators in the chapter opening, for the engineer, the designer, and the drafter, engineering-type drawings are more than "a bunch of drawn lines." No doubt lines are important; there can be little argument with this point. Moreover, to correctly interpret the blueprint in servicing a part or assembly, the maintenance operator and technician must recognize and understand the meaning of ten kinds of lines that are commonly used in engineering and technical type drawings.

These lines, known as the *alphabet of lines*—a list of line symbols, are universally used throughout the industry. Each line has a definite form and shape (width—thick, medium, or thin), and when combined in a drawing they convey information essential to understanding the blueprint (i.e., shape and size of an object).

Important Point: Each line on a technical drawing has a definite meaning and is drawn in a certain way. We use the line conventions, together with illustrations showing various applications, recommended by the American National Standards Institute (ANSI, 1973a) throughout this text.

In regard to Case Study 6.2, whereby the seasoned maintenance operator stated that technical drawings are "nothing more than a bunch of drawn lines," we agree to a point but feel it would be more accurate to say that the *line* is the basis of all technical drawings. The point is that by combining lines of different thicknesses, types, and lengths, just about anything can be described graphically and in sufficient detail so that persons with a basic understanding of blueprint reading can accurately visualize the shape of the component.

Therefore, to understand the blueprint we must know and understand the alphabet of lines. The following sections explain and describe each of these lines.

VISIBLE LINES

The *visible line* (also called *object line*) is a thick (dark), continuous line that represents all edges and surfaces of an object that are visible in the view. A visible line (see Figure 6.9) is always drawn thick (dark) and solid so that the outline or shape of the object is clearly emphasized in the drawing.

Important Point: The visible line represents the outline of an object. The thickness of the line may vary according to the size and complexity of the part being described (Olivo and Olivo, 1999).

HIDDEN LINES

Hidden lines are thin, dark, medium-weight, short dashes used to show edges, surfaces, and corners that are not visible in a particular view (see Figure 6.10). Many of these lines are invisible to the observer because they are covered by other portions of the object. They are used when their presence helps to clarify a drawing and are sometimes omitted when the drawing seems to be clearer without them.

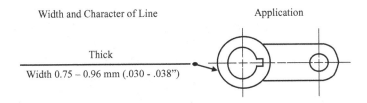

FIGURE 6.9 Visible lines. Adaptation from Brown, W.C., *Blueprint Reading for Industry*. South Holland, IL: The Goodheart-Wilcox Co., pp. 14–18, 82, 1989.

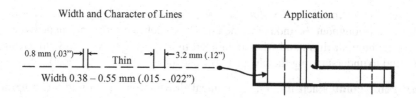

FIGURE 6.10   Hidden line. Source: Adaptation from Brown, W.C., Blueprint Reading for Industry. South Holland, Illinois: The Goodheart-Wilcox Co., pp. 14–18, 83, 1989.

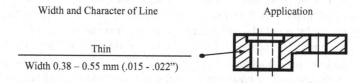

FIGURE 6.11   Section line. Source: Adaptation from Brown, W.C., Blueprint Reading for Industry. South Holland, Illinois: The Goodheart-Wilcox Co., pp. 14–18, 1989.

SECTION LINES

Usually drawn at an angle of 45 degrees, *section lines* are thin lines used to indicate the cut surface of an object in a sectional view. In Figure 6.11, the section lining is composed of cast iron. This particular section lining is commonly used for other materials in the section unless the draftsperson wants to indicate the specific material in the section. For example, Figure 6.12 shows symbols for other specific materials.

CENTERLINES

Centerlines are thin (light), broken lines of long and short dashes, spaced alternately, used to designate the centers of a whole circle or a part of a circle, of holes, arcs, and symmetrical objects (see Figure 6.13). The symbol L is often used with a centerline. On some drawings, only one side of a part is drawn and the letters *SYM* are added to indicate the other side is identical in dimension and shape. Centerlines are also used to indicate paths of motion.

DIMENSION AND EXTENSION LINES

Dimension lines are thin, dark, solid lines broken at the dimension and terminated by arrowheads, which indicate the direction and extent of a dimension (see Figure 6.14). Fractional, decimal, and metric dimensions are used on drawings to give size dimensions. On machine drawings, the dimension line is broken, usually near the middle, to provide an open space for the dimension figure.

Important Point: The tips of arrowheads used on dimension lines indicate the exact distance referred to by a dimension placed at a break in the line. The tip of the

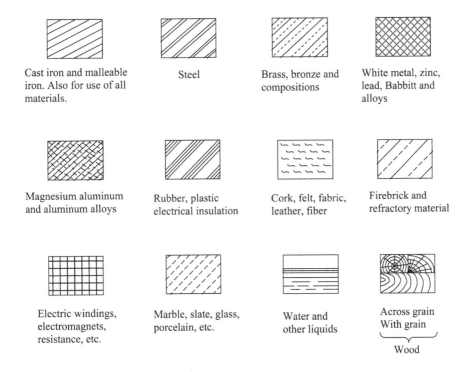

FIGURE 6.12   Symbols for materials in section.

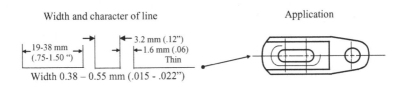

FIGURE 6.13   Centerline. Source: Adaptation from Brown, W.C., Blueprint Reading for Industry. South Holland, Illinois: The Goodheart-Wilcox Co., pp. 83, 1989.

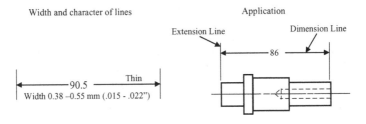

FIGURE 6.14   Dimension line, extension line, and leaders. Source: Adaptation from Brown, W.C., Blueprint Reading for Industry. South Holland, Illinois: The Goodheart-Wilcox Co., p. 84, 1989.

arrowhead touches the extension line. The size of the arrow is determined by the thickness of the dimension line and the size of the drawing.

Extension lines are thin, dark, solid lines that "extend" from a point on the drawing to which a dimension refers. Simply, extension lines are used in dimensioning to show the size of an object (see Figure 6.14).

Important Point: A space of one-sixteenth inch is usually allowed between the object and the beginning of the extension line.

Leaders

Leaders are thin inclined solid lines leading from a note or a dimension (see Figure 6.14) and terminated by an arrowhead or a dot touching the part to which attention is directed.

CUTTING PLANE OR VIEWING PLANE LINES

To obtain a sectional view, an imaginary cutting plane is passed through the object as shown in Figure 6.15. This *cutting plane line* or *viewing plane line* is either a thick (heavy) line with one long and two short dashes or a series of thick (heavy), equally spaced long dashes (see Figure 6.15).

BREAK LINES

To break out sections for clarity (e.g., from behind a hidden surface) or shorten parts of objects that are constant in detail and would be too long to place on a blueprint, *break lines* are used. Typically, three types of break lines are used. When the part to be broken requires a short line, the thick, wavy *short-break line* is used (see Figure 6.16). If the part to be broken is longer, the thin *long-break line* is used (see Figure 6.17). In round stock such as shafts or pipes, the thick "*S*" *break* is used (see Figure 6.18).

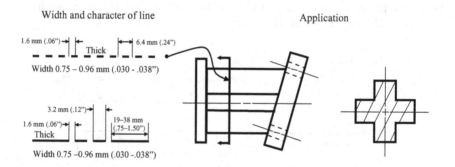

FIGURE 6.15 Cutting plane or viewing plane lines. Source: Adaptation from Brown, W.C., Blueprint Reading for Industry. South Holland, Illinois: The Goodheart-Wilcox Co., pp. 10–11, 85, 1989.

Width and character of line   Application

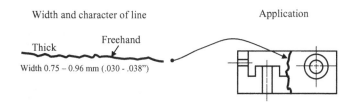

FIGURE 6.16 Short-break line. Source: Adaptation from Brown, W.C., Blueprint Reading for Industry. South Holland, Illinois: The Goodheart-Wilcox Co., pp. 14–16, 86, 1989.

Width and character of line   Application

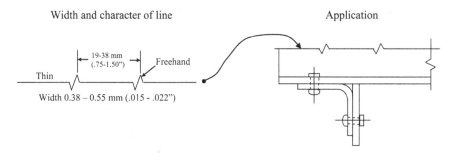

FIGURE 6.17 Long-break line. Source: Adaptation from Brown, W.C., Blueprint Reading for Industry. South Holland, Illinois: The Goodheart-Wilcox Co., p. 80, 1989.

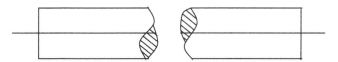

FIGURE 6.18 Cylindrical S-break.

PHANTOM LINES

Limited almost entirely to detail drawings, *phantom lines* are thin lines composed of long dashes alternating with pairs of short dashes. They are used primarily to indicate: (1) alternate positions of moving parts such as those shown in Figure 6.19 (right end); (2) adjacent positions of related parts such as an existing column (see Figure 6.20); and (3) in representing objects having a series of identical features (repeated detail) as in screwed shafts and long springs (see Figure 6.21A and 6.21B).

Line Gage

The *line gage*, used by draftspersons and shown in Figure 6.22, is convenient when referring to lines of various widths.

VIEWS

In the context of reading blueprints, when we speak of various views, they speak for themselves.

Width and Character   Application

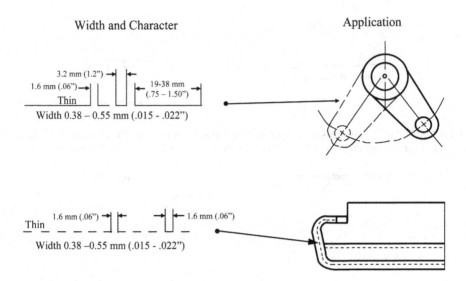

FIGURE 6.19 Phantom line. Source: Adaptation from Brown, W.C., Blueprint Reading for Industry. South Holland, Illinois: The Goodheart-Wilcox Co., p. 79, 1989.

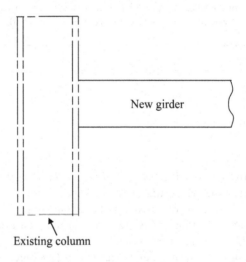

New girder

Existing column

FIGURE 6.20 Phantom line.

ORTHOGRAPHIC PROJECTIONS

When a draftsperson sets pencil to paper to draw a particular object such as a machine part, a basic problem is faced because such objects are three-dimensional. That is, they have height, width, and depth. No matter the skill of the draftsperson, he/she can only show the object drawn in two dimensions on a flat two-dimensional (2-D) sheet of paper—height and width.

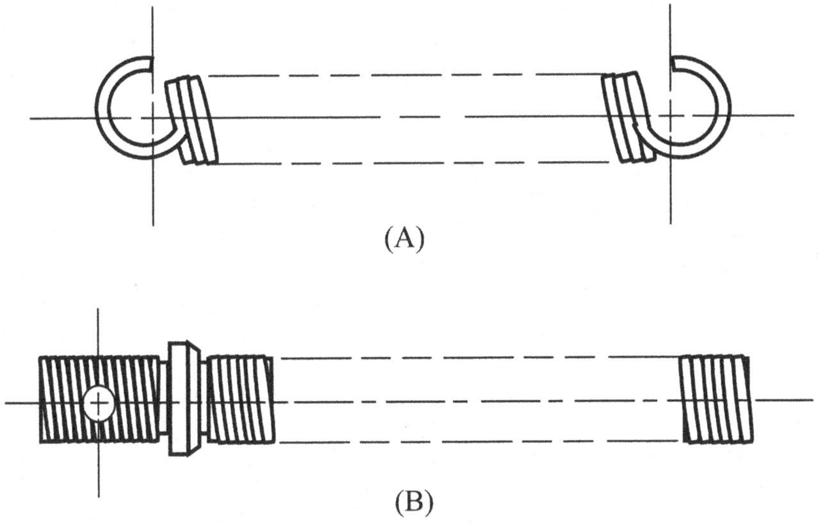

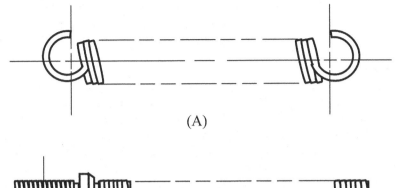

(A)

(B)

FIGURE 6.21   Phantom lines: (A) Spring; (B) Screw shaft.

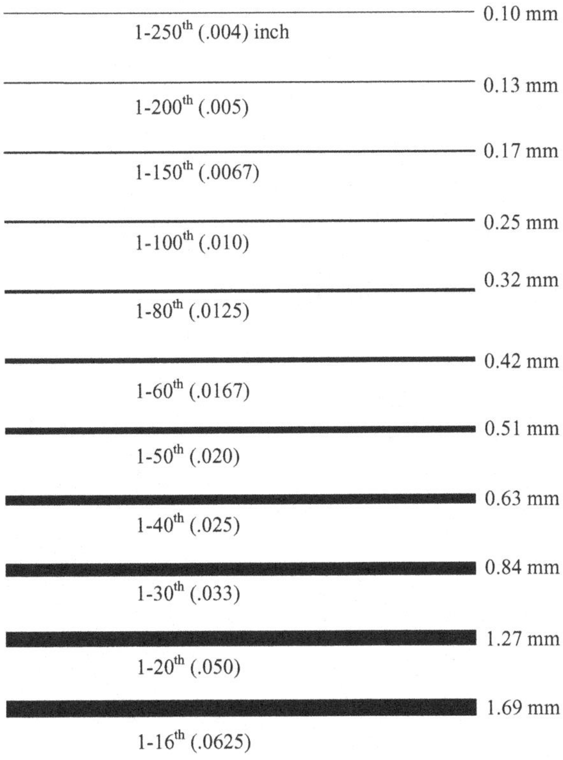

1-250th (.004) inch	0.10 mm
1-200th (.005)	0.13 mm
1-150th (.0067)	0.17 mm
1-100th (.010)	0.25 mm
1-80th (.0125)	0.32 mm
1-60th (.0167)	0.42 mm
1-50th (.020)	0.51 mm
1-40th (.025)	0.63 mm
1-30th (.033)	0.84 mm
1-20th (.050)	1.27 mm
1-16th (.0625)	1.69 mm

FIGURE 6.22   Line gauge.

Again, the draftsperson typically works with three-dimensional (3-D) objects. How does a draftsperson represent a 3-D object on a 2-D sheet of paper?

One way to do this is with a *three-dimensional pictorial*. A 3-D pictorial is a drawing that displays three sides of an object. A pictorial view of a 3-D object is shown in Figure 6.23.

In a pictorial view, the object is seen in such a way that three of the six sides of the object are visible. In this case, the *top, front*, and *right* sides of the object are visible. The other sides (*bottom, rear*, and *left*) are not visible in this view.

As mentioned, in addition to addressing the "sides" of a 3-D object, we also refer to the three "dimensions" of an object: *length, width*, and *height*. The dimensions of an object are shown in Figure 6.24.

To get around the basic problem of drawing 3-D objects in such a manner as to make them usable in industry, *orthographic views* are used. Simply, it is often useful to choose the position from which an object is seen, or *viewpoint*, so that only one side and two dimensions of the object are visible. This is called an orthographic view of an object.

A *top* (or *plan*) view shows the top side of the object, with the object's length and width displayed. A *front* view (or *front elevation*) shows the front side of the object,

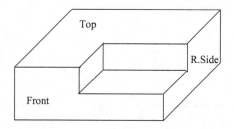

FIGURE 6.23 Pictorial view of a 3-D object.

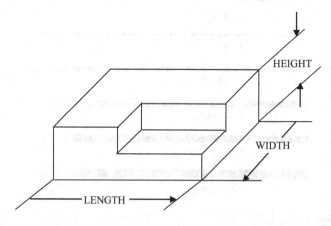

FIGURE 6.24 Dimension of an object.

with the object's length and height displayed. A *right-side* view (or *right elevation*) shows the right side of the object, with the object's width and height displayed.

Important Point: The object is usually drawn so that its most important feature appears in the front view.

To create an orthographic view, imagine that the object shown in Figure 6.25 is inside a glass box. The edges of the object are **projected** onto the glass sides.

Next, imagine that the sides of the glass box are hinged so that when opened the views are as shown in Figure 6.26A, 6.26B, and 6.26C. These are the orthographic projections of the 3-D object.

Important Point: Notice that the views shown in Figure 6.26A, B, and C are arranged so that the top and front projections are in vertical alignment, while the front and side views are in horizontal alignment—again, shown in Figure 6.27.

ONE-VIEW DRAWINGS

Frequently a single view supplemented by a note or lettered symbols is sufficient to clearly describe the shape of a relatively simple object (i.e., simple meaning parts that are uniform in shape). For example, cylindrical objects (shafts, bolts, screws, and similar parts) require only one view to describe them adequately.

According to ANSI standards, when a one-view drawing of a cylindrical part is used (see Figure 6.28), the dimension for the diameter must be preceded by the symbol ø. In many cases, the older, but widely used, practice for dimensioning diameters is to place the letters *DIA* after the dimension.

The main advantage of one-view drawings is the saving in drafting time; moreover, they simplify blueprint reading.

Important Point: In both applications, the symbol ø, or the letters **DIA**, and the use of a centerline indicate that the part is cylindrical.

The one-view drawing is also used extensively for flat parts. With the addition of notes to supplement the dimensions on the view, the one view furnishes all the necessary information for accurately describing the part (see Figure 6.29). (Note that in Figure 6.29, a note indicates the thickness as 3/8″).

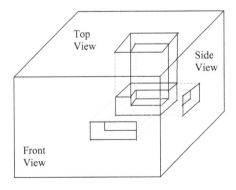

FIGURE 6.25   3-D object inside a glass box.

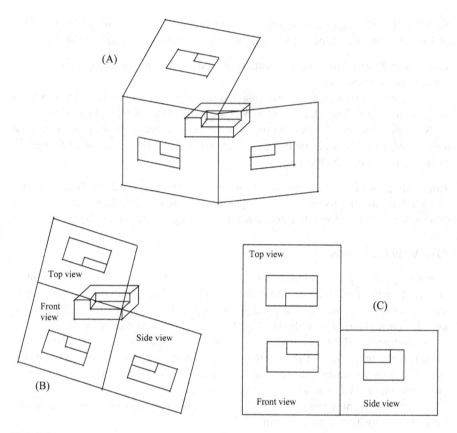

FIGURE 6.26 (A), (B), (C) Various orthographic projections of the 3-D object.

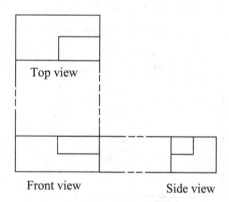

FIGURE 6.27 Proper alignment of orthographic projections.

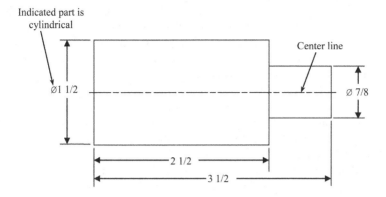

FIGURE 6.28 One-view drawing of a cylindrical shaft. Ø or DIA indicates the part is cylindrical. Centerline shows part is symmetrical.

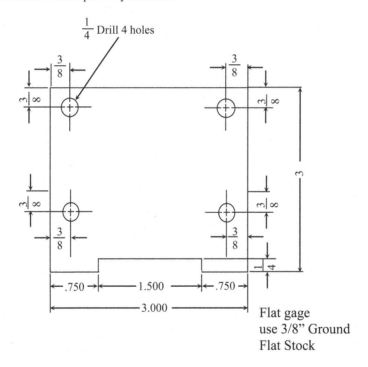

FIGURE 6.29 One-view drawing of a flat machine cover.

Two-View Drawings

Often only two views are needed to clearly describe the shape of simple, symmetrical flat objects and cylindrical parts. For example, sleeves, shafts, rods, and studs only require two views to give the full details of construction (see Figure 6.30A and 6.30B). The two views usually include the front view and a right-side or left-side view, or a top or bottom view.

Important Point: If an object requires only two views and the left-side and right-side views are equally descriptive, the right-side view is customarily chosen. Similarly, if an object requires only two views and the top and bottom views are equally descriptive, the top view is customarily chosen. Finally, if only two views are necessary and the top view and right-side view are equally descriptive, the combination chosen is the one that spaces best on the paper.

In the front view shown in Figure 6.30A and 6.30B, the centerlines run through the axis of the part as a horizontal centerline. If the rotor shaft is in a vertical position, the centerline runs through the axis as a vertical centerline.

The second view of the two-view drawing shown in Figure 6.30A and 6.30B contain a horizontal and a vertical centerline intersecting at the center of the circles which make up the part in the view.

Some of the two-view combinations commonly used in industrial blueprints are shown in Figure 6.31.

(A)   Horizontal Position

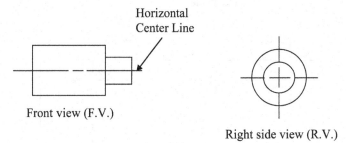

Front view (F.V.)

Right side view (R.V.)

(B)   Vertical Position

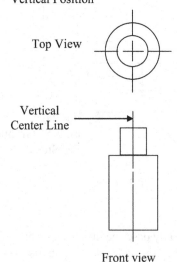

Front view

FIGURE 6.30   Examples of two-view drawings of a rotor shaft.

(A)   Horizontal Position

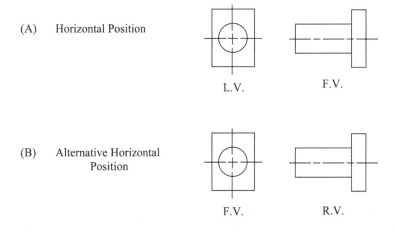

L.V.   F.V.

(B)   Alternative Horizontal
        Position

F.V.   R.V.

(C)   Alternate Vertical Position Views

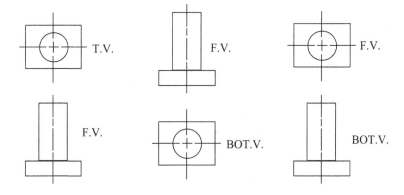

T.V.   F.V.   F.V.

F.V.   BOT.V.   BOT.V.

FIGURE 6.31   Several views for a two-view drawing of the same object.

In many two-view drawings, hidden edge or invisible edge lines, as shown in Figure 6.32, are common. A hidden detail may be straight, curved, or cylindrical.

THREE-VIEW DRAWINGS

Regularly shaped flat objects which require only simple machining operations are often adequately described with notes on a one-view drawing. Moreover, any two related views will show all three dimensions. But two views may not show enough detail to make the intentions of the designer completely clear. In addition, when the shape of the object changes, portions are cut away or relieved, or complex machine or fabrication processes must be represented on a drawing, the one view may not be sufficient to describe the part accurately. Therefore, a set of three related views has been established as the usual standard for technical drawings.

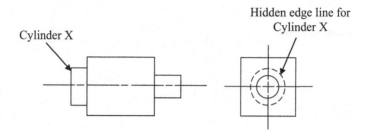

FIGURE 6.32   Two-view object with invisible edge lines.

Important Point: The number and selection of views are governed by the shape or complexity of the object. A view should not be drawn unless it makes a drawing easier to read or furnishes other information needed to describe the part clearly.

The combination of front, top, and right-side views represents the method most commonly used by draftspersons to describe simple objects (see Figure 6.33). The object is usually drawn so that its most important feature appears in the front view.

AUXILIARY VIEWS

The purpose of the technical drawing is to show the size and shape of each surface. As long as all the surfaces of an object are parallel or at right angles to one another, they may be represented in one or more views. However, on occasion, even three views are not enough. To overcome this problem, draftspersons sometimes find it necessary to use *auxiliary* views of an object to show the shape and size of the surfaces which can't be shown in the regular view (i.e., many objects are of such shape that their principal faces cannot always be assumed parallel to the regular planes of projection). For example, if an object has a surface that is not at a 90 degree angle from the other surfaces, the drawing will not show its true size and shape; thus, an auxiliary view is drawn to overcome this problem.

Figure 6.34(A) shows an object with an inclined surface (surface cut at an angle). Figure 6.34(B) shows the three standard views of the same object. However, because of the object's inclined surface, it is impossible to determine its true size and shape. To show its true size and shape, the draftsperson draws an auxiliary view of the inclined surface (see Figure 6.35). In Figure 6.35, the auxiliary view shows the inclined surface from a position perpendicular to the surface.

Important Point: Auxiliary views may be projected from any view in which the inclined surface appears as a line.

DIMENSIONS AND SHOP NOTES

Early on, dimensioning (or measuring using basic units of measurement) was rather simple and straightforward. For example, in the time of Noah and the Ark, a *cubit* was the length of a man's forearm, or about 18″. In pre-industrialized England, an inch used to be "three barleycorns, round and dry." More recently, we have all heard

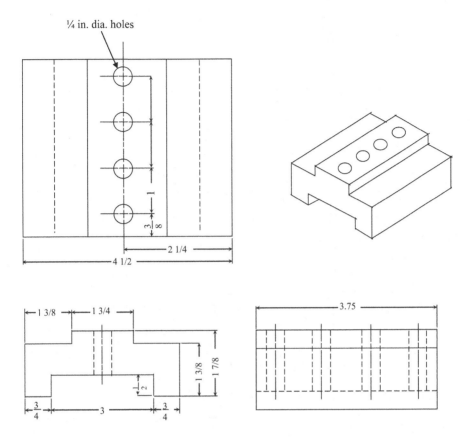

FIGURE 6.33 Three-view drawing.

of "rule of thumb." Actually, at one time, an inch was defined as the width of a thumb, and a foot was simply the length of a man's foot.

Though it is still somewhat common to hear some of the terms and sayings stated above, today, dimensions are stated somewhat differently. One major difference is in the adoption of the standardized dimensioning units we currently use. This use came about because of the relatively recent rapid growth of worldwide science, technology, and commerce—all of which has fostered an international system of units (SI) we use today.

DIMENSIONING

As mentioned, technical drawings consist of several types of lines that are used singly or in combination with each other to describe the shape and internal construction of an object or mechanism. However, to rebuild a machine or re-machine or reproduce a part, the blueprint or drawing must include dimensions which indicate exact sizes and locations of surfaces, indentations, holes, and other details. Stated

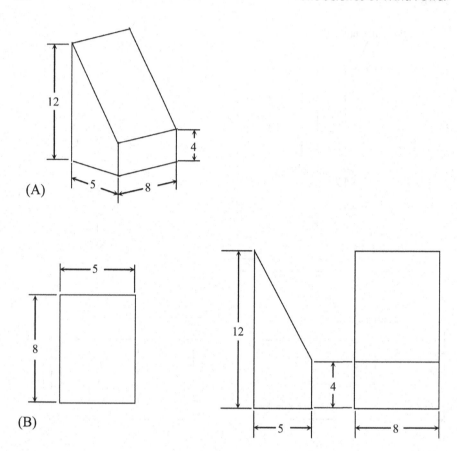

FIGURE 6.34 (A) Object with an inclined surface; (B) Three views of an object with the inclined surface.

differently, in addition to a complete *shape description* of an object, a technical drawing of the object must also give a complete size description; that is, it must be *dimensioned*.

In the early days of industrial manufacturing, products were typically produced under one roof, often by one individual, using parts and subassemblies manufactured on the premises. Today, most major industries do not manufacture all of the parts and subassemblies in their products. Frequently, the parts are manufactured by specialty industries to standard specifications or the specifications provided by the major industry.

Important Point: The key to the successful operation of the various parts and subassemblies in the major product is the ability of two or more nearly identical duplicate parts to be used individually in an assembly and function satisfactorily.

The modern practice of *interchanging* parts (i.e., their interchangeability) is the basis for the development of widely accepted methods for size description. Drawings

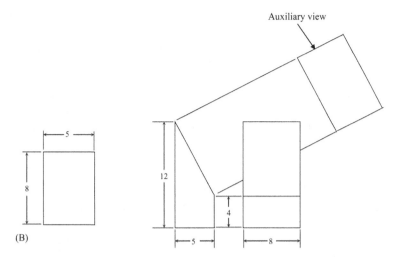

FIGURE 6.35 Object with the inclined view shown in Figure 3.32 with auxiliary view added to give size and shape of the inclined surface.

today are dimensioned so that machinists in widely separated places can make mating parts that will fit properly when brought together for final assembly in the factory or when replacement parts are used to make repairs to plant equipment.

Technical drawings show the object in its completed condition and contain all necessary information to bring it to its final state. A properly dimensioned drawing considers the shop processes required to finish a piece and the function of the part in the assembly. Moreover, shop drawings are dimensioned for convenience for the shopworker or wind turbine maintenance operator or repair technician. These dimensions are given so that it is not necessary to calculate, scale, or assume any dimensions.

Designers and draftspersons provide dimensions that are neither duplicitous nor superfluous. Only those dimensions are given that are needed to produce and inspect the part exactly as intended by the designer. More importantly, only those dimensions are given that are needed by the maintenance operator, who may have to rely on the blueprint (usually as a last resort) to determine the exact dimensions of the replacement part.

The meaning of various terms and symbols and conventions used in shop notes, as well as procedures and techniques relating to dimensioning, are presented here to assist the wind turbine technician in accurately interpreting plant blueprints. However, before defining these important terms we first discuss decimal and size dimensions.

DECIMAL AND SIZE DIMENSIONS

Dimensions may appear on blueprints as decimals, usually two-place decimals (normally given in even hundredths of an inch). In fact, it is common practice to use two-place decimals when the range of dimensional accuracy of a part is between 0.01″ larger or smaller than nominal size (specified dimension).

For more precise dimensions (i.e., dimensions requiring machining accuracies in thousandths or ten-thousandths of an inch), three- and four-place decimal dimensions are used.

Every solid object has three size dimensions: depth (or thickness), length (or width), and height. In the case of the object shown in Figure 6.36, two of the dimensions are placed on the principal view and the third dimension is placed on one of the other views.

DEFINITION OF DIMENSIONING TERMS

An old Chinese proverb states: "The beginning of wisdom is to call things by their right names." This statement is quite fitting because to satisfactorily read and interpret blueprints, it is necessary to understand the terms relating to conditions and applications of dimensioning.

1. Nominal Size

Nominal size is the designation that is used for the purpose of general identification. It may or may not express the true numerical size of the part or material. For example, the standard 2×4 stud used in building construction has an actual size of 1-1/2×3-1/2 inches (see Figure 6.37). However, in the case of the hole and shaft shown in Figure 6.38, the nominal size of both hole and shaft is 1-1/4″, which would be 1.25″ in a decimal system of dimensioning. So, again, it may be seen that the nominal size may or may not be the true numerical size of a material.

Important Point: When the term "nominal size" is used synonymously with basic size, we are to assume the exact or theoretical size from which all limiting variations are made.

2. Basic Size

Basic size (or *dimension*) is the size of a part determined by engineering and design requirements. More specifically, it is the theoretical size from which limits of size

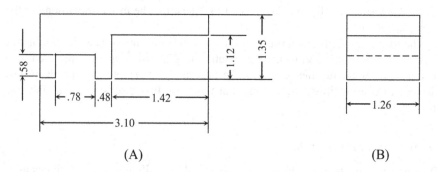

(A)   (B)

FIGURE 6.36   Size dimensions: (A) View with two-size dimensions; (B) third-side dimension on this view.

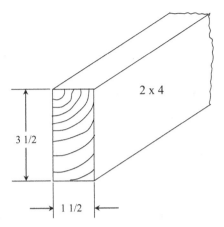

FIGURE 6.37   Nominal size of a construction 2×4.

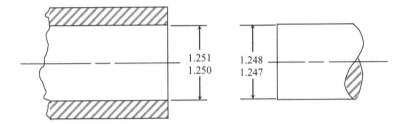

FIGURE 6.38   Nominal size—1.25″.

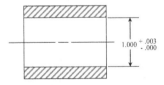

FIGURE 6.39   Basic size.

are derived by the application of allowances and tolerances. That is, it is the size from which limits are determined for the size, shape, or location of a feature. For example, strength and stiffness may require a 1-inch diameter shaft. The basic 1-inch size (with tolerance) will most likely be applied to the hole size since allowance is usually applied to the shaft (see Figure 6.39).

3.   Allowance

Allowance is the designed difference in the dimensions of mating parts to provide for different classes of fit. Simply, it is the minimum clearance space (or maximum interference) of mating parts. Consequently, it represents the tightest permissible fit and is simply the smallest hole minus the largest shaft. For example, recall that in Figure 6.39 we allowed .003 on the shaft for clearance (1.000 − .003 = .997; see Figure 6.39).

4. Design Size

Design size is the size of a part after an allowance for clearance has been applied and tolerances have been assigned. The design size of the shaft (see Figure 6.40) is .997 after the allowance of .003 has been made. A tolerance of ±.003 is assigned after the allowance is applied (see Figure 6.41).

Important Point: After defining basic and design size, the reader may be curious as to what the definition of "actual size" is. *Actual size* is simply the measured size.

5. Limits

Limits are the maximum and minimum sizes indicated by a tolerance dimension. For example, the design size of a part may be 1.435. If tolerance of plus or minus two thousandths (±.002) is applied, then the two limit dimensions are that the maximum limit is 1.437 and the minimum limit is 1.433 (see Figure 6.42).

6. Tolerance

Tolerance is the total amount by which a given dimension may vary or the difference (variation) between limits (as shown in Figure 6.42). Tolerance should always be as large as possible, other factors considered, to reduce manufacturing costs. It can also be expressed as the design size followed by the tolerance (see Figure 6.43). Moreover, tolerance can be expressed when only one tolerance value is given; the other value is assumed to be zero (see Figure 6.44). Tolerance is also applied to *location dimensions* for other features (holes, slots, surfaces, etc.) of a part (see Figure 6.45).

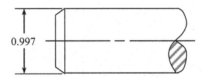

0.997

FIGURE 6.40   Design size (after application of allowance).

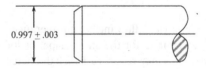

0.997 ± .003

FIGURE 6.41   Design size (after allowance and tolerance are applied).

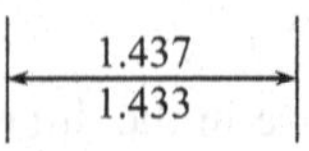

1.437
1.433

FIGURE 6.42   Tolerance expressed by limits.

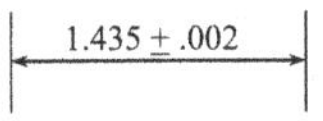

FIGURE 6.43   Design size with tolerance.

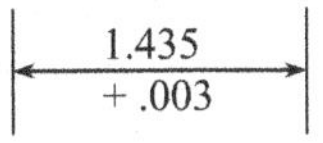

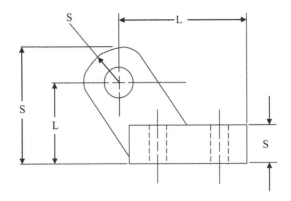

FIGURE 6.44   One tolerance value given.

FIGURE 6.45   Size and location dimensions.

(**Note:** Location dimensions are usually made from either a centerline or a finished surface—this practice is followed to overcome inaccuracies due to variations caused by surface irregularities).

Because the size of the shaft shown in Figure 6.40 is 0.997 after the allowance has been applied, the tolerance applied must be below this size in order to assure the minimum clearance (allowance) of .003. If tolerance of ±.003 is permitted on the shaft, the total variation of .006 (+.003 and −.003) must occur between .997 and .994. Then, the design of the shaft is given a *bilateral tolerance* (i.e., variation is permitted in both directions from the design size, as shown in Figure 6.43) vs. *unilateral tolerance* (variation is permitted only in one direction from the design size, as shown in Figure 6.44).

Important Point: Tolerances may be specific and given with the dimension value or, general and given by means of a printed note in or just above the title block.

7.   Datums

A *datum* (a point, axis, or plane) identifies the origin of a dimensional relationship between a particular (designated) point or surface and a measurement; that is, it is assumed to be exact for the purpose of reference and is the origin from which the location or geometric characteristic of features or a part are established. The datum is indicated by the assigned letter preceded and followed by a dash, enclosed in a small rectangle or box (see Figure 6.46).

Types of Dimensions

The types of dimensions include linear, angular, reference, tabular, and arrowless. Each of these is discussed in the following sections.

Linear Dimensions

Linear dimensions are typically used in aerospace, automotive, machine tool, sheet metal, electrical and electronic, and similar industries. Linear dimensions are usually given in inches for measurements of 72″ and under, and in feet and inches if greater than 72″.

Important Point: In the construction and structural industries, linear dimensions are given in feet, inches, and common fractional parts of an inch.

Angular Dimensions

Angular dimensions are used on blueprints to indicate the size of angles in degrees (°) and fractional parts of a degree, minutes (′) and seconds (″). Each degree is one three hundred sixtieth of a circle (1/360). There are 60 minutes (′) in each degree. As mentioned, each minute may be divided into smaller units called seconds. There are 60 seconds (″) in each minute. To simplify the dimensioning of angles, these symbols are used. For example, fifteen degrees, twelve minutes and forty-five seconds can be written 15° 12′ 45″.

The current practice is to use decimalized angles. To convert angles given in whole degrees, minutes, and seconds, the following example should be followed.

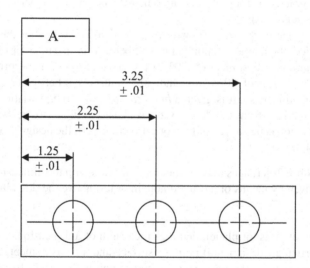

FIGURE 6.46 Dimensioning datum surface.

Example 6.1

PROBLEM:

Convert 15° 12′ 45″ into decimal degrees.

SOLUTION:

Step 1: Convert minutes into degrees by dividing by 60 (60′ = 1°)

$$12 \div 60 = .20°$$

Step 2: Convert seconds into degrees by dividing seconds by 3,600

$$3600° = 1$$

$$45″ \div 3600 = .01°$$

Step 3: Add whole degrees plus decimal degrees

$$15° + .20° + .01° = 15.21°$$

$$\text{therefore } 15°12′.45″ = 15.21° \text{ decimal degrees}$$

The size of an angle with the tolerance may be shown on the angular dimension itself. The tolerance may also be given in a note on the drawing, as shown in Figure 6.47.

Reference Dimensions

Reference dimensions are occasionally given on drawings for reference and checking purposes; they are given for information only. They are not intended to be measured and do not govern the shop operations. They represent the calculated dimensions and are often useful in showing the intended design size. Reference dimensions are marked by parentheses or followed by REF (see Figure 6.48).

Tabular Dimensions

Tabular dimensions are used when a series of objects having the same features but varying in dimensions may be represented by one drawing. Letters are substituted

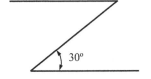

Upper limit: 30° + 30' = 30°30'
Lower limit: 30° - 30' = 29°30'

30°

| UNLESS OTHERWISE SPECIFIED: |
| TOLERANCE ON ANGULAR DIMENSIONS |
| ARE ± .5 |

FIGURE 6.47   Tolerance specified as a note.

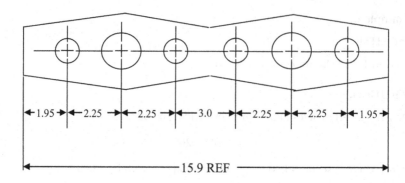

FIGURE 6.48   Dimensions for reference only.

PART NO.	A	B	C	D	E

FIGURE 6.49   Example table used for dimensioning a series of sizes.

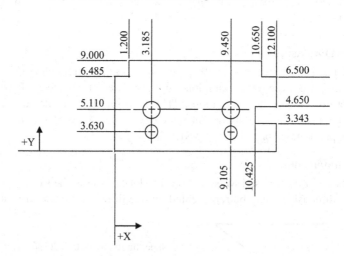

FIGURE 6.50   Application of arrowless dimensioning.

for dimension figures on the drawing, and the varying dimensions are given in a tabular form (see Figure 6.49).

Arrowless Dimensions

Arrowless dimensions are frequently used on drawings that contain datum lines or planes (see Figure 6.50). This practice improves the clarity of the drawing by eliminating numerous dimensions and extension lines.

Shop Notes

To convey the information the machinist needs to make a part, the draftsperson typically uses notes. Notes such as those used for reaming, counterboring, drilling, or countersinking holes are added to ordinary dimensions. The order of items in a note corresponds to the order of procedure in the shop in producing the hole. Two or more holes are dimensioned by a single note, the leader pointing to one of the holes.

Important Point: Notes should always be lettered horizontally on the drawing paper, and guidelines should always be used.

A note may consist of a very brief statement at the end of a leader, or it may be a complete sentence that gives an adequate picture of machining processes and all necessary dimensions. On drawings of parts to be produced in large quantity for interchangeable assembly, dimensions and notes may be given without specification of the shop process to be used. A note is placed on a drawing near the part to which it refers.

MACHINE DRAWINGS

Your ability to work with machines like wind turbines depends on your ability to comprehend them. If wind turbine maintenance operators know how the parts of a machine fit together and how they are intended to work together, they are better able to operate the machine, perform proper preventive maintenance on it, and repair it when it breaks down.

At larger wind farms, no one can possibly know all the details of every machine and/or machine tool. There are too many variations among machines of the same type.

For example, the *submersible pump* is another familiar machine to most maintenance operators; it is used extensively in many industries and may be used at a wind turbine farm on occasion to pump out sumps and on occasion the bottom level of the turbine tower. The submersible pump is, as the name suggests, placed directly in deep wells and pumping station wet wells. In some cases, only the pump is submerged, while in other cases, the entire pump-motor assembly is placed in the well or wet well. A simplified drawing of a typical submersible pump is shown in Figure 6.51; each numbered component is identified as follows:

1. Electrical connection
2. Drop pipe
3. Inlet screen
4. Electric motor
5. Check valve
6. Inlet screen
7. Bowls and impellers

HYDRAULIC AND PNEUMATIC DRAWINGS

Hydraulic and pneumatic power systems are widely used in wind turbine operations and discussed in detail in Chapter 7; however, for now, it is important to know that they operate small tools and large machines. As a wind turbine maintenance

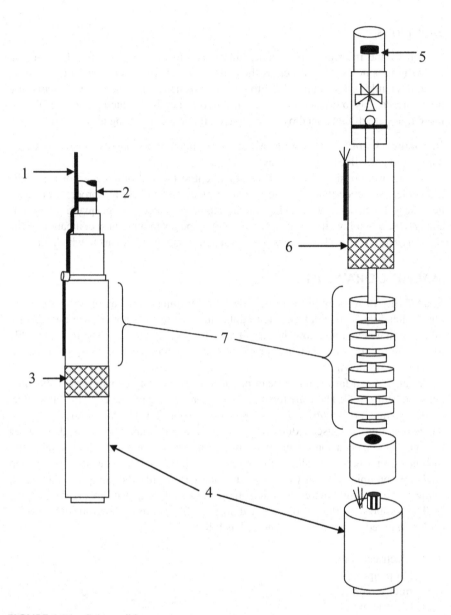

FIGURE 6.51 Submersible pump.

operator, you must know how these fluid power systems work in order to repair and maintain them. However, before you can understand and service fluid power systems, you must also know how to read hydraulic and pneumatic drawings.

Although it is not the author's intention in this section to make fluid mechanics out of anyone (i.e., to discuss such fluid principles as Pascal's Law and multiplying forces), it is my intention to explain the basics of hydraulic and pneumatic systems.

Many wind turbine machines are operated by *hydraulic* (uses noncompressible fluids) and *pneumatic* (is a compressible fluid) power systems. These systems transmit forces through a *fluid* (defined as either a gas or a liquid)—also called *fluid power*. Air-powered drills and grinders are tools that operate on *compressed air*. Larger pneumatic devices are used on larger machines. Because all these mechanisms must be maintained and repaired, we need to know how they operate. Moreover, we must also be able to read and interpret the drawings that show the construction of these systems.

STANDARD HYDRAULIC SYSTEM

A standard hydraulic system operates by means of a *liquid* (hydraulic fluid) under pressure. A basic hydraulic system has five components.

- *Reservoir*—provides storage space for the liquid.
- *Pump*—provides pressure to the system.
- *Piping*—directs fluid through the system.
- *Control valve*—controls the flow of fluid.
- *Actuating unit*—device that reacts to the pressure and does some kind of useful work.

Because the hydraulic fluid never leaves the system, it is a *closed system*. The hydraulic system on a forklift is an example of a hydraulic system.

STANDARD PNEUMATIC SYSTEM

A standard pneumatic system operates by means of a *gas* under pressure; it is not used in the wind turbine itself but supplies power to many of the tools used in maintenance activities. The gas is usually dry air. A basic pneumatic system is very much like the hydraulic system described above and includes the following main components.

- *Atmosphere*—serves the same function as the reservoir of the hydraulic system.
- *Intake pipe and filter*—provides a passage for air to enter the system
- *Compressor*—compresses the air, putting it under pressure. Its counterpart in the hydraulic system is the pump.
- *Receiver*—stores pressurized air until it is needed. It helps provide a constant flow of pressurized air in situations where air demand is high or varies.
- *Relief valve*—is set to open and "bleed off" some of the air if the pressure becomes too high.
- *Pressure-regulating valve*—assures that the air delivered to the actuating unit is at the proper pressure.
- *Control valve*—provides a path for the air to the actuating unit.

Pneumatic systems are usually *open* systems (i.e., the air leaves the system after it is used).

HYDRAULIC/PNEUMATIC SYSTEMS: SIMILARITIES AND DIFFERENCES

In regard to their similarities, both hydraulic and pneumatic systems use a pressure-building source. This can be either a pump or a compressor. They also need either a reservoir or a receiver to store the fluid. In addition, they need valves and actuators, and lines to connect these components in a system. The motion that results from the actuator may be either straight line (linear) or circular (rotary).

In regard to their differences, there is an important difference between a liquid (for hydraulic systems) and a gas (for pneumatic systems). A liquid is difficult to compress. Water, for example, cannot be compressed into a space that is noticeably smaller in size. On the other hand, gas is easy to compress. For example, a large volume of air from the atmosphere can be compressed into a much smaller volume.

TYPE OF HYDRAULIC AND PNEUMATIC DRAWINGS

Several types of drawings are used in showing instrumentation and control circuits and/or hydraulic/pneumatic systems. These include graphic, pictorial, cutaway, and combination drawings.

> *Graphic drawings* consist of graphic symbols joined by lines that provide an easy method of emphasizing the functions of the system and its components (see Figure 6.52).
> *Pictorial drawings* are used when piping is to be shown between components (see Figure 6.53).
> *Cutaway drawings* consist of cutaway symbols of components and emphasize component function and piping between components (see Figure 6.54).
> *Combination drawings* utilize (in one drawing), the type of component illustration that best suits the purpose of the drawing (see Figure 6.55).

Note: The emphasis in this section will be graphic diagrams since these are the most widely used in wind turbine operations, and since the graphic symbols have been standardized.

GRAPHIC SYMBOLS FOR FLUID POWER SYSTEMS

When we reviewed Figures 6.52 through 6.55, you may or may not have had difficulty reading any of these four types of drawings. On the other hand, unless you were familiar with the basic symbols used in the drawings, you probably did have some difficulty understanding them.

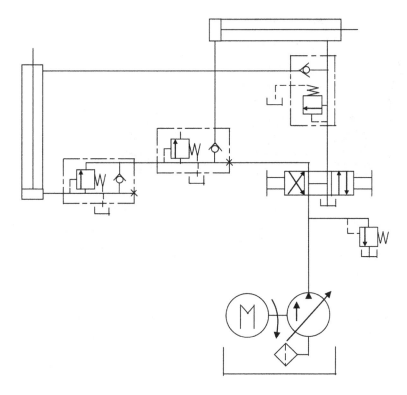

FIGURE 6.52   Graphic drawing for a fluid power system.

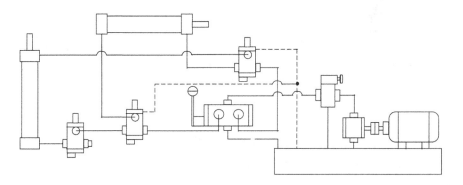

FIGURE 6.53   Pictorial drawing for a fluid power system.

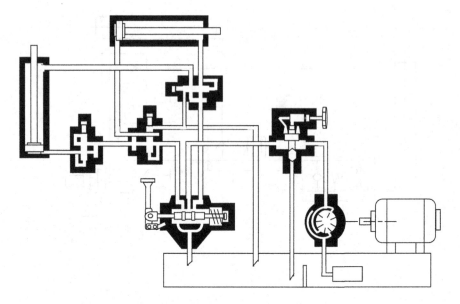

FIGURE 6.54   Cutaway drawing of a fluid power system.

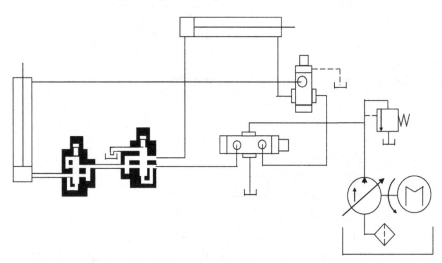

FIGURE 6.55   Combination drawing of a fluid power system.

SYMBOLS FOR METHODS OF OPERATION (CONTROLS)

Figure 6.56 shows the standard graphic symbols used in hydraulic and pneumatic system diagrams for methods of operation (controls).

SYMBOLS FOR ROTARY DEVICES

Figure 6.57 shows the standard graphic symbols used in hydraulic and pneumatic system diagrams for rotary devices, such as pumps, motors, oscillators, and internal combustion engines.

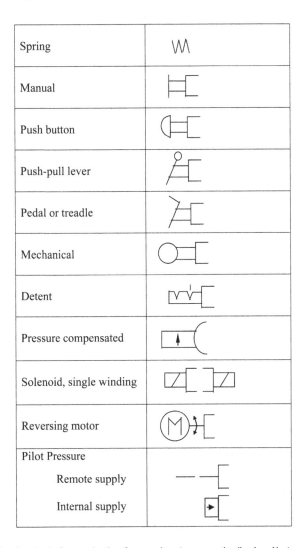

Spring	
Manual	
Push button	
Push-pull lever	
Pedal or treadle	
Mechanical	
Detent	
Pressure compensated	
Solenoid, single winding	
Reversing motor	

Pilot Pressure
 	
    Remote supply
 	
    Internal supply	

FIGURE 6.56   Symbols for methods of operation (pneumatics/hydraulics).

SYMBOLS FOR LINES

Figure 6.58 shows the standard graphic symbols used in hydraulic and pneumatic system diagrams for system lines.

SYMBOLS FOR VALVES

Figure 6.59 shows the standard graphic symbols used in hydraulic and pneumatic system diagrams for valves.

SYMBOLS FOR MISCELLANEOUS UNITS

Figure 6.60 shows the standard graphic symbols used in hydraulic and pneumatic system diagrams for miscellaneous units, such as energy storage and fluid storage devices.

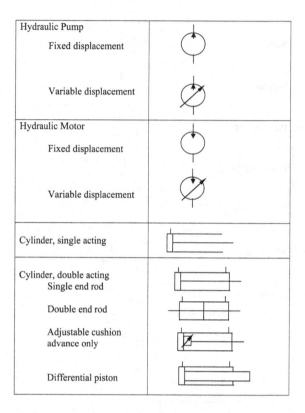

Hydraulic Pump	
Fixed displacement	
Variable displacement	
Hydraulic Motor	
Fixed displacement	
Variable displacement	
Cylinder, single acting	
Cylinder, double acting Single end rod	
Double end rod	
Adjustable cushion advance only	
Differential piston	

FIGURE 6.57 Symbols for rotary devices (pneumatics/hydraulics).

SUPPLEMENTARY INFORMATION ACCOMPANYING GRAPHIC DRAWINGS

Once we have become familiar with the symbols used in graphic drawings, the drawings are relatively easy to read and understand. In addition to the graphic drawing, prints of hydraulic and pneumatic systems usually include a listing of the sequence of operations, solenoid chart, and parts used to facilitate understanding of the function and purpose of the system and its components.

SEQUENCE OF OPERATIONS

A listing of the sequence of operations is an explanation of the various functions of the system explained in order of occurrence. Each phase of the operation is numbered or lettered, and a brief description is given of the initiating and resulting action.

Important Note: A listing of the sequence of operations is usually given in the upper part of the print or on an attached sheet.

Line, working (main)	 ———————	Station, testing, measurement or power take-off	————✗
Line, pilot (for control)	 – – – – – – –	Variable component (run arrow through symbol at 45°)	
Line, liquid drain	 – – – – – –		

Flow, direction of
    Hydraulic
    Pneumatic
	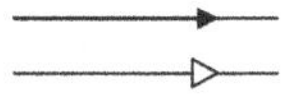		

Lines crossing	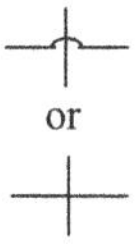 or	Pressure compensated units (arrow parallel to short side of symbol)	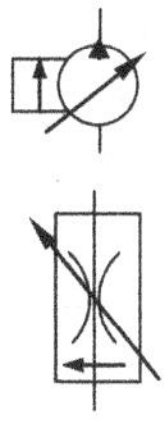
Lines joining		Temperature cause or effect	

Line with fixed restrictions		Reservoir
		    Vented
		
		    Pressurized
			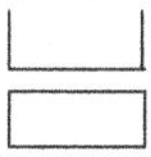

Line, flexible		Line, to reservoir above fluid level	
Vented manifold		below fluid level	

FIGURE 6.58 Symbols for lines (pneumatics/hydraulics).

Solenoid Chart

If solenoids are used in the instrumentation and/or control circuits of a hydraulic or pneumatic system, a chart is normally located in the lower-left corner of the print to help explain the operation of the electrically controlled circuit.

Important Point: Solenoids are usually given a letter on the drawings, and the chart shows where the solenoids are energized (+) or de-energized (–) at each phase of system operation.

BILL OF MATERIALS

Bill of materials, sometimes called a Component or Parts List, includes an itemized list of the several parts of a structure or device shown on a graphic detail drawing or a graphic assembly drawing. This parts list usually appears right above the title block. However, this list is also often given on a separate sheet.

Important Point: The title strip alone is sufficient on graphic detail drawings of only one part, but a parts list is necessary on graphic detail drawings of several parts.

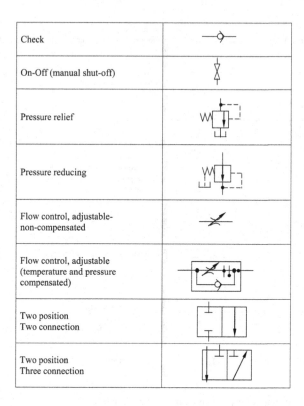

Check	
On-Off (manual shut-off)	
Pressure relief	
Pressure reducing	
Flow control, adjustable-non-compensated	
Flow control, adjustable (temperature and pressure compensated)	
Two position Two connection	
Two position Three connection	

FIGURE 6.59 Symbols for valves.

Parts lists on machine drawings contain the part numbers or symbols, a descriptive title of each part, the quantity required, the material specified, and frequently other information, such as pattern numbers, stock sizes of materials, and weights of parts.

Important Point: Parts are listed in general order of size or importance. For example, the main castings or forgings are listed first, parts cut from cold-rolled stock second, and standard parts such as bushings and roller bearings third.

ELECTRICAL DRAWINGS

Working drawings for the repair, troubleshooting, and installation of replacement components for electrical machinery, switching devices, chassis for electronic equipment, cabinets, housings, wind turbines, and other mechanical elements associated with electrical equipment are based on the same principles as given in the earlier sections.

To operate, maintain, and repair electrical equipment in a wind turbine and/or wind farm, the wind turbine maintenance operator/technician (qualified in electrical work) must understand electrical systems. The wind turbine electrical system

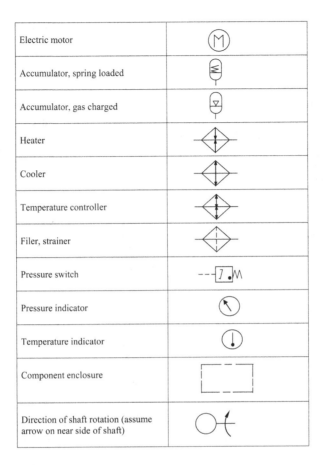

Electric motor	
Accumulator, spring loaded	
Accumulator, gas charged	
Heater	
Cooler	
Temperature controller	
Filer, strainer	
Pressure switch	
Pressure indicator	
Temperature indicator	
Component enclosure	
Direction of shaft rotation (assume arrow on near side of shaft)	

FIGURE 6.60   Miscellaneous symbols.

troubleshooter must be able to read electrical drawings and determine what is wrong when electrical equipment fails to run properly.

This section introduces electrical drawings and the functions of important electrical components; how they are shown on drawings is also explained.

TROUBLESHOOTING AND ELECTRICAL DRAWINGS

The wind turbine maintenance operator/technician—qualified to perform electrical work—is often assigned to repair or replace components of a wind turbine electrical system. To repair anything, the first thing that must be done is to find the problem. The technician may solve the problem by simply restoring electrical power (i.e., resetting a circuit breaker or replacing a fuse). At other times, however, a good deal of troubleshooting may be required. *Troubleshooting* is like detective work: Find the culprit (the problem or what happened) and remedy (fix) the situation. Troubleshooting is a skill, but even the best troubleshooter would have some difficulty in troubleshooting many complex electrical machines without the proper electrical blueprint or wiring diagram to lead the way—a roadmap of sorts.

Electrical Symbols

Figure 6.61 shows some of the most common symbols used on electrical drawings. It is not necessary to memorize these symbols, but the maintenance operator should be familiar with them (i.e., be able to recognize them) as an aid to reading electrical drawings.

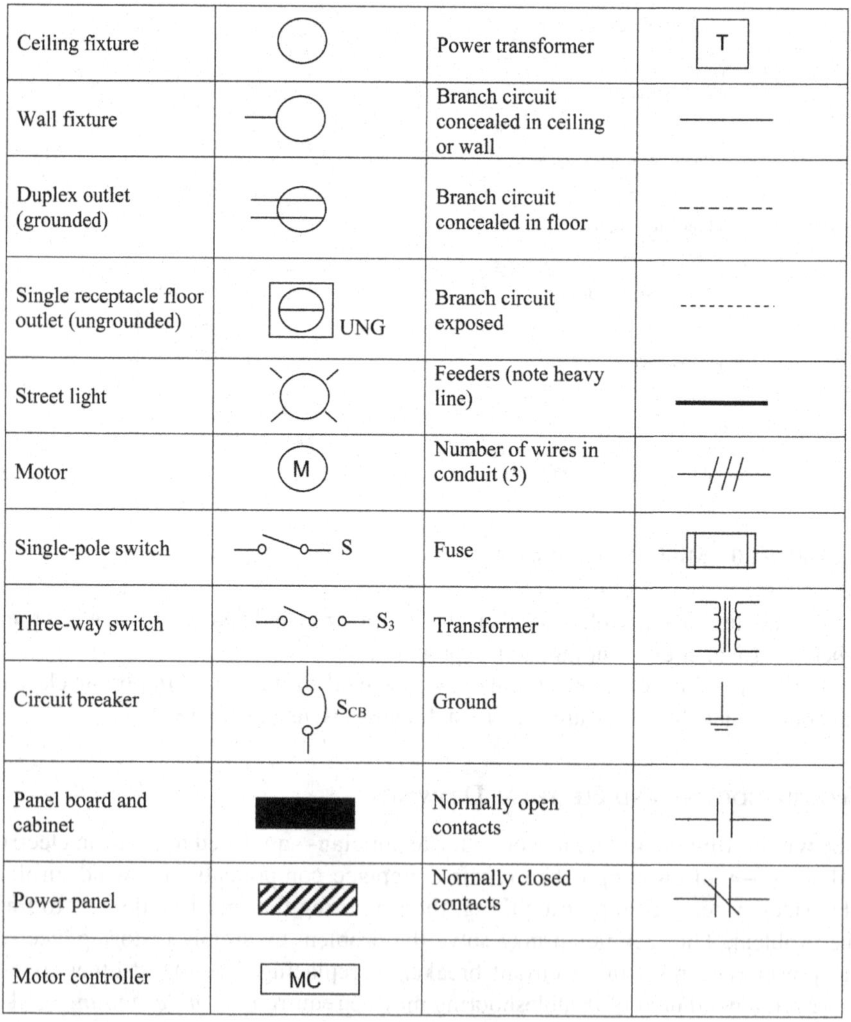

Ceiling fixture		Power transformer	T
Wall fixture		Branch circuit concealed in ceiling or wall	—————
Duplex outlet (grounded)		Branch circuit concealed in floor	– – – – –
Single receptacle floor outlet (ungrounded)	UNG	Branch circuit exposed	· · · · ·
Street light		Feeders (note heavy line)	▬▬▬
Motor	M	Number of wires in conduit (3)	
Single-pole switch	S	Fuse	
Three-way switch	S₃	Transformer	
Circuit breaker	S_CB	Ground	
Panel board and cabinet	▬▬▬	Normally open contacts	
Power panel	▨▨▨	Normally closed contacts	
Motor controller	MC		

FIGURE 6.61   Common electrical symbols.

Electrical Drawings

There are two kinds of electrical drawings used for troubleshooting in wind turbine/farm operations. They are called *architectural drawings* and *circuit drawings*. An architectural drawing shows the physical locations of the electric lines in a wind farm building or between buildings. A circuit drawing shows the electrical loads served by each circuit.

Important Point: A circuit drawing does not indicate the physical location of any load or circuit.

TYPES OF ARCHITECTURAL DRAWINGS

Figure 6.62 indicates that there are three types of architectural drawings. One type of architectural drawing is called a *plot plan*; it shows electric distribution to all the plant buildings.

Another kind of architectural drawing is called a *floor plan*. A floor plan shows where branch circuits are located in one building or pumping station. The floor plan shows where equipment is located and where outside-inside tie-ins to water, heat, and electric power are located.

The third type of architectural drawing shown in Figure 6.62 is called a *riser diagram*. The riser diagram shows how the wiring goes to each floor of the building.

Circuit Drawings

A *circuit drawing* shows how a single circuit distributes electricity to various loads (e.g., pumps, motors, etc.). Unlike an architectural drawing, a circuit drawing does not show the location of these loads.

Figure 6.63 depicts a typical single-line circuit drawing; it shows power distribution to 11 loads. The number in each circle indicates the power rating of the loads in horsepower. (**Note**: Electrical loads in all plants can be divided into two

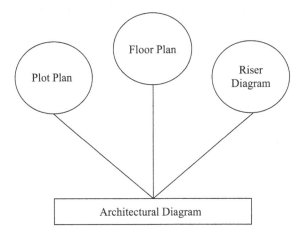

FIGURE 6.62 Types of architectural diagrams.

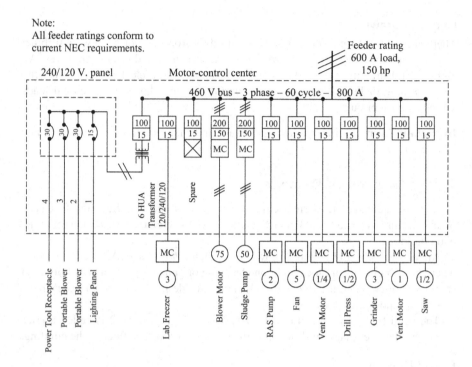

FIGURE 6.63 Single-line circuit diagram.

categories: critical and noncritical. Critical loads are those that are essential to the operation of the plant and cannot be turned off (e.g., critical unit processes). Noncritical loads include those pieces of equipment that would not disrupt the operation of the plant or pumping station or compromise safety if they were turned off for a short period of time (e.g., air conditioners, fan systems, electric water heaters, and certain lighting systems)). The numbers in the rectangles show the current ratings of circuit breakers. The upper number is the current in amps the circuit breaker will allow as a momentary surge. (**Note:** Circuit breakers are typically equipped with surge protection for three-phase motors and other devices. When a three-phase motor is started, current demand is six to ten times the normal value. After the start, the current flow decreases to its normal rated value. Surge protection is also provided to allow slight increases in current flow when the load varies or increases slightly). The lower number is the maximum current the circuit breaker will allow to flow continuously. (**Note:** Most circuit breakers are also equipped with an instantaneous trip valve for protection against short circuits).

Ladder Drawing

Another type of electrical drawing is called a *ladder drawing*. It is a type of schematic diagram that shows a control circuit. The parts of the control circuit lie on horizontal lines, like the rungs of a ladder. Figure 6.64 is an example of a ladder diagram.

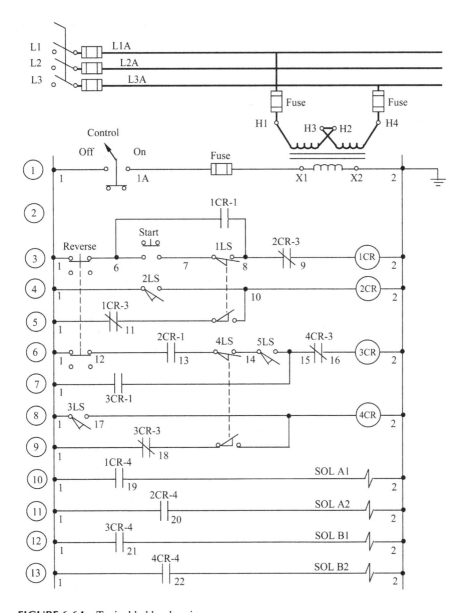

FIGURE 6.64 Typical ladder drawing.

Important Point: The size of electrical drawings is important. This can be understood first in the problem of storing drawings. If every size and shape were allowed, the task of systematic and protective filing of drawings could be tremendous. Page sizes of 8-1/2″ × 11″ or 9″ × 12″ and multiples thereof are generally accepted. The drawing size can also be a problem for the troubleshooter or maintenance operator. If the drawing is too large, it is unwieldy to handle at the machine. If it is too small, it is hard to read the schematic.

The purpose of a ladder drawing, such as the one shown in Figure 6.64, is to cut maintenance and troubleshooting time. This is accomplished when the designer uses certain guidelines in making electrical drawings and layouts.

Let's take a closer look at the ladder drawing for the control circuit shown in Figure 6.64. Note the numbering of elementary circuit lines. Normally closed contacts are indicated by a bar under the line number. Moreover, note that the line numbers are enclosed in a geometric figure to prevent mistaking the line numbers for circuit numbers.

All contacts and the conductors connected to them are properly numbered. Typically, numbering is carried throughout the entire electrical system. This may involve going through one or more terminal blocks. The incoming and outgoing conductors, as well as the terminal blocks, carry the proper electrical circuit numbers. When possible, connections to all electrical components are taken back to one common checkpoint.

All electrical elements on a machine should be correctly identified with the same markings as shown on the ladder drawing in Figure 6.64. For example, if a given solenoid is marked "solenoid A2" on the drawing, the actual solenoid on the machine should carry the marking "solenoid A2."

The bottom line: An electrical drawing is made to show the relative location of each electrical component on the machine. The drawing is not normally drawn to scale, and it need not be. However, usually, it is reasonably accurate in showing the location of parts relative to each other and in relative size.

SCHEMATICS AND SYMBOLS

Because of the complexity of many electrical/mechanical systems, it would be almost impossible to show these systems in a full-scale detailed drawing. Instead, we use symbols and connecting lines to represent the parts of a system.

Figure 6.65 shows a voltage divider containing resistance and capacitance connected in a circuit by means of a switch. Such a series arrangement is called an *RC series circuit*. Note that unless the reader is an electrician or electronics technician, it is not important to understand this circuit. However, it is important for the reader

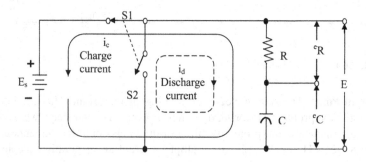

FIGURE 6.65 Schematic of an RC series circuit.

to understand that Figure 6.65 depicts a *schematic* representation formed by the use of symbols and connecting lines for a technical purpose.

Important Point: In the preceding paragraph, we defined what a schematic is; namely, it is a line drawing made for a technical purpose that uses symbols and connecting lines to show how a system operates.

How to Use Schematic Diagrams

(**Note**: To explain how to use a basic schematic diagram, Figure 6.66 is provided with an explanation below).

Learning to read and to use a schematic diagram (any schematic diagram) is a little bit like map reading. In a schematic for an electrical circuit, for example, we need to know which wires connect to which component and where each wire starts and finishes. With a map book, this would be equivalent to knowing the origin and destination points, which roads connect to the highway network, etc. However, schematics are a little more complicated as components need to be identified and some are polarity conscious (must be wired up in the circuit the correct way) in order to work. The reader does not need to understand what the circuit does, or how it works, in order to read it, but the reader does need to correctly interpret the schematic. Here are some basic rules that will help with reading a simple diagram. Refer to Figure 6.66.

The heavy lines represent wires, and for simplicity, we have labeled them as A, B, and C. There are just three components here, and it is easy to see where each wire starts and ends and which components a wire is connected to. As long as the wire labeled A connects to the switch and negative terminal of the battery, wire B connects to the switch and lamp, and C connects to the lamp and the battery positive terminal, then this circuit should operate.

Before we move on and describe how to read the diagram, it is important to point out that any schematic may be drawn in a number of different ways. For example,

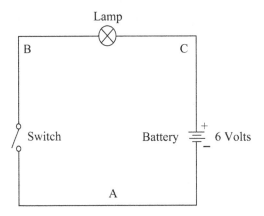

FIGURE 6.66   A single schematic diagram.

Figures 6.67 and 6.68 show two electrically equivalent lamp dimmer circuits; they may look very different, but in fact, if we mentally label the wires and trace them, we will see that in both diagrams each wire starts and finishes at the same components. The components have been labeled and so have the three terminals of the transistor (i.e., NPN is the transistor).

In Figure 6.68 there are two wire junctions as indicated by a "dot." A wire connects from battery positive to the C (transistor collector) terminal, and also a wire runs from the collector terminal to one end of the potentiometer, VR1. The wires could be joined at the transistor collector, battery positive, or even one end of the potentiometer; it does not matter as long as both wires exist. Similarly, a wire runs from the battery negative to the lamp and also from the lamp to the other end of VR1. The wires could be joined at the negative terminal of the battery, the lamp, or the opposite leg of VR1. In Figure 6.68, we could have drawn the wires from the lamp and bottom terminal of VR1 back to the battery negative terminal and placed the dot

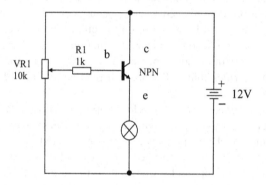

FIGURE 6.67   Schematic of simple lamp dimmer circuit.

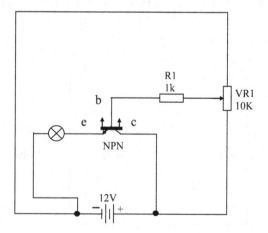

FIGURE 6.68   Schematic of a simple dimmer circuit.

there; it would be the same. Looking at Figure 6.68, we see that one wire junction appears at the negative battery terminal; the other junction is a similar place.

SCHEMATIC CIRCUIT LAYOUT

Sometimes, the way a circuit is wired may compromise its performance. This is particularly important for high-frequency and radio circuits, and some high-gain audio circuits.

Consider the audio circuit shown in Figure 6.69 (**Note**: For clearer understanding, the following explanation is simplified). Although this circuit has a voltage gain of less than one, wires to and from the transistor should be kept as short as possible. This will prevent a long wire from picking up radio interference or hum from a transformer. Moreover, in this circuit, input and output terminals have been labeled, and a common reference point or earth (ground) is indicated. The ground terminal would be connected to the chassis (metal framework of the enclosure) in which the circuit is built. Many schematics contain a chassis or ground point. Generally, it is just to indicate the common reference terminal of the circuit, but in radio work, the ground symbol usually requires a physical connection to a cold-water pipe or a length of pipe or earth spike buried in the soil.

SCHEMATIC SYMBOLS

Wind turbine operators/technicians are generally expected to be Jacks and Jills of many trades and skills. Simply, a good maintenance operator must be able to do many different kinds of jobs. To become a fully qualified "Jack" or "Jill," the

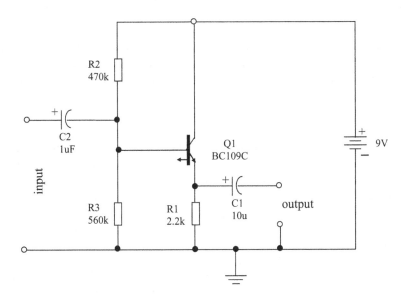

FIGURE 6.69 Schematic of a simple audio circuit.

maintenance operator must learn to perform electrical tasks, mechanical tasks, piping tasks, fluid power tasks, hotwork tasks, and many other special tasks. Moreover, maintenance operators must be flexible; they must be able to work on both familiar and new equipment and systems.

Undoubtedly you have heard the seasoned maintenance operator state that he or she can "fix" anything and everything using nothing more than their own intuition (i.e., "seat of the pants" troubleshooting). However, in the real world, to troubleshoot systems, maintenance operators must be able to read and understand schematics. By learning this skill, operators will have little difficulty understanding, maintaining, and repairing almost any equipment or unit process in the plant—old or new.

LINES ON A SCHEMATIC

As mentioned, symbols are used instead of pictures on schematics. Moreover, as mentioned, a schematic is a line diagram. Lines on a schematic show the connections between the symbols (devices) in a system. Each line has meaning; thus, we can say that schematic lines are part of the symbology employed. The meaning of certain lines, however, depends on the kind of system the schematic portrays. For example, a simple solid line can have different meanings. On an electrical diagram, it probably represents wiring. On a fluid power diagram, it stands for a working line. On a piping diagram, it could mean a low-pressure steam line. Figure 6.70 shows some other common lines used in schematics.

A schematic diagram is not necessarily limited to one kind of line. In fact, several kinds of lines may appear on a single schematic. Following applicable ANSI standards, most schematics use only one thickness, but they may use various combinations of solid and broken lines.

Important Point: Not all schematics adhere to standards set by national organizations as an aid in providing uniform drawings. Some designers prefer to use their own line symbols. These symbols are usually identified in a legend.

LINES CONNECT SYMBOLS

If we look at a diagram filled with lines, we may simply have nothing more than a diagram filled with lines. Likewise, if we look at a diagram with assorted symbols, we may simply have a diagram filled with various symbols. Such diagrams may have meaning to someone but probably have little meaning to most of us. In order to make a schematic readable (understandable) to a wide audience, we must have a diagram that uses a combination of recognizable lines and symbols.

When symbols are combined with lines in schematic form, we must also understand the meaning of the symbols used. The meaning of certain symbols depends on the kind of system the schematic shows. For example, the symbols used in electrical systems differ from those used in piping and fluid power systems.

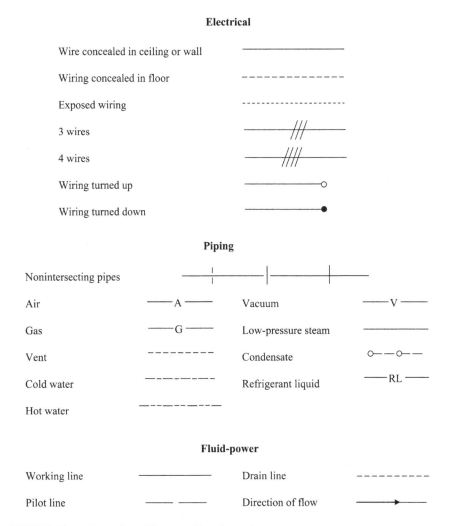

FIGURE 6.70   Examples of lines used in schematics.

The bottom line: To understand and properly use a schematic diagram, we must understand the meaning of both the lines and the symbols used.

SCHEMATIC DIAGRAM: AN EXAMPLE (ANSI, 1982)

Note that Figure 6.71 shows a schematic diagram used in electronics and communications. The layout of this schematic involves the same principles and procedures (except for lesser detail) suggested for more complex schematics. Although less complex than most schematics, Figure 6.71 serves our intended purpose: to provide a simplified schematic diagram for basic explanation and for easier understanding of a few key points—essential to understanding schematics and on how to use them.

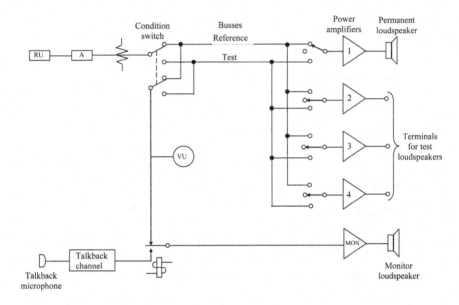

FIGURE 6.71   Single-line diagram. (From ANSI Y14.15 Dimensioning and Tolerancing. New York: American National Standards, text, 1982.)

A Schematic by Any Other Name Is a Line Diagram

The schematic (or line) diagram is intended to describe the basic functions of a circuit or system. As such, the individual lines connecting the symbols may represent single conductors or multiple conductors. The emphasis is on the function of each stage of a device and the composition of the stage.

The various parts or symbols used in a schematic (or line) diagram are typically arranged to provide a pleasing balance between blank areas and lines (see Figure 6.71). Sufficient blank spaces are provided adjacent to symbols for the insertion of reference designations and notes.

It is standard practice to arrange schematic and line diagrams so that the signal or transmission path from input to output proceeds from left to right (see Figure 6.71) and from top to bottom for a diagram in successive layers. Supplementary circuits, such as a power supply and an oscillator circuit, are usually shown below the main circuit.

Stages of an electronic device, such as shown in Figure 6.71, are groups of components, usually associated with a transistor or other semiconductor, which together perform one function of the device.

Connecting lines (for conductors) are drawn horizontally or vertically, for the most part, minimizing bends and crossovers. Typically, long interconnecting lines are avoided. Instead, *interrupted paths* are used in place of long, awkward interconnecting lines or where a diagram occupies more than one sheet. When parallel connecting lines are drawn close together, the spacing between lines is not less than .06″ after reduction. As a further visual aid, parallel lines are grouped with consideration of function and with double spacing between groups.

Crossovers are usually necessary for schematic diagrams. The looped crossovers shown in Figure 6.72(A) have been used for several years to avoid confusion. However, this method is not approved by the American National Standard. A simpler practice recognized by ANSI is shown in Figure 6.72(B). Connection of more than three lines at one point, shown at A, is not recommended and can usually be avoided by moving or staggering one or more lines as at B.

ANSI Y14.15 (cited earlier) recommends crossovers, as shown in Figure 6.72(C). In this system, it is understood that the termination of a line signifies a connection. If more than three lines come together, as at C, the dot symbol becomes necessary.

Interrupted paths, either for a single line or for groups of lines, may be used where desirable for the overall simplification of a diagram.

SCHEMATICS AND TROUBLESHOOTING

As mentioned, one of the primary purposes of schematic diagrams is to assist the maintenance operator in troubleshooting system, components, or unit process faults. While it is true that a basic schematic can be the troubleshooter's best friend, experience has shown that many mistakes and false starts can be avoided by taking a step-by-step approach to troubleshooting.

Seasoned wind turbine maintenance operators usually develop a standard trouble-shooting protocol or step-by-step procedure to assist them in their troubleshooting activities. No single protocol is the same; each troubleshooter proceeds based on intuition and experience (not on "seat of the pants" solutions). However, the simple

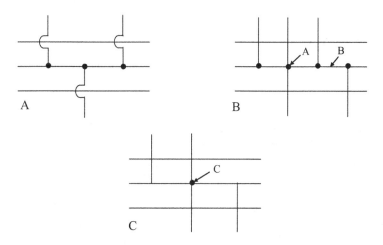

FIGURE 6.72 Crossovers.

15-step protocol (along with an accurate system schematic) described below has worked well for those of us who have used it (**Note:** Recognize that several steps may occur at the same time):

(1) Recognize a problem exists (figure out what it is designed to do and how it should work).
(2) Review all available data.
(3) Find the part of the schematic that shows the troubled area, and study it in detail.
(4) Evaluate the current plant operation.
(5) Decide what additional information is needed.
(6) Collect the additional data.
(7) Test the process by making modifications and observing the results.
(8) Develop an initial opinion as to the cause of the problem and potential solutions.
(9) Fine-tune your opinion.
(10) Develop alternative actions to be taken.
(11) Prioritize alternatives (i.e., prioritize based on its chances of success, how much it will cost, etc.).
(12) Confirm your opinion.
(13) Implement the alternative actions (this step may be repeated several times).
(14) Observe the results of the alternative actions implemented (i.e., observe the impact on effluent quality; impact on individual unit process performance; changes (trends) in the results of process control tests and calculations; impact on operational costs).
(15) During project completion, evaluate other, more permanent long-term solutions to the problem (such as chemical addition, improved preventive maintenance, design changes). Continue to monitor results. Document the actions taken and the results produced for use in future problems.

ELECTRICAL SCHEMATICS

A good deal of standard wind turbine equipment in use today produces electricity. Moreover, certain components within the wind turbine are operated by electricity. Wind turbine operators/technicians are required to keep turbine electrical equipment working. When a machine fails or a system stops working, the wind turbine maintenance operator must find the problem and solve it quickly.

No maintenance person can be expected to remember every detail of a wind farm's electrical operation and generation equipment; this is especially the case where wind farms use wind turbine models of different types. The information wind turbine technicians need must be stored in diagrams or drawings in a format that can be readily understood by trained and qualified maintenance operators. Electrical schematics store the information in a user-friendly form.

In this section, the basics of electrical schematics and wiring diagrams are discussed. Typical symbols and circuits are used as examples.

In describing electrical systems, three kinds of drawings are typically used: pictorial, wiring, and schematic. *Pictorial drawings* show an object or system much as it would appear in a photograph—as if we were viewing the actual object. Several sides of the object are visible in the one pictorial view. Pictorial drawings are quite easy to understand. They can be used in making or servicing simple objects but are usually not adequate for complicated parts or systems, such as electrical components and systems. A *wiring diagram* shows the connections of an installation or its component devices or parts. It may cover internal or external connections, or both, and contain such details as are needed to make or trace connections that are involved. The wiring diagram usually shows the general physical arrangement of the component devices or parts. A *schematic diagram* uses symbols instead of pictures for the working parts of the circuit. These symbols are used in an effort to make the diagrams easier to draw and easier to understand. In this respect, schematic symbols aid the maintenance operator in the same way that shorthand aids the stenographer.

Important Point: A schematic diagram emphasizes the flow in a system. It shows how a circuit functions rather than how each part actually looks. Stated differently, a schematic represents the *electrical*, not the physical, situation in a circuit.

ELECTRICAL SYMBOLS

Electrical and electronic circuits are indicated by very simple drawings, called *schematic symbols*, which are standardized throughout the world, with minor variations. Some of these symbols look like the components they represent. Some look like key parts of the components they represent. As a maintenance operator, you must know these symbols so that you can read the diagrams and keep the plant equipment in working order. The more schematics are used, the easier it becomes to remember what these symbols mean.

SCHEMATIC LINES

In electrical and electronic schematics, lines symbolize (or stand for) wires connecting various components. Different kinds of lines have different meanings in schematic diagrams. Figure 6.73 shows examples of some lines and their meanings; other lines are usually identified by a diagram legend.

To understand any schematic diagram, we must observe how the lines intersect. These intersections show that two or more wires are connected or that the wires pass over or under each other without connecting.

Note, as mentioned, Figure 6.73 shows some of the connections and crossings of lines. When wires intersect in a connection, a dot is used to indicate this. If it is clear that the wires connect, the dot is not used.

Important Point: Wires and how they intersect are important. Maintenance operators must be able to tell the difference between wires that connect and those that do not to be able to properly read the schematic and determine the flow of current in a circuit.

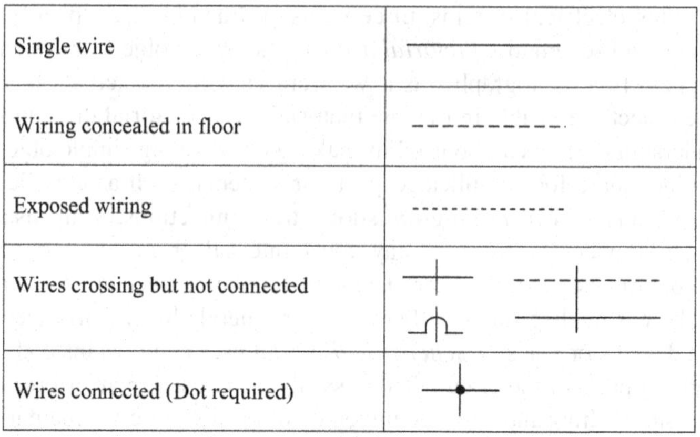

Single wire	![single wire symbol]
Wiring concealed in floor	![dashed line]
Exposed wiring	![dotted line]
Wires crossing but not connected	![crossing not connected symbols]
Wires connected (Dot required)	![connected dot symbol]

FIGURE 6.73 Symbols for wires.

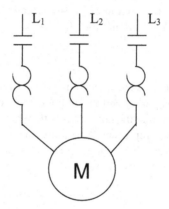

FIGURE 6.74 Electrical power supply lines.

POWER SUPPLIES: ELECTRICAL SYSTEMS

Most wind turbine farms receive electrical power from the transmission lines of a utility company and in some applications from the wind turbine farm and associated farm storage system. On a schematic, the entry of power lines into the plant's electrical system can be shown in several ways. Electrical power supply lines to a motor are shown in Figure 6.74.

Another source of electrical power is a battery. A battery consists of two or more cells. Each cell is a unit that produces electricity by chemical means. The cells can be connected together to produce the necessary voltage and current. For example, a 12-V storage battery might consist of six 2-V cells.

On a schematic, a battery power supply is represented by a symbol. Consider the symbol in Figure 6.76. The symbol is rather simple and straightforward—but is also very important. For example, by convention, the shorter line in the symbol for a battery represents the *negative* terminal. It is important to remember this, because it is sometimes necessary to note the direction of the current flow, which is from negative to positive, when you examine the schematic. The battery symbol shown in Figure 6.75 has a single cell, so only one short and one long line are used. The number of lines used to represent a battery varies (and they are not necessarily equivalent to the number of cells), but they are always in pairs, with long and short lines alternating. In the circuit shown in Figure 6.76, the current would flow in a *counterclockwise* direction, that is, in the opposite direction that a clock's hands move. If the long and short lines of the battery symbol (symbol shown in Figure 6.76) were reversed, the current in the circuit shown in Figure 6.76 would flow *clockwise*, that is, in the direction of a clock's hands.

Important Point: Current flows from the negative (–) terminal of the battery, shown in Figure 6.76, through the switch, fuse, and resistor (R) to positive (+) battery terminal and continues going through the battery from the positive (+) terminal to the negative (–) terminal. As long as the pathway is unbroken, it is a closed circuit, and the current will flow. However, if the path is broken (e.g., the switch is in an open position), it is an open circuit, and no current flows.

Important Point: In Figure 6.76, a fuse is placed directly into the circuit. A fuse will open the circuit whenever a dangerous large current starts to flow (i.e., a short-circuit condition occurs, caused by an accidental connection between two points in a circuit that offer very little resistance). A fuse will permit currents smaller than the fuse value to flow but will melt and therefore break or open the circuit if a larger current flows.

Figure 6.77 shows a simple diagram for a basic power supply.

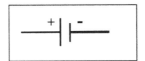

FIGURE 6.75   Schematic symbol for a battery.

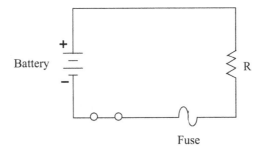

Fuse

FIGURE 6.76   Schematic of a simple fused circuit.

In wind turbine operations, maintenance operators/technicians are most likely to maintain or troubleshoot circuits connected to a wind farm distribution system or electrical circuits fed to turbine towers, control buildings, and so forth for lighting or other purposes. However, on occasion, they may also work on some circuits that are battery-powered—battery-powered emergency lighting, for example.

ELECTRICAL LOADS

An electric circuit, which provides a complete path for electric current, includes an energy source (source of voltage; that is, a battery or utility line), a conductor (wire), a means of control (switch), and a load. As shown in the schematic representation in Figure 6.78, the energy source is a battery. The battery is connected to the circuit by conductors (wire). The circuit includes a switch for control. The circuit also consists of a load (resistive component). The *load* that dissipates battery-stored energy could be a lamp, a motor, heater, a resistor, or some other device (or devices) that does useful work, such as an electric toaster, a power drill, radio, or soldering iron.

Figure 6.79 shows some symbols for common loads and other components in electrical circuits. The maintenance operator should become familiar with these symbols (and others not shown here) because they are widely used.

Important Point: Every complete electrical/electronic circuit has at least one load.

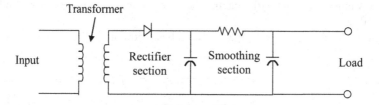

FIGURE 6.77   Basic power supply.

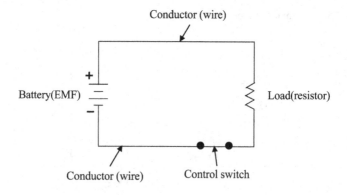

FIGURE 6.78   Schematic of a simple closed circuit.

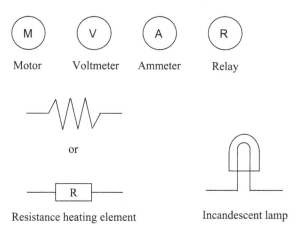

FIGURE 6.79   Symbols for electrical loads.

Type of switch	Abbreviation	Symbol
Single pole, single throw	SPST	Toggle switch Knife switch
Double pole, single throw	DPST	
Single pole, double throw	SPDT	
Double pole, double throw	DPDT	
Normally open	NO	
Normally closed	NC	
Two-position	NC-NO	

FIGURE 6.80   Types of switches.

SWITCHES

In Figure 6.78, the schematic shows a simple circuit with switches. A *switch* is a device for making or breaking the electrical connection at one point in a wire. A switch allows the starting, stopping, or changing of the direction of current flow in a circuit. Figure 6.80 shows some common switches and their symbols.

INDUCTORS (COILS)

Simply put, an *inductor* is a coil of wire, usually many turns (of wire) around a piece of soft iron (magnetic core). In some cases, the wire is wound around a nonconducting material. Inductors are used as ballasts in fluorescent lamps and for magnets and solenoids. When electric current flows through a coil, it creates a magnetic field (an electromagnetic field). The magnetism causes certain effects needed in electric circuits (e.g., in an alarm circuit, the magnetic field in a coil can cause the alarm bell to ring). It is not important to understand these effects in order to read schematic diagrams. It is, however, important to recognize the symbols for inductors (or coils). These symbols are shown in Figure 6.81.

TRANSFORMERS

Transformers are used to increase or decrease AC voltages and currents in circuits. The operation of transformers is based on the principle of *mutual inductance*. A transformer usually consists of two coils of wire wound on the same core. The primary coil is the input coil of the transformer, and the secondary coil is the output coil. Mutual induction causes the voltage to be induced in the secondary coil. If the output voltage of a transformer is greater than the input voltage, it is called a *step-up* transformer. If the output voltage of a transformer is less than the input voltage, it is called a *step-down* transformer. Figure 6.82 shows some of the basic symbols that are used to designate transformers on schematic diagrams.

FUSES

A *fuse* is a device that automatically opens a circuit when the current rises above a certain limit. When the current becomes too high, part of the fuse melts. Melting

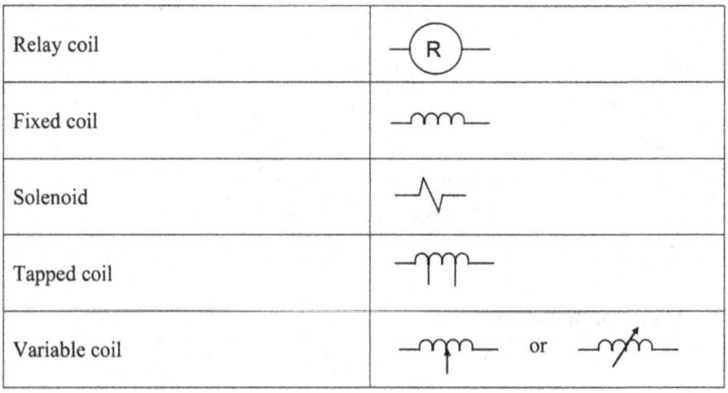

Relay coil	—(R)—
Fixed coil	⌐ⵣⵣ⌐
Solenoid	—⌐Λ⌐—
Tapped coil	—⌐ⵣⵣ⌐—
Variable coil	—⌐ⵣ⌐— or —⌐ⵣ⌐—

FIGURE 6.81   Symbols for coils and inductors.

Transformer with iron core   Transformer with taps   Autotransformer

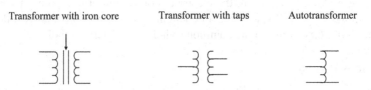

FIGURE 6.82   Transformer symbols.

opens the electrical path, stopping the flow of electricity. To restore the flow of electricity, the fuse must be replaced. Figure 6.83 shows some of the basic symbols that are used to designate fuses on schematic diagrams.

CIRCUIT BREAKERS

A *circuit breaker* is an electric device (similar to a switch) that, like a fuse, interrupts an electric current in a circuit when the current becomes too high. The advantage of a circuit breaker is that it can be reset after it has been tripped; a fuse must be replaced after it has been used once. When a current supplies enough energy to operate a trigger device in a breaker, a pair of contacts conducting the current are separated by preloaded springs or some similar mechanism. Generally, a circuit breaker registers the current either by the current's heating effect or by the magnetism it creates in passing through a small coil. Figure 6.84 shows some of the basic symbols that are used to designate circuit breakers on schematic diagrams.

ELECTRICAL CONTACTS

Electrical *contacts* (usually wires) join two conductors in an electrical circuit. *Normally closed* (NC) contacts allow the current to flow when the switching device is at rest. *Normally open* (NO) contacts prevent current from flowing when the switching device is at rest. Figure 6.85 shows some of the basic symbols that are used to designate contacts on schematic diagrams.

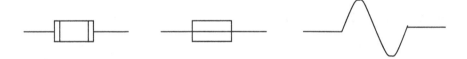

FIGURE 6.83   Fuse symbols.

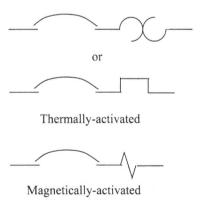

or

Thermally-activated

Magnetically-activated

FIGURE 6.84   Circuit breaker symbols.

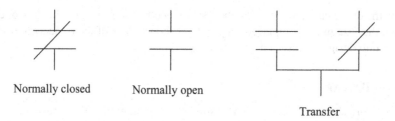

Normally closed   Normally open

Transfer

FIGURE 6.85   Electrical contacts symbols.

RESISTORS

Electricity travels through a conductor (wire) easily and efficiently, with almost no other energy released as it passes. On the other hand, electricity cannot travel through a resistor easily. When electricity is forced through a resistor, often the energy in the electricity is changed into another form of energy, such as light or heat. The reason a light bulb glows is that electricity is forced through the tungsten filament, which is a resistor.

Resistors are commonly used for controlling the current flowing in a circuit. A *fixed resistor* provides a constant amount of resistance in a circuit. A *variable resistor* (also called a potentiometer) can be adjusted to provide different amounts of resistance, such as in a dimmer switch for lighting systems. A resistor also acts as a load in a circuit in that there is always a voltage drop across it. Figure 6.86 shows some of the basic symbols that are used to designate resistors on schematic diagrams. A summary of electrical symbols is shown in Figure 6.87.

READING TURBINE POWER SCHEMATICS

With the information provided in the preceding sections on electrical schematic symbols and an explanation of their function(s), you should be able to read simple schematic diagrams. Many of the schematics used in wind turbine operations are of simple motor circuits, such as the one shown in Figure 6.88, which is for a reversing motor starter.

Important Point: In the *reversing starter*, there are two starters of equal size for a given horsepower motor application. The reversing of a three-phase, squirrel-cage induction motor is accomplished by interchanging any two-line connections to the motor. The concern is to properly connect the two starters to the motor so that the line feed from one starter is different from the other. Both mechanical and electrical interlocks are used to prevent both starters from closing their line contacts at the same time. Only one set of overloads is required as the same load current is available for both directions of rotation.

From the schematic shown in Figure 6.88, it can be seen that the motor is connected to the plant's power source by the three power lines (line leads) L_1, L_2, and L_3. The circuits for the forward and reverse drive are also shown.

For forward drive, lead L_1 is connected to terminal T_1 (known as a T lead) on the motor. Likewise, L_2 is connected to T_2 and L_3 to T_3. When the three normally open

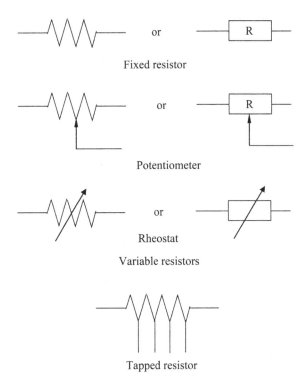

Fixed resistor

Potentiometer

Rheostat

Variable resistors

Tapped resistor

FIGURE 6.86   Resistor symbols.

F contacts (F for forward) are closed, these connections are made, current flows between the power source and the motor, and the motor rotor turns in the forward direction.

In the reverse drive condition, the three leads (L_1, L_2, L_3) connect to a set of R contacts. The R contacts reverse the connections of terminals T_1 and T_3, which reverses the rotation of the motor rotor. To reverse the motor, the three normally open R contacts must close, and the F contacts must be open.

The three fuses located on lines L_1, L_2, and L_3 protect the circuit from overloads. Moreover, three thermal overload cutouts protect the motor from damage.

Important Point: In actual operation, the circuit shown in Figure 6.88 utilizes a separate control to open or close the forward and reverse contacts. It has a mechanical interlock to make sure the R contacts stay open when the F contacts are closed and vice versa.

GENERAL PIPING SYSTEMS AND SYSTEM SCHEMATICS

It would be difficult to imagine any modern wind turbine without pipes. Pipes convey all the fluids—that is, liquids and gases. Within the turbine most of the pipes and valves are visible.

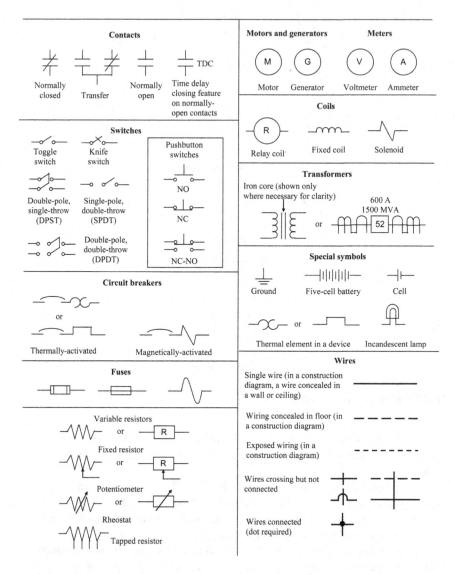

FIGURE 6.87 Summary of electrical symbols.

However, they may be arranged in complex ways that look confusing to the untrained eye. Notwithstanding a piping network's complexity, the wind turbine maintenance operator can trace through a piping system, no matter how complex the system, by reading a piping schematic of the system.

In this section, piping systems, piping system schematic diagrams, and schematic symbols typical of wind turbine and industrial operations are described. Also described are how piping system symbols are used to represent various connections and fittings used in piping arrangements.

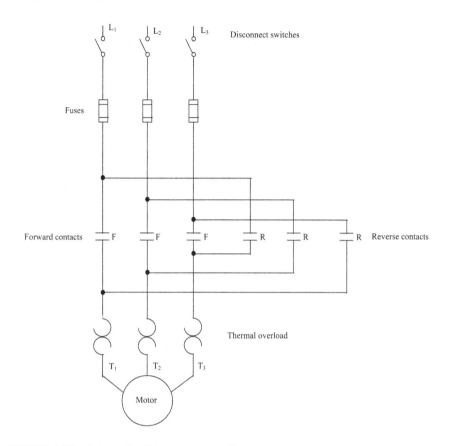

FIGURE 6.88   Schematic of the motor controller.

PIPING SYSTEMS

Pipe is used for conveying fluids (liquid and gases) and chemicals and for structural elements such as columns and handrails. The choice of the type of pipe is determined by the purpose for which it is to be used.

A *pipe* is defined as an enclosed, stationary device that conducts a fluid or a semi-solid from one place to another in a controlled way. A *piping system* is a set of pipes and control devices that work together to deliver a fluid where it is needed, in the right amount, and at the proper rate.

Pipes are made of several kinds of solid materials. They can be made of wood, glass, porcelain, cast iron, lead, aluminum, stainless steel, brass, copper, plastic, clay, concrete, plastic, lead, and many other materials. Cast iron, steel, wrought iron, brass, copper, plastic, and lead pipes are most commonly used for conveying fluids.

When conveying a substance in a piping system, the flow of the substance must be controlled. The substance must be directed to the place where it is needed. The amount of substance that flows, and how fast it flows, must be regulated.

The flow of a substance in a piping system is controlled, adjusted, and regulated by *valves*. In this section, the symbols that depict valves on different kinds of piping schematics are described. Also, basic *joints* and other fittings used in piping systems are described.

PIPING SYMBOLS: GENERAL

In order to read piping schematics correctly, maintenance operators must identify and understand the symbols used. It is not necessary to memorize them, but maintenance operators should keep a table of basic symbols handy and refer to it whenever the need arises. In the following sections, the most common symbols used in general piping systems are described.

Piping Joints

The joints between pipes, fittings, and valves may be screwed (or threaded), flanged, welded, or, for nonferrous materials, soldered. A *joint* is a connection between two elements in a piping system. There are five major types of joints—screwed (or threaded joints), welded, flanged, bell-and-faucet, and soldered.

Screwed Joints

Screwed joints are threaded together. That is, screwed joints can be made up tightly by simply screwing the threads together. Figure 6.89 shows the schematic symbol for a screwed joint. Screwed joints are usually used with pipes ranging from 1/4 inches to about 6 inches in diameter.

Important Point: In most screwed joints, the threads are on the inside of the fitting and on the outside of the pipe.

Welded Joints

Piping construction employing *welded joints* is almost the universal practice today. This kind of joint is used when the coupling must be permanent, particularly for higher pressure and temperature conditions. Figure 6.90 shows the schematic symbol for welded joints.

Important Point: Welded joints may either be socket-welded or be butt-welded.

FIGURE 6.89   Screwed joint symbol.

FIGURE 6.90   Welded joint symbol.

Flanged Joints

Flanged joints are made by bolting two flanges together with a gasket between the flange faces. Flanges may be attached to the pipe, fitting, or appliance by means of a screwed joint, by welding, by lapping the pipe, or by being cast integrally with the pipe, fitting, or appliance. A flanged joint is shown in Figure 6.91, along with its symbol used on schematics.

Bell-and-Spigot Joints

Cast-iron pipes for handling water, in particular, usually fit together in a special way. When each fitting and section of pipe is cast, one end is made large enough to fit loosely around the opposite end of another fitting or pipe. When this type of fitting is connected, it forms a joint called the *bell-and-faucet joint*. Figure 6.92 shows a bell-and-faucet joint and the symbol used on schematics.

Soldered Joints

Nonferrous fittings, such as copper piping and fittings, are often joined by *soldering* them with a torch or soldering iron and then melting the solder on the joint. The solder flows into the narrow space between the two mating parts and seals the joint. Figure 6.93 shows the schematic symbol for a soldered joint.

Symbols for Joints and Fittings

Figure 6.94 shows most of the common schematic piping joint and fitting symbols (for elbows only) used today. However, it does not show all the symbols used on all piping schematics.

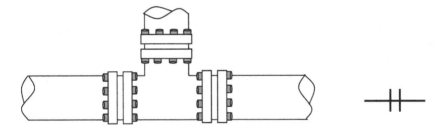

FIGURE 6.91   Flanged joint.

FIGURE 6.92   Symbol for the bell-and-faucet joint.

FIGURE 6.93   Symbol for the soldered joint.

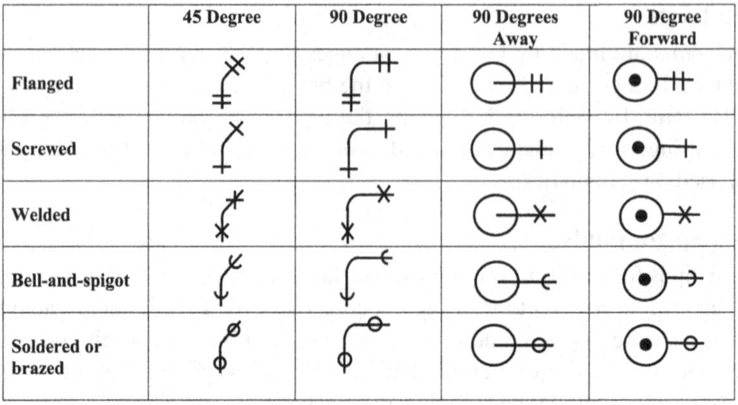

	45 Degree	90 Degree	90 Degrees Away	90 Degree Forward
Flanged				
Screwed				
Welded				
Bell-and-spigot				
Soldered or brazed				

FIGURE 6.94   Symbols for elbow fittings.

VALVES

Wind turbine systems have valves that are important components of different piping systems. Simply as a matter of routine, a wind turbine maintenance operator must be able to identify and locate valves in order to inspect them, adjust them, and repair or replace them. For this reason, the maintenance operator should be familiar with schematic diagrams that include valves and with schematic symbols used to designate valves.

VALVES: DEFINITION AND FUNCTION (VALVES, 1998)

A *valve* is defined as any device by which the flow of fluid may be started, stopped, or regulated by a movable part that opens or obstructs passage. As applied in fluid power systems, valves are used for controlling the flow, the pressure, and the direction of the fluid flow through a piping system. The fluid may be a liquid, a gas, or some loose material in bulk (like a biosolids slurry). Designs of valves vary, but all valves have two features in common:

- A passageway through which fluid can flow
- Some kind of movable (usually machined) part that opens and closes the passageway

Important Point: It is all but impossible, obviously, to operate a practical fluid power system without some means of controlling the volume and pressure of the fluid and directing the flow of fluid to the operating units. This is accomplished by the incorporation of different types of valves.

Whatever type of valve is used in a system, it must be accurate in the control of fluid flow and pressure and the sequence of operation. Leakage between the valve element and the valve seat is reduced to a negligible quantity by precision-machined surfaces, resulting in carefully controlled clearances. This is, of course, one of the very important reasons for minimizing contamination in fluid power systems. Contamination causes valves to stick, plugs small orifices, and causes abrasions of the valve seating surfaces, which results in leakage between the valve element and valve seat when the valve is in the closed position. Any of these can result in inefficient operation or complete stoppage of the equipment. Valves may be controlled manually, electrically, pneumatically, mechanically, hydraulically, or by combinations of two or more of these methods. Factors that determine the method of control include the purpose of the valve, the design and purpose of the system, the location of the valve within the system, and the availability of the source of power.

Valves are made from bronze, cast iron, steel, Monel®, stainless steel, and other metals. They are also made from plastic and glass. Special valve trim is used where seating and sealing materials are different from the basic material of construction. (**Note**: *Valve trim* usually means those internal parts of a valve controlling the flow and in physical contact with the line fluid). Valves are made in a full range of sizes, which match pipe and tubing sizes. Actual valve size is based upon the internationally agreed definition of nominal size. *Nominal size* (DN) is a numerical designation of size which is common to all components in a piping system other than components designated by outside diameters. It is a convenient number for reference purposes and is only loosely related to manufacturing dimensions. Valves are made for service at the same pressures and temperatures that piping and tubing are subject to. Valve pressures are based upon the internationally agreed definition of nominal pressure. *Nominal pressure* (PN) is a pressure that is conventionally accepted or used for reference purposes. All equipment of the same nominal size (DN) designated by the same nominal pressure (PN) number must have the same mating dimensions appropriate to the type of end connections. The permissible working pressure depends upon materials, design, and working temperature and should be selected from the (relevant) pressure/temperature tables. The pressure rating of many valves is designated under the American (ANSI) class system. The equivalent class rating to PN ratings is based upon international agreement.

Usually, valve end connections are classified as flanged, threaded, or other.

Valves are also covered by various codes and standards, as are the other components of piping and tubing systems.

Many valve manufacturers offer valves with special features. Table 6.4 lists a few of these special features; however, this is not an exhaustive list, and for more details of other features the manufacturer should be consulted.

Valve Construction

Figure 6.95 shows the basic construction and principle of operation of a common valve type. Fluid flows into the valve through the inlet. The fluid flows through passages in the body and past the opened element that closes the valve. It then flows out of the valve through the outlet or discharge.

If the closing element is in the closed position, the passageway is blocked. Fluid flow is stopped at that point. The closing element keeps the flow blocked until the

TABLE 6.4
Valve Special Features

High temperature	Valves are those usually able to operate continuously on services above 250°C.
Cryogenic	Valves are those that will operate continuously on services in the range −50°C to 196°C.
Bellows sealed	Valves are glandless designs having metal bellows for stem sealing.
Actuated	Valves may be operated by a gearbox, pneumatic or hydraulic cylinder (including diaphragm actuator), or electric motor and gearbox.
Fire-tested design	Refers to a valve that has passed a fire test procedure specified in an appropriate inspection standard.

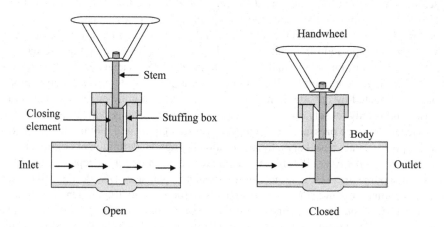

FIGURE 6.95   Basic valve operation.

valve is opened again. Some valves are opened automatically, and others are controlled by manually operated handwheels. Other valves, such as check valves, operate in response to pressure or the direction of flow.

TYPES OF VALVES

The types of valves covered in this text include the following:

- Ball valve
- Cock valve
- Gate valve
- Globe valve
- Check valve

Each of these valves is designed to perform either control of the flow, the pressure, the direction of fluid flow or some other special application. With a few exceptions, these valves take their names from the type of internal element that controls the passageway. The exception is the check valve.

Ball Valve

A *Ball* valve, as the name implies, is a stop valve that uses a ball to stop or start a flow of fluid. The ball performs the same function as the disk in other valves. As the valve handle is turned to open the valve, the ball rotates to a point where part or all of the hole through the ball is in line with the valve body inlet and outlet, allowing fluid to flow through the valve. When the ball is rotated so the hole is perpendicular to the flow openings of the valve body, the flow of fluid stops.

Most ball valves are the quick-acting type. They require only a 90-degree turn to either completely open or close the valve. However, many are operated by planetary gears. This type of gearing allows the use of a relatively small handwheel and operating force to operate a fairly large valve. The gearing does, however, increase the operating time for the valve. Some ball valves also contain a swing check located within the ball to give the valve a check valve feature.

The two main advantages of using ball valves are (1) the fluid can flow through it in either direction, as desired, and (2) when closed, pressure in the line helps to keep it closed.

Symbols that represent the ball valve in schematic diagrams are shown in Figure 6.96.

Cock Valves

The *cock valve*, like the gate valve, has only two positions—on and off. The difference is in the speed of operation. The gate valve, for example, opens and closes gradually. The cock valve opens and closes quickly. The cock valve is used when the flow must be started quickly and stopped quickly. The schematic symbols for a cock valve appear in Figure 6.97.

Gate Valves

Gate valves are used when a straight-line flow of fluid and minimum flow restriction are needed. Gate valves are so-named because the part that either stops or allows

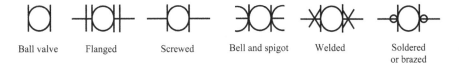

| Ball valve | Flanged | Screwed | Bell and spigot | Welded | Soldered or brazed |

FIGURE 6.96   Symbols for ball valves.

| Cock valve | Flanged | Screwed | Bell and spigot | Welded | Soldered or brazed |

FIGURE 6.97   Symbols for cock valves.

flow through the valve acts somewhat like a gate. The gate is usually wedge-shaped. When the valve is wide open the gate is fully drawn up into the valve bonnet. This leaves an opening for flow through the valve the same size as the pipe in which the valve is installed. For these reasons, the pressure loss (pressure drop) through these types of valves is about equal to the loss in a piece of pipe of the same length. Gate valves are not suitable for throttling (i.e., to control the flow as desired, by means of intermediate steps between fully open and fully closed) purposes. The control of flow is difficult because of the valve's design, and the flow of fluid slapping against a partially open gate can cause extensive damage to the valve. The schematic symbols for a gate valve appear in Figure 6.98.

Important Point: Gate valves are well-suited to service on equipment in distant locations, where they may remain in the open or closed position for a long time. Generally, gate valves are not installed where they will need to be operated frequently because they require too much time to operate from fully open to closed (AWWA, 1998).

Globe Valves

Probably the most common valve type in existence, the *globe* valve principle is commonly used for water faucets and other household plumbing. As illustrated in Figure 6.99, the valves have a circular disk—the "globe"—that presses against the valve seat to close the valve. The disk is the part of the globe valve that controls the flow. The disk is attached to the valve stem. Fluid flow through a globe valve is at right angles to the direction of flow in the conduits. Globe valves are seated very tightly and can be adjusted with fewer turns of the wheel than gate valves; thus, they are preferred for applications that call for frequent opening and closing. On the other hand, globe valves create high head loss when fully open; thus, they are not suited in systems where head loss is critical. The schematic symbols that represent the globe valve are also shown in Figure 6.99.

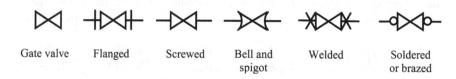

| Gate valve | Flanged | Screwed | Bell and spigot | Welded | Soldered or brazed |

FIGURE 6.98 Symbols for gate valves.

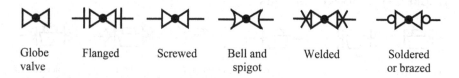

| Globe valve | Flanged | Screwed | Bell and spigot | Welded | Soldered or brazed |

FIGURE 6.99 Symbols for globe valves.

Important Point: The globe valve should never be jammed in the open position. After a valve is fully opened, the handwheel should be turned toward the closed position, approximately one-half turn. Unless this is done, the valve is likely to seize in the open position, making it difficult, if not impossible, to close the valve. Another reason for not leaving globe valves in the fully open position is that it is sometimes difficult to determine if the valve is open or closed (Globe Valve, 1998).

Check Valves

Check valves are usually self-acting and designed to allow the flow of fluid in one direction only. They are commonly used at the discharge of a pump to prevent back-flow when the power is turned off. When the direction of flow is moving in the proper direction, the valve remains open. When the direction of flow reverses, the valve closes automatically from the fluid pressure against it.

There are several types of check valves used in fluid systems, including

- Slanting disk check valves
- Cushioned swing check valves
- Rubber flapper swing check valves
- Double door check valves
- Ball check valves
- Foot valves
- Backflow prevention devices

In each case, pressure from the flow in the proper direction pushes the valve element to an open position. Flow in the reverse direction pushes the valve element to a closed position. Symbols that represent check valves are shown in Figure 6.100. (Note: Figure 6.101 shows the standard symbols for valves discussed in this text).

Hydraulic Solenoid Valves

Figure 6.102 illustrates the common symbol for a solenoid valve. Solenoid valves are commonly used in industrial wind turbines. The wind turbine's hydraulic system is the driving system of the braking system, which carries out the start-stop job of the turbine. It is customarily comprised of two pressure holding circuits: one is to supply the impeller blade system with an accumulator (maintains a fixed volume of fluid at system pressure); the other supplies the yaw brake system with an accumulator. The function of these two hydraulic circuits is to keep the wind turbine brake system under certain pressure during the normal operation of the wind turbine. When the pressure measured by the pressure sensor is less than the set value of the system, the

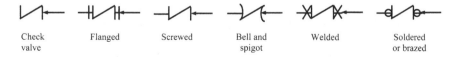

| Check valve | Flanged | Screwed | Bell and spigot | Welded | Soldered or brazed |

FIGURE 6.100 Symbols for check valves.

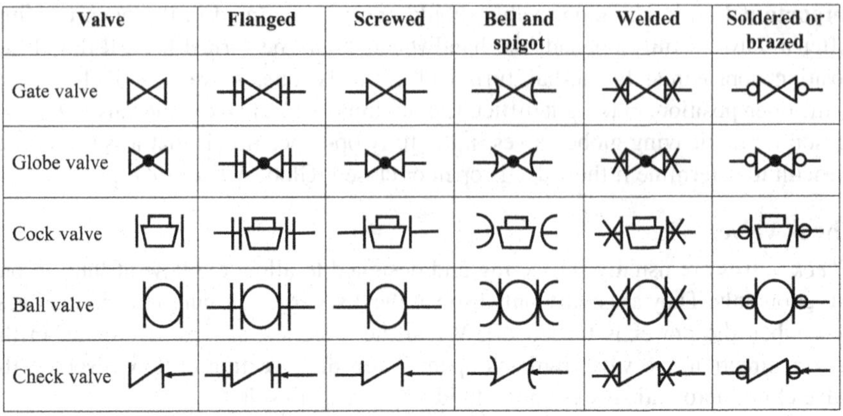

Valve	Flanged	Screwed	Bell and spigot	Welded	Soldered or brazed
Gate valve					
Globe valve					
Cock valve					
Ball valve					
Check valve					

FIGURE 6.101   Standard symbols for valves.

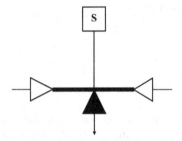

FIGURE 6.102   Symbol of the solenoid valve.

programmable logic controller (PLC) will control the start of the hydraulic station motor to compensate for the loss of pressure so that the pressure valve is always above the set value.

During wind turbine energy conversion, the pitch system is the main braking system. It is designed to have an independent electric drive, with a backup emergency battery supply, for each blade. A third braking system of the wind energy conversion system consists of a mechanical brake mounted on the high-speed side of the drive chain. This third braking system is not designed to prevent the wind energy conversion system from exceeding the allowable speed limit whenever the first and second braking systems malfunction.

Wind turbine brake operation is important. The variable pitch system decelerates the impeller blade rotation, and the rotor stops completely driven by the hydraulic pressure in the brake system. In an emergency shutdown, the mechanical brakes are used to slow down the wind energy conversion system as quickly as possible.

In operation, the opening and setting of the brake and the setting or closing of the brake make the solenoid valve energize, and the rotor brake is released. In an emergency situation, brake opening and closing de-energize the solenoid valves, the accumulator pressure oil enters brake calipers through the brake opening valves, braking the rotor. The pressure sensor disconnects the signal when the brake oil chamber is below a certain pressure parameter (set point). The accumulator's charging condition is monitored by pressure sensor 2. When the pressure value is less than the set value, the PLC initiates action for the oil motor to begin charging.

Yaw system brakes, located on the yaw system, help to position the nacelle into wind direction for optimum performance. The yaw brake mechanism consists of 8 to 10 hydraulic control brake discs. The yaw system, consisting of two pressures, provides damping at the yaw and braking force at the end.

The hydraulic system includes ball valves on one side of the brake; this enables the screw piston to be used for braking the solenoid values when hydraulic pressure can't be applied.

THE BOTTOM LINE

Whenever a road trip away from home is planned, away from familiar surroundings, it is common practice to check out the desired trip route. Roadmaps can be consulted in a road atlas or online using the Internet. From the map, one can see the principal cities and towns of a state or area, the main roads, tourist attractions, and places of historical significance to check out. Moreover, using a road map enables the user to determine various routes to follow and the mileage to those locations that are the preferred stopover spots or the desired destination.

A blueprint, wiring diagram, schematic, and/or a technical drawing is very much like a road map; simply, it is a guide. If one understands the symbols and notes on a blueprint and is capable of following the lines as shown, he or she can find whatever it is they are attempting to find for operational and maintenance purposes. Having a detailed drawing of the wiring and piping inside a wind turbine and for its associated equipment is not only desirable but necessary to ensure optimum maintenance activities and operation.

REFERENCES

ANSI. 1980. *Blueprint Sheet Size.* New York: American National Standards Institute.
ANSI. 1982. *Dimensioning and Tolerancing.* New York: American National Standards Institute.
AWWA. 1998. *Water Transmission & Distribution*, 2nd Edition. Denver, CO: American Water Works Association.
Brown, W.C. 1989. *Blueprint Reading for Industry.* South Holland, Il: The Goodheart-Wilcox C.
Globe Valve 1998. Integrated Publications @ Pub web page tpub.com.
Olivo, C. and Olivo, T. 1999. Visible lines in basic blueprint reading and sketching. Accessed 11/22/2021 @ https://cc.thelaternativedaily.com/ba/basic.
Valves 1998. Integrated Publishing web page @ tpub.com.

7 Basic Hydraulics and Accessories
Earth's Blood (An Analogy)

The watery environment in which single-cell organisms lives provides them food and removes their wastes, a function that the human circulatory system provides for the 60–100 trillion cells in a human body. The circulatory system brings each cell its daily supply of nutritive amino acids and glucose, and carries away waste carbon dioxide and ammonia, to be filtered out of our systems and flushed away through micturition and excretory functions. The heart, the center of our circulatory system, keeps the blood moving on its predetermined circular path, so essential that if the pump fails, we quickly fail as well—we die.

As single-celled organisms no longer, humans sometimes assume they no longer need a watery environment in which to live—but they aren't paying close attention to the world around them. Actually, those of us who live on earth are as dependent upon the earth's circulatory system as we are on our own circulatory system. As our human hearts pump blood, circulating it through a series of vessels, and as our lives are dependent upon that flow of blood, so too life on the earth is, and our own lives are dependent on the earth's water cycle, and on water, in every aspect of our lives.

This cycle is so automatic that we generally ignore it until we are slapped in the face by it. Just as we don't control or pay attention to the beating of our hearts unless the beat skips or falters, unless we are confronted by flood or drought, unless our plans are disrupted by rain, we ignore the water cycle, preferring to believe that the water we drink comes out of the faucet, and not from deep within the belly of the earth, placed there by a process we only dimly comprehend. However, water is as essential to us and to the earth as blood is in our bodies, and the constant cycle water travels makes our lives possible.

The earth's blood, water, is pumped, not by a heart, but by the hydrological cycle—the water cycle. A titanic force of nature, the water cycle is beyond our control—a fact that we ignore until weather patterns shift and suddenly inundated rivers flow where they will and not within human-engineered banks, floodwalls, dikes, and levees. In the water cycle, the water evaporated from the oceans falls as rain, hail, sleet, or snow and it strikes the earth again; the cycle continues.

In cities, in summer, rain strikes hot cement and asphalt and swiftly evaporate, or run into storm drains, swiftly rejoining the cycle. In fields, rain brings essential moisture to crops, and sinking deeper into the earth, ends as groundwater. If water strikes a forested area, the forest canopy breaks the force of the falling drops. The forest floor, carpeted in twigs, leaves, moss, dead, and decaying vegetation, keeps the

DOI: 10.1201/9781003288947-9

soil from splashing away in erosion as the water returns to the depths of the earth, or runs over the land to join a stream.

Whenever water strikes the earth, it flows along four pathways, which carry water through the cycle as our veins, arteries, and capillaries carry our blood to our cells.

It may evaporate directly back into the air.
It may flow overland into a stream as runoff.
It may soak into the ground and be taken up by plants for evapotranspiration.
It may seep down to groundwater.
Whatever pathway water takes, one fact is certain: water is dynamic, vital, constantly on the move. And like human blood, which sustains our lives, earth's blood, to sustain us as well, must continue to flow.

– Frank R. Spellman (2007)

THE 411 ON WIND TURBINE HYDRAULICS

The reader may wonder: Why is this chapter opened with an epigraph describing the earth's blood, the water cycle, and the human heart? Moreover, one might ask what does the hydrological system have to do with wind turbine hydraulics? Well, this analogous comparison is done for good reason. Consider, for example, that hydraulics is an engineering discipline which applies the theoretical science of fluid mechanics in order to generate control and transmit power through the use of pressurized liquids. So, again, what does all this have to do with the functioning of the human heart? Hydraulics (plural but commonly used as singular) is/are at work in the human body—in the human heart and circulatory system (Figure 7.1).

Hydraulics is all about fluid—hydraulic fluid; moreover, the human heart is also about fluid—blood. The hydraulic fluid in an industrial hydraulic system such as a wind turbine consists of mineral oil, a phosphate ester, or water glycol or skydrol (i.e., a fire-resistant aviation hydraulic fluid), depending on the application. Along with hydraulic fluid the industrial hydraulic system (aka power pack or power unit—similar to the human heart) is comprised of and forms the basics of a hydraulic system:

- A prime mover
- A reservoir of fluid
- Pipework and hoses
- Hydraulic pump
- Actuators and valves
- Hydraulic power units
- Hydraulic cylinder
- Hydraulic pitch components

In the human body it is the heart that provides strong, regular contractions utilizing the heart's pacemaker—the sinoatrial node, the heart's prime mover, which applies pressure. In industrial hydraulic and wind turbine systems, it is an electric motor that powers the hydraulic pump.

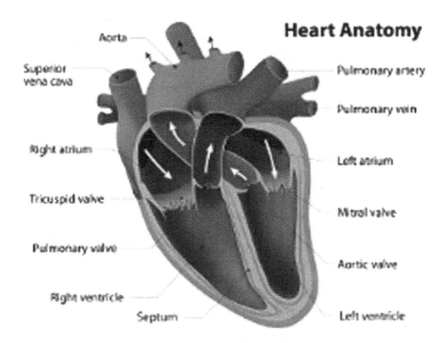

Heart Anatomy

Aorta

Superior vena cava

Pulmonary artery

Pulmonary vein

Right atrium

Left atrium

Tricuspid valve

Mitral valve

Pulmonary valve

Aortic valve

Right ventricle

Septum

Left ventricle

FIGURE 7.1 Human heart. Source: National Institutes of Health. Accessed 09/17/21 @ https://www.nig.nih.gov/health/heart.

Besides electrical motor power, a modern industrial wind turbine can use hydraulic power, which is actually the preferred power source. This is the case because hydraulics are better for heavy-duty uses, as they can produce more power and energy. Hydraulics are applicable in all forms of wind technologies, and it does not have the disadvantages of power by electricity compared to hydraulics. For example, hydraulics systems have proven to be more dependable, more cost effective, and lighter-weight than their electric counterparts—the weight of the total turbine system is a consideration that must be taken seriously, and hydraulics are hundreds of pounds lighter than electrical power systems. Moreover, the mechanical gearbox used in a hydraulic-powered system is much lighter; again, reducing weight is a good thing. Also, a hydraulic-powered system has the advantage or minimizing torsional vibrations produced in the rotor hub. The bottom line on utilizing a hydraulic power system instead of an electrical system is cost, shelf-life (longevity), efficiency, reliability, and versatility.

With regard to versatility, hydraulics in modern industrial wind turbines is used in many applications. For instance, brake control for regulating blade rotation and setting, turning the blades for more speed are important hydraulic uses. Hydraulic tubing or hoses and connections for hose assemblies create the arterial flow, the hydraulic drivetrain with rotor and blades using a simple hose fitting. In most modern wind turbines, hoses are the preferred hydraulic fluid conveyance arteries and tubing. Tubing, while used in some applications, generally takes a back seat with regard to actual usage as compared to hoses.

The truth be told, a modern industrial wind turbine can't operate and produce electricity as per design without hydraulics. As explained in this chapter opening, hydraulics within the huge machine is like the human heart in its function to power turbine operation. This operation would not be possible without the proper circulation of hydraulic fluid—basically arterial flow within the wind turbine control system—without hoses. Therefore, at this point it is important to discuss hoses and connections.

PIPING BY ANOTHER NAME MIGHT BE HOSES

Industrial hoses, the type used in wind turbine hydraulic systems, are at the core of the operation. Industrial hydraulic hoses are classified as a slightly different tubular product and the best ones must be able to handle pressure capacity, durable, flexible, abrasion resistant, and able to deal with varying temperature exposure. Their basic function is the same as piping and tubing, and that is to carry fluids (liquids and gases) from one point to another. The outstanding feature of the industrial hose is its flexibility, which allows it to be used in applications where vibrations would make the use of rigid pipe impossible. Most industrial operations use industrial hoses to convey steam, water, air, and hydraulic fluids over short distances. It is important to point out that each application must be analyzed individually, and an industrial hose must be selected which is compatible with the system specification.

In this section, we study industrial hoses—what they are, how they are classified and constructed, and the ways in which sections of hoses are connected to one another and to components. The maintenance requirements for industrial hoses and what the turbine maintenance operator should look for during routine inspections or checks for specific problems are also discussed.

Industrial hoses, piping, and tubing all are used to convey a variety of materials under a variety of circumstances. Beyond this similar ability to convey a variety of materials, however, there are differences between industrial hoses and piping and tubing. For example, in their construction and in their advantages, industrial hoses *are* different from piping and tubing. As mentioned, one of the requirements and outstanding advantages of the hose is its flexibility; its ability to bend means that the hose can meet the requirements of numerous applications that can't be met by rigid piping and some tubing systems. Two examples of this flexibility are in turbine pitch control and in the hose that supplies hydraulic fluids used to provide a pathway for liquids to safely stop the turbines from rotating at high speeds. Hoses must remain tough under extraordinary circumstances with quality that is apparent and undeniable. Rigid piping would be cumbersome and impractical to use in both situations.

Industrial hose is not only flexible but also has a dampening effect on vibration. Certain components within an industrial wind turbine vibrate while performing their function(s). Obviously, the built-in rigidity of piping and tubing will not allow vibrating occasions to stand up for very long under such conditions. Many of these devices are equipped with moving members that require the hydraulic fluid supply to move with them. In such circumstances, of course, rigid piping cannot be used.

It is important to note that the flexibility of industrial hose is not the only consideration that must be taken to account when selecting hose over either piping or tubing. That is, the hose must be selected according to the potential damaging conditions of an application. These conditions include the effects of pressure, temperature, and corrosion. Hose applications within the industrial wind turbine are exposed to the effects of all three of these. To meet such service requirements, hoses are manufactured from a number of different materials for different applications.

HOSE NOMENCLATURE

To gain a fuller understanding of industrial hoses and their applications, it is important to be familiar with the nomenclature or terminology normally associated with industrial hoses. Accordingly, in this section, the hose terminology wind turbine maintenance operators should be familiar with is discussed. Figure 7.2 is a cutaway view of a high-pressure air hose of the kind that supplies portable air hammers and drills and other pneumatic tools commonly used in turbine maintenance and repair operations. The hose is the most common type of reinforced nonmetallic hose in general use. Many of the terms given have already been mentioned. The I.D., which designates the hose size, refers to the inside diameter throughout the length of the hose body, unless the hose has enlarged ends. The O.D. is the diameter of the outside wall of the hose.

Important Point: If the ends of an industrial hose are enlarged, as shown in Figure 7.3, the letters E.E. are used (meaning *expanded* or *enlarged end*). Some hoses have enlarged ends to fit a fixed end of piping tightly (e.g., an automobile engine).

As shown in Figure 7.2, the *tube* is the inner section (i.e., the core) of the hose, through which the fluid flows. Surrounding the tube is the *reinforcement* material, which provides resistance to pressure—either from the inside or outside. Notice that the hose shown in Figure 7.2 has two layers of reinforcement *braid* (this braid is fashioned from a high-strength synthetic cord). The hose is said to be *mandrel-braided* because a spindle or core (the mandrel) is inserted into the tube before the reinforcing materials are put on. The mandrel provides a firm foundation over which the

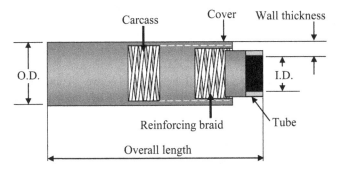

FIGURE 7.2 Common hose nomenclature.

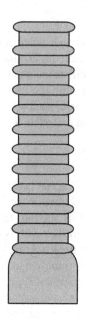

FIGURE 7.3 Expanded-end hose.

cords are evenly and tightly braided. The *cover* of the hose is an outer, protective covering. The hose in Figure 7.2 has a cover of tough, abrasion-resistant material.

The *overall length* is the true length of a straight piece of hose. Hose, which is not too flexible, is formed or molded in a curve (e.g., automobile hose used in heating systems; see Figure 7.4). As shown in Figure 7.4, the *arm* is the section of a curved hose that extends from the end of the hose to the nearest centerline intersection. The *body* is the middle section or sections of the curved hose. Figure 7.5 shows the *bend radius* (i.e., is the radius of the bend measured to the centerline) of the curved hose and designated as the radius *R*. In a straight hose, bent on the job, the radius of the bend is measured to the surface of the hose (i.e., the radius *r* in Figure 7.5).

Important Point: Much of the nomenclature used above does not apply to nonmetallic hose that is not reinforced. However, non-reinforced nonmetallic hose is not very common in wind turbine operations.

FACTORS GOVERNING HOSE SELECTION

The amount of *pressure* that a hose will be required conveys is one of the important factors governing hose selection. Typically, pressure range falls in any of three general groups:

- <250 psi (low-pressure applications)
- 250–3000 psi (medium-pressure applications)
- 3,000–6,000+ psi (high-pressure applications)

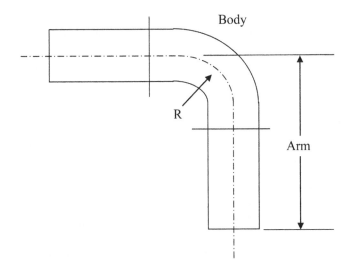

FIGURE 7.4   Bend radius.

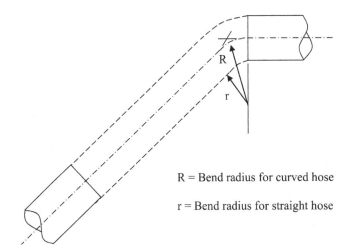

R = Bend radius for curved hose

r = Bend radius for straight hose

FIGURE 7.5   Bend radius: Measurement.

Important Point: Note that some manufacturers have their own distinct hose pressure rating scheme; we can't assume that a hose rated as "low-pressure" hose will automatically be useful at 100 or 200 psi. It may, in fact, be built for pressures not to exceed 50 psi, for example. Therefore, whenever we replace a particular hose, we must ensure that the same type of hose with the same pressure rating as the original hose is used. In high-pressure applications, this precaution is of particular importance.

In addition to the pressure rating of a hose, we must also consider, for some applications, the *vacuum rating* of a hose refers to suction hose applications, in which the pressure outside the hose is greater than the pressure inside the hose. It is important, obviously, to know the degree of vacuum that can be created before a hose begins to collapse. A drinking straw, for example, collapses rather easily if too much vacuum is applied. Thus, it has a low vacuum rating. In contrast, the lower automobile radiator hose (also works under vacuum) has a relatively higher vacuum rating.

STANDARDS, CODES, AND SIZES

Just as they have for piping and tubing, authoritative standards organizations have devised standards and codes for hoses.

Standards and Codes

Standards and codes are safety measures designed to protect personnel and equipment. For example, specifications are provided for working pressures, sizes, and material requirements. The working pressure of a hose, for example, is typically limited to one-fourth, or 25%, of the amount of pressure needed to burst the hose. Let's look at an example. If we have a hose that has a maximum rated working pressure of 200 psi, it should not rupture until 800 psi has been reached, and possibly not even then. Thus, the use of hoses that meet specified standards or codes is quite evident and necessary.

HOSE SIZE

The parameter typically used to designate hose size is its inside diameter (I.D.). In regard to the classification of hose, ordinarily, a *dash numbering* system is used. Current practice by most manufacturers is to use the dash system to identify both hose and fittings. In determining the size of a hose, we simply convert the size into 16ths. For example, a hose size of 1/2 in. (a hose with a 1/2-in. I.D.) is the same as 8/16 in. The numerator of the fraction (the top number, or "8" in this case) is the dash size of the hose. In the same way, a 1 1/2-in. size can be converted to 24/16 in. and so is identified as a –24 (pronounced "dash 24") hose. By using the dash system, we can match a hose line to the tubing or piping section and be sure the I.D. of both will be the same. This means, of course, that the non-turbulent flow of fluid will not be interrupted. Based on I.D., hoses range in size from 3/16 in. to as large as 24 in.

HOSE CLASSIFICATIONS

Hose is classified in a number of ways; for example, hose can be classified by type of service (hydraulic, pneumatic, corrosion-resistant), by material, by pressure, and by type of construction. Hose may also be classified by type. The three types include metallic, nonmetallic, and reinforced nonmetallic. Generally, the terminology is the same for each type.

METALLIC HOSE

The construction of a braided, flexible all-metal hose includes a tube of corrugated bronze. The tube is covered with the woven metallic braid to protect against abrasion and to provide increased resistance to pressure. Metal hose is also available in steel, aluminum, Monel®, stainless steel, and other corrosion-resistant metals in diameters up to 3 in. and in lengths of 24 in. In addition to providing protection against abrasion and resistance to pressure, a flexible metal hose also dampens vibration. For example, the wind turbine can experience a considerable amount of vibration. The use of flexible hoses in such machines increases their portability equipment and dampens vibrations. Other considerations such as constant bending at high temperatures and pressures are extremely detrimental on most other types of hoses.

Other common uses for metallic hoses include serving as steam lines, lubricating lines, gas and oil lines, and exhaust hose for diesel engines. The corrugated type, for example, is used for high-temperature, high-pressure leak-proof service. Another type of construction is the *interlocked* flexible metal hose, used mainly for low-pressure applications. The standard shop oil can use a flexible hose for the flexible spout. Other metal hoses, with a liner of flexible, corrosion-resistant material, are available in diameters of up to 24 in.

Another type of metallic hose is used in the ductwork. This type of hose is usually made of aluminum, galvanized steel, and stainless steel and is used to protect against corrosive fumes, as well as gases at extreme hot or cold temperatures. This hose is fire resistant because it usually does not burn.

NONMETALLIC HOSE

Relatively speaking, the use of hose is not a recent development. Hoses, in fact, have been used for one application or another for hundreds of years. Approximately 100 years ago, after new developments in the processing of rubber, layering rubber around mandrels was usually used to make hoses. Later, the mandrel was removed, leaving a flexible rubber hose. However, these flexible hoses tended to collapse easily. Even so, they were an improvement over the earlier types. Manufacturers later added layers of rubberized canvas. This improvement gave hoses more strength and gave them the ability to handle higher pressures. Later, after the development of synthetic materials, manufacturers had more rugged and more corrosion-resistant rubber-type materials to work with. Today, neoprene, nitrile rubber, and butyl rubber are commonly used in hose.

The current manufacturing practice is not to make hoses from a single material. Instead, different materials form layers in the hose, reinforcing it in various ways for strength and resistance to pressure. Hose manufactured today usually has a rubber-type inner tube or a synthetic (e.g., plastic) lining surrounded by a *carcass* (usually braided) and cover. The type of carcass braiding used is determined by the requirements of the application.

To reinforce a hose, two types of braiding are used, *vertical* braiding and *horizontal* braiding. Vertical braiding strengthens the hose against pressure applied at right angles to the centerline of the hose. Horizontal braiding strengthens the hose along its length, giving it greater resistance to expansion and contraction.

Types of Nonmetallic Hose

Descriptions of the types of nonmetallic hose follow, with references to their general applications.

Vertical-Braided Hose—It has an inner tube of seamless rubber (see Figure 7.6). The reinforcing wrapping (carcass) around the tube is made of one or more layers of braided yarn. This type of hose is usually made in lengths of up to 100 ft with I.D.s of up to 1.5 in. Considered a small hose, it is used in low-pressure applications to carry fuel oil, acetylene gas and oxygen for welding, water for lawns, gardens, and other household uses, and paint for spraying.

Horizontal-Braided Hose—It is mandrel built; it is used to make hose with an I.D. of up to 3 in. Used in high-pressure applications, the seamless rubber tube is reinforced by one or more layers of braided fibers or wire. This hose is used to carry propane and butane gas and steam, and for various hydraulic applications that require high working pressures.

Reinforced Horizontal-Braided-Wire Hose—In this type of hose, the carcasses around the seamless tube are made up of two or more layers of fiber braid with steel wire reinforcement between them. The I.D. may be up to 4 in. Mechanically very strong, this hose is used where there are high working pressures and/or strong suction (vacuum) forces, such as in chemical transfer and petroleum applications.

Wrapped Hose—Made in diameters up to 24 in., wrapped hose is primarily used for pressure service rather than suction. The hose is constructed of mandrels, and to close tolerances (see Figure 7.7). It also has a smooth bore, which encourages laminar flow and avoids turbulence. Several plies (layers) of woven cotton or synthetic fabric make up the reinforcement. Selected for their resistance to corrosive fluids, the tube is made from a number of synthetic rubbers. It is also used in sandblasting applications.

Wire-Reinforced Hose—In this type of hose, wires wound in a spiral around the tube, or inside the carcass, in addition to a number of layers of wrapped fabrics, provide the reinforcement (see Figure 7.8). With I.D.s of 16–24 in. common, this

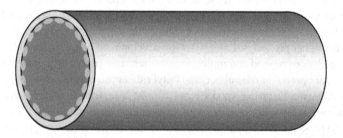

FIGURE 7.6   Vertical-braided hose.

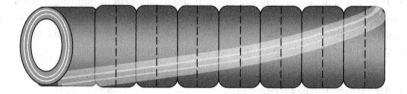

FIGURE 7.7   Wrapped hose.

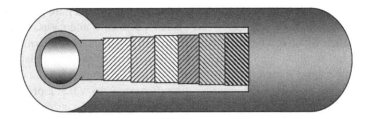

FIGURE 7.8   Wire-reinforced hose.

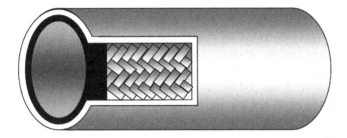

FIGURE 7.9   Wire-woven hose.

type of hose is used in oil-suction and discharge situations that require special hose ends, maximum suction (without collapsing), or special flexing characteristics (must be able to bend in a small radius without collapsing), or a combination of these three requirements.

Wire-Woven Hose—It (see Figure 7.9) has cords interwoven with a wire running spirally around the tube and is highly flexible, low in weight, and resistant to collapse even under suction conditions. This kind of hose is well suited for such negative pressure applications.

Other Types of Nonmetallic Hose—Hoses are also made of other nonmetallic materials, many of them non-reinforced. For example, materials like Teflon®, Dacron®, polyethylene, and nylon have been developed. Dacron remains flexible at very low temperatures, even as low as −34°C (up to −350°F), nearly the temperature of liquid nitrogen. Consequently, these hoses are used to carry liquefied gas in cryogenic applications. Where corrosive fluids and fluids up to 230+°C (up to 450°F) are to be carried, Teflon is often used. Teflon can also be used at temperatures as low as −55°C (−65°F). Usually sheathed in a flexible, braided metal covering, Teflon hoses are well protected against abrasion; they also have added resistance to pressure.

Nylon hoses (small diameter) are commonly used as air hoses, supplying compressed air to small pneumatic tools. The large plastic hoses (up to 24 in.) we use to ventilate utility access holes are made of such neoprene-coated materials as nylon fabric, glass fabric, and cotton duck. The cotton duck variety is for light-duty applications. The glass fabric type is used with portable heaters and for other applications involving hot air and fumes.

Various hoses made from natural latex, silicone rubber, and pure gum are available. The pure gum hose will safely carry acids, chemicals, and gases. Small hoses of natural latex, which can be sterilized, are used in hospitals, with pharmaceuticals, blood, and intravenous solutions, and in food-handling operations and laboratories. Silicone rubber hose is used in situations where extreme temperatures and chemical reactions are possible. It is also used for aircraft starters, to which it provides compressed air in very large volumes. Silicone rubber hose works successfully over a temperature range from −57°C (−70°F) to 232°C (450°F).

HOSE COUPLINGS

The methods of connecting or coupling hoses vary. Hose couplings may be either permanent or reusable. They can also be manufactured for the obvious advantage of quick-connect or quick-disconnect. Probably the best example of the need for quick-connect is fire hose—quick-disconnect couplings permit a rapid connection between separate lengths of hose, and between hose ends and hydrants or nozzles. Another good example of where the quick-connect, quick-disconnect feature is user-friendly is in plant or mobile compressed air systems—a single line may have a number of uses. Changes involve disconnecting one section and connecting another. In plant shops, for example, the compressed air supply from a single source is used to power pneumatic tools, cleaning units, paint sprayers, and so on. Each unit has a hose that is equipped for rapid connecting and disconnecting at the fixed airline.

CAUTION: Before connections are broken, unless quick-acting, self-closing connectors are used, pressure must be released first.

For general low-pressure applications, a coupling like that shown in Figure 7.10 is used. To place this coupling on the hose by hand, first cut the hose to the proper length, then oil the inside of the hose and outside of the coupling stem. Force hose over stem into protective cap until it seats against the bottom of the cap. No brazing is involved, and the coupling can be used repeatedly. After the coupling has been inserted in the hose, a yoke is placed over it in such a way that its arms are positioned along opposite sides of the hose behind the fitting. The arms are then tightly strapped or banded.

CAUTION: Where the pressure demands are greater, such a coupling can be blown out of the tube. Hose couplings designed to meet high-pressure applications must be used.

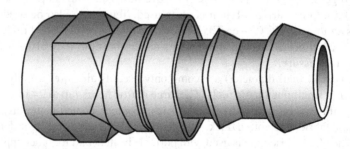

FIGURE 7.10   Low-pressure hose coupling.

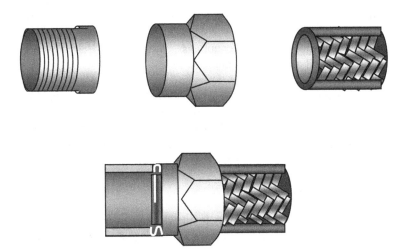

FIGURE 7.11 Coupling installation for all-metal hose.

A variation of this type uses a clamp that is put over the inner end of the fitting and is then tightly bolted, thus holding the hose firmly. In other cases, a plain clamp is used. Each size of clamp is designed for a hose of a specified size (diameter). The clamp slides snugly over the hose and is then crimped tight by means of a special hand tool or air-powered tool.

Coupling for all-metal hose, described earlier, involves two brazing operations, as shown in Figure 7.11. The sleeve is slipped over hose end and brazed to it, and then the nipple is brazed to the sleeve.

Important Point: For large hoses of rugged wall construction, it is not possible to insert push-on fittings by hand. Special bench tools are required.

Quick-connect, quick-disconnect hose couplings provide flexibility in many plant process lines where a number of different fluids or dry chemicals from a single source are either to be blended or routed to different vats or other containers. Quick-connect couplings can be used to pump out excavations, utility access holes, and so forth. They would not be used, however, where highly corrosive materials are involved.

HOSE MAINTENANCE

All types of equipment and machinery require proper care and maintenance, including hoses. Depending on the hose type and its application, some require more frequent checking than others. The maintenance procedures required for most hose are typical and are outlined here as an example. To maintain a hose, we should:

1. Examine for cracks in the cover caused by weather, heat, oil, or usage
2. Look for a restricted bore because of tube-swelling or foreign objects
3. Look for cover blisters, which permit material pockets to form between carcass and cover

4. Look for leaking materials, which is usually caused by improper couplings or faulty fastenings of couplings
5. Look for corrosion damage to couplings
6. Look for kinked or otherwise damaged hose

CAUTION: Because any of the faults listed above can result in a dangerous hose failure, regular inspection is necessary. At the first sign of weakness or failure, replace the hose. System pressure and temperature gauges should be checked regularly. Do not allow the system to operate above design conditions—especially when the hose is a component of the system.

PIPE AND TUBE FITTINGS

The term *piping* refers to the overall network of pipes or tubing, fittings, flanges, valves, and other components that comprise a conduit system used to convey fluids. Whether a piping system is used to simply convey fluids from one point to another or to process and condition the fluid, piping components serve an important role in the composition and operation of the system. A system used solely to convey fluids may consist of relatively few components, such as valves and fittings, whereas a complex chemical processing system may consist of a variety of components used to measure, control, condition, and convey the fluids. In this section, the characteristics and functions of various piping and tubing fittings are described (Geiger, 2000).

FITTINGS

The primary function of *fittings* is to connect sections of piping and tubing and to change the direction of flow. Whether used in piping or tubing, fittings are similar in shape and type, even though pipefittings are usually heavier than tubing fittings. Several methods can be used to connect fittings to piping and tubing systems. However, most tubing is threadless because it does not have the wall thickness needed to carry threads. Most pipes, on the other hand, because they have heavier walls, are threaded.

In regard to changing direction of flow, the simplest way would be simply to bend the conduit, which, of course, is not always practical or possible. When piping is bent, it is usually accomplished by the manufacturer in the production process (in larger shops equipped with their own pipe-bending machines), but not by the maintenance operator on the job. Tube bending, on the other hand, is a common practice. Generally, a tubing line requires fewer fittings than a pipeline; however, in actual practice, many tube fittings are used.

Important Point: Recall that improperly made bends can restrict fluid flow by changing the shape of the pipe and weakening the pipe wall.

Fittings are basically made from the same materials (and in the same broad ranges of sizes) as piping and tubing, including bronze, steel, cast iron, glass, and plastic. Various established standards are in place to ensure that fittings are made from the proper materials and are able to withstand the pressures required; they are also made to specific tolerances, so that they will properly match the piping or tubing that they join. A fitting stamped "200 lb," for example, is suitable (and safe) for use up to 200 psi.

FUNCTIONS OF FITTINGS

Fittings in piping and tubing systems have five main functions:

- Changing the direction of flow
- Providing branch connections
- Changing the sizes of lines
- Closing lines
- Connecting lines

Changing the Direction of Flow

Usually, a 45° or 90° *elbow* (or "ell") fitting is used to change the direction of flow. Elbows are among the most commonly used fittings in piping and are occasionally used in tubing systems. Two types of 90° elbows are shown in Figure 7.12. From the figure, it is apparent that the *long-radius* fitting (the most preferred elbow) has the more gradual curve of the two. This type of elbow is used in applications where the rate of flow is critical and space presents no problem. Flow loss caused by turbulence is minimized by the gradual curve. The *short-radius* elbow (see Figure 7.12) should not be used in a system made up of long lines and has many changes in direction. Because of the greater frictional loss in the short-radius elbow, heavier more expensive pumping equipment may be required. Figure 7.13 shows a *return bend* fitting that carries fluid through a 180° ("hairpin") turn. This type of fitting is used for piping in heat exchangers and heater coils. Note that tubing, which can be bent into this form, does not require any fittings in this kind of application.

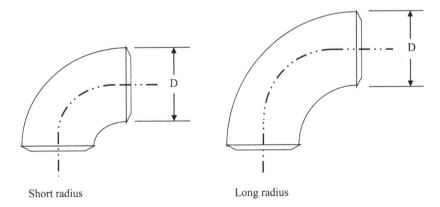

Short radius   Long radius

FIGURE 7.12   Short- and long-radius elbows.

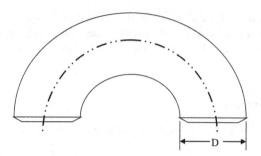

FIGURE 7.13   Long-radius return bend.

Providing Branch Connections

Because they are often more than single lines running from one point to another point, piping and tubing systems usually have a number of intersections. In fact, many complex piping and tubing systems resemble the layout of a town or city.

Changing the Sizes of Lines

For certain applications, it is important to reduce the volume of fluid flow or to increase flow pressure in a piping or tubing system. To accomplish this, a *reducer* (which reduces a line to a smaller pipe size) is commonly used.

Important Point: Reducing is also sometimes accomplished by means of a bushing inserted into the fitting.

Sealing Lines

Pipe *caps* are used to seal or close off (similar to "corking" a bottle) the end of a pipe or tube. Usually, caps are used in a part of the system that has been dismantled. To seal off openings in fittings, *plugs* are used. Plugs also provide a means of access into the piping or tubing system in case the line becomes clogged.

Connecting Lines

To connect two lengths of piping or tubing together, a coupling or union is used. A *coupling* is simply a threaded sleeve. A *union* is a three-piece device that includes a threaded end, an internally threaded bottom end, and a ring. A union does not change the direction of flow, or close off the pipe, or provide for a branch line. Unions make it easy to connect or disconnect pipes without disturbing the position of the pipes.

Figure 7.14 is a diagram of a shortened piping system, which illustrates how some fittings are used in a piping system. (Figure 7.14 is only for illustrative purposes; it is unlikely that such a system with so many fittings would actually be used.)

TYPES OF CONNECTIONS

Pipe connections may be screwed, flanged, or welded. Each method is widely used, and each has its own advantages and disadvantages.

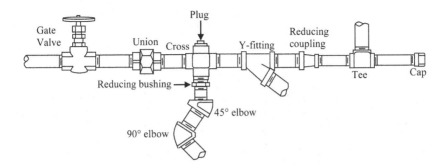

FIGURE 7.14   Diagram of a hypothetical shortened piping system.

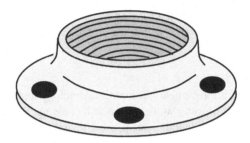

FIGURE 7.15   Flanged fitting.

Screwed Fittings

Screwed fittings are joined to the pipe by means of threads. The main advantage of using threaded pipe fittings is that they can be easily replaced. The actual threading of a section of replacement pipe can be accomplished on the job. The threading process itself, however, which cuts right into the pipe material, may weaken the pipe in the joint area. The weakest links in a piping system are the connection points. Because threaded joints can be potential problem areas, especially where higher pressures are involved, the threads must be properly cut to ensure the "weakest" link is not further compromised. Typically, the method used to ensure a good seal in a threaded fitting is to coat the threads with a paste dope. Another method is to wind the threads with Teflon® tape.

Flanged Connections

Figure 7.15 shows a flanged fitting. *Flanged* fittings are forged or cast-iron pipes. The *flange* is a rim at the end of the fitting, which mates with another section. Pipe sections are also made with flanged ends. Flanges are joined either by being bolted or welded together. The flange faces may be ground and lapped to provide smooth, flat mating surfaces. Obviously, a tight joint must be provided to prevent leakage of fluid and pressure.

Figure 7.16 shows a typical example of a flanged joint. The mating parts are bolted together with a gasket inserted between their faces to ensure a tight seal. The procedure requires proper alignment of clean parts and tightening of bolts.

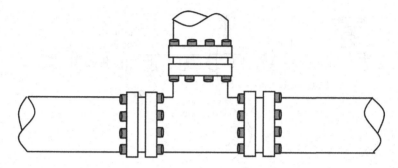

FIGURE 7.16   Flanged joint.

Important Point: Some flanges have raised faces and others have plain faces. Like faces must be matched—a flange with a raised face should never be joined to one with a plain face.

Welded Connections

Currently, because of improvements in piping technology and welding techniques/equipment, the practice of using welded joints is increasing. When properly welded, a piping system forms a continuous system, which combines piping, valves, flanges, and other fittings. Along with providing a long leak-proof and maintenance-free life, the smooth joints simplify insulation and take up less room.

TUBING FITTINGS AND CONNECTIONS

Tubing is connected by brazed or welded flange fittings, compression fittings, and flare fittings.

The *welded flange* connection is a reliable means of connecting tubing components. The flange welded to the tube end fits against the end of the fitting. The lock-nut of the flange is then tightened securely onto the fitting. The *compression fitting* connection uses a ferrule that pinches the tube as the locknut is tightened on the body of the fitting. The *flare fitting* connection uses tubing flared on one end of the tubing that matches the angle of the fitting. The tube's flared end is butted against the fitting, and a locknut is screwed tightly onto the fitting, sealing the tube connection properly.

MISCELLANEOUS FITTINGS

Other fittings used for flanged connections include expansion joints and vibration dampeners. *Expansion joints* function to compensate for slight changes in the length of pipe by allowing joined sections of rigid pipe to expand and contract with changes in temperature. They also allow pipe motion, either along the length of the pipe or to the side, as the pipe shifts around slightly after installation. Finally, expansion joints also help dampen vibration and noise carried along the pipe from distant equipment (e.g., pumps).

One type of expansion joint has a leak-proof tube that extends through the bore and forms the outside surfaces of the flanges. Natural or synthetic rubber compounds are normally used, depending on the application.

Other types of expansion joints include metal corrugated types, slip-joint types, and spiral-wound types. In addition, high-temperature lines are usually made up with a large bend or loop to allow for expansion.

Vibration dampeners absorb vibrations that, unless reduced, could shorten the life of the pipe and the service life of the operating equipment. They also eliminate line humming and hammering (water hammer) carried by the pipes.

THE BOTTOM LINE

The chapter opened with an analogy comparing the human heart and its importance and function with the heart of the hydraulics system within an industrial wind turbine. Wind turbine systems, like the human heart, can't function without pressurized fluid and accessories such as conveyance lines such as arteries, veins, and valves.

REFERENCES

Geiger, E.L. 2000. Piping components, in *Piping Handbook*, 7th Edition. Nayyar, M.L., Ed., New York: McGraw-Hill.
Spellman, F.R. 2007. *The Science of Water*, 2nd Edition. Boca Raton, FL: CRC Press.

THE BOTTOM LINE

REFERENCES

8 Electrical Aspects

INTRODUCTION

At the beginning of my lectures on wind power in one of my environmental engineering classes for upper level and graduate students, I always ask the students the same introductory question:

What do you people think about wind turbines and wind energy production? I have never had a problem receiving answers to these questions—and the answers are wide-ranging, down-to-earth, so to speak and in many cases downright humorous. The nature of the answers I receive is always interesting and more importantly gets the ball rolling in my classes dealing with renewable energy and especially wind turbines for energy production. Engaging the students in conversation with their input, their discussion of opinions, views, thoughts—getting them to think (isn't that the essence of college instruction?) is when teaching is not only fulfilling but educating for all involved, including the teacher.

Anyway, some of the responses I have elicited from the students are listed as follows:

"Well, watching them monsters twirl in the wind is hypnotizing, for sure."
"I like the swoop of the blades as they cut through the air."
"I think they are ugly."
"They are loud … noisy."
"I lived by a wind farm and hated that strobe-like effect those blade lights give off at night."
"For me, they give off instant vertigo."
"They are monsters."
"They are bird killers … I like birds and killing them is not right!"
"They are a good way to produce electricity … clean electricity."

Now, this last comment is my bottom line. I always wait until someone; anyone in the class makes a statement similar to this last one about wind turbines producing electricity. It is at this point when I take over the class and present my lecture, much of which is included in the material presented in this chapter.

ELECTRICAL FACTORS IMPORTANT TO WIND ENERGY

"They are a good way to produce electricity … clean electricity." This certainly is, as far as it goes, a true statement. However, when I take over my classes I immediately expand this statement whereby I list electrical factors important to wind energy on my large flipchart, and flip the sheets one after the other, to show students that there

DOI: 10.1201/9781003288947-10

is more to a wind turbine's interface with electricity than just its production. My list includes:

Power generation—generators and power converters
Instrumentation and distribution—power cables, switch gear, circuit breakers, transformers, and power quality
Control—sensors, controller, yaw and pitch motors, and solenoids
Site monitoring—data measurement, recording, and analysis
Storage—batteries, rectifiers, and converters
Lightning protection—grounding and lightning rods

ELECTRICAL POWER CONCEPTS

In order to understand electrical power use generated by wind turbines, the wind turbine's internal use of electricity, and the subsequent distribution of generated electricity, it is important to be knowledgeable in the fundamentals of electricity. Specific topics that are introduced in this chapter include:

- Simple circuits
- Voltage
- Current
- Resistance
- Impedance
- Power
- True power
- Apparent power
- Conductors
- Series circuits
- Parallel circuits
- Rectifiers
- Inverters
- Series-parallel circuits
- Switches
- Batteries
- Electromagnets
- Inductance
- Capacitance
- Inductive and capacitance reactance
- ac current/voltage
- dc current/voltage
- Reactance
- Power factor
- ac motors/generators
- dc motors/generators
- Transformers
- Power distribution

ELECTRICITY: WHAT IS IT?

Wind turbine maintenance operators and technicians generally have little difficulty in recognizing electrical equipment. Electrical equipment is everywhere within and is integral to wind turbines and electricity output. For example, typical wind turbines are outfitted with electrical equipment that

- Generates electricity (a generator—or emergency generator)
- Stores electricity (batteries)
- Steps up or steps down electricity levels (transformers)
- Transports or transmits and distributes electricity throughout the plant site (wiring distribution systems)
- Measures electricity (meters)
- Converts electricity into other forms of energy (rotating shafts—mechanical energy, heat energy, light energy, chemical energy, or radio energy)
- Protects other electrical equipment (fuses, circuit breakers, or relays)
- Operates and controls other electrical equipment (motor controllers)
- Converts some condition or occurrence into an electric signal (sensors)
- Converts some measured variable to a representative electrical signal (transducers or transmitters)

A question that does not always result in a resounding note of assurance, for the untrained wind turbine operator/maintenance technician, is for him or her to explain to us in very basic terms how electricity works to make their turbine equipment operate and generate; the answers we would receive would be varied, jumbled, disjointed—and probably not all that accurate. The question? Hold on, we are getting to that. Even on a basic level, how many wind turbine operators (or anyone else, for that matter) are able to accurately answer the question: What is electricity?

Probably very few. Why do a few maintenance operators and technicians in wind turbine operations know so little about electricity? Part of the answer resides in the fact that operators are expected to know so much (and they are—and they do) but are given so little opportunity to be properly trained.

We all know that experience is the great trainer. As an example, let us look at what an operator assigned to change the bearings on an a-c three-phase motor would need to know to accomplish this task. The wind turbine maintenance operator would have to know:

1. How to deenergize the equipment (i.e., proper lockout/tagout procedures)
2. Once deenergized, how to properly disassemble the motor coupling from the device it operates (e.g., a motor coupling from a pump shaft) and the proper tools to use
3. Once uncoupled, how to properly disassemble the motor end-bells (preferably without damaging the rotor shaft)
4. Once disassembled, how to recognize if the bearings are really in need of replacement (though once removed from the end-bells, the bearings are typically replaced)

Questions the operator needs to know how to find answers to include the following:

1. If the bearings need replacement, how are they to be removed without causing damage to the rotor shaft?
2. Once removed, what bearings should be used to replace the old bearings?
3. When the proper bearings are identified and obtained, how are they to be installed properly?
4. When the bearings are replaced properly, how is the motor to be reassembled properly?
5. Once the motor is correctly put back together, how is it properly aligned to the pump and then reconnected?
6. What is the test procedure to ensure that the motor has been restored properly to full operational status?

Each of the steps and questions on the above procedures is important—obviously, errors at any point in the procedure could cause damage (maybe more damage than occurred in the first place). Here's another question to go along with that one—does the wind turbine maintenance operator need to know electricity to perform the sequence of tasks described above?

The short answer is no, not exactly. Fully competent operators (who received most of their training via on-the-job experience) are usually qualified to perform the bearing-change-out activity on most site motors with little difficulty.

The long answer is yes. Consider the motor mechanic who tunes your automobile engine. Then ask yourself: is it important that the mechanic have some understanding of internal combustion engines? It is important. You probably do, too. It is important for the operator who changes bearings on an electrical motor to have some understanding of how the electric motor operates.

This chapter is all about the how and the where—the here and the now.

NATURE OF ELECTRICITY*

The word *electricity* is derived from the Greek word "electron" (meaning amber). Amber is a translucent (semitransparent) yellowish fossilized mineral resin. The ancient Greeks used the words "electric force" in referring to the mysterious forces of attraction and repulsion exhibited by amber when it was rubbed with a cloth. They did not understand the nature of this force. They could not answer the question, "What is electricity?" The fact is this question remains unanswered. Today, we often attempt to answer this question by describing the effect and not the force. That is, the standard answer given is, "the force which moves electrons" is electricity, which is about the same as defining a sail as "that force which moves a sailboat."

At the present time, little more is known than the ancient Greeks knew about the fundamental nature of electricity, but we have made tremendous strides in harnessing

---

* The material in this section is taken from F. Spellman (2001), *Basic Electricity for Maintenance Operators*. Boca Raton, FL: CRC Press.

and using it. As with many other unknown (or unexplainable) phenomena, elaborate theories concerning the nature and behavior of electricity have been advanced and have gained wide acceptance because of their apparent truth—and because they work. When something works and you think you know why it is difficult to argue against.

Scientists have determined that electricity seems to behave in a constant and predictable manner in given situations or when subjected to given conditions. Scientists, like Faraday, Ohm, Lenz, and Kirchhoff, have described the predictable characteristics of electricity and electric current in the form of certain rules. These rules are often referred to as laws. Thus, though electricity itself has never been clearly defined, its predictable nature and easily used form of energy have made it one of the most widely used power sources in modern times.

The bottom line: You can "learn" about electricity by learning the rules, or laws, applying to the behavior of electricity, and by understanding the methods of producing, controlling, and using it. Thus, this learning can be accomplished without ever having determined its fundamental identity.

You are probably scratching your head—puzzled, mystified, baffled, confused, or just plain bewildered.

Welcome to a long-standing and ancient club.

We understand the main question running through your brain at this exact moment: "This is a chapter about basic electricity and the author can't even explain what electricity is?"

That is correct; I cannot provide a definitive, an absolute answer. Who can? The point is no one can definitively define electricity. Electricity is one of those subject areas where the old saying "we don't know what we don't know about it" fits perfectly.

Again, thankfully, there are a few theories about electricity that have so far stood the test of extensive analysis and much time (relatively speaking, of course). One of the oldest and most generally accepted theories concerning electric current flow (or electricity) is known as the *Electron Theory*.

The electron theory states that electricity or current flow is the result of the flow of free electrons in a conductor. Thus, electricity is the flow of free electrons or simply electron flow. In addition, in this text, this is how we define electricity; that is, again, electricity is the flow of free electrons.

Electrons are extremely tiny particles of matter. To gain an understanding of electrons and exactly what is meant by "electron flow," it is necessary to briefly discuss the structure of matter.

The Structure of Matter

Matter is anything that has mass and occupies space. To study the fundamental structure or composition of any type of matter, it must be reduced to its fundamental components. All matter is made of *molecules*, or combinations of *atoms* (Greek: not able to be divided), that are bound together to produce a given substance, such as salt, glass, or water. For example, if you keep dividing water into smaller and smaller drops, you will eventually arrive at the smallest particle that was still water. That particle is the molecule, which is defined as the *smallest bit of a substance that retains the characteristics of that substance.*

Note: Molecules are made up of atoms, which are bound together to produce a given substance.

Atoms are composed, in various combinations, of subatomic particles of *electrons*, *protons*, and *neutrons*. These particles differ in weight (a proton is much heavier than the electron) and charge. We are not concerned with the weights of particles in this text, but the **charge** is extremely important in electricity. The electron is the fundamental negative charge (−) of electricity. Electrons revolve about the nucleus or center of the atom in paths of concentric *orbits*, or shells (see Figure 8.1). The proton is the fundamental positive (+) charge of electricity. Protons are found in the nucleus. The number of protons within the nucleus of any particular atom specifies the atomic number of that atom. For example, the helium atom has two protons in its nucleus, so the atomic number is 2. The neutron, which is the fundamental neutral charge of electricity, is also found in the nucleus.

Most of the weight of the atom is in the protons and neutrons of the nucleus. Whirling around the nucleus are one or more negatively charged electrons. Normally, there is one proton for each electron in the entire atom so that the net positive charge of the nucleus is balanced by the net negative charge of the electrons rotating around the nucleus (see Figure 8.2).

Note: Most batteries are marked with the symbols + and − or with the abbreviations POS (positive) and NEG (negative). The concept of a positive or negative polarity and its importance in electricity will become clear later. However, for the moment, you need to remember that an electron has a negative charge and that a proton has a positive charge.

It was stated earlier that in an atom the number of protons is usually the same as the number of electrons. This is an important point because this relationship determines the kind of element (the atom is the smallest particle that makes up an

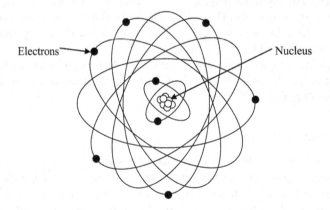

FIGURE 8.1   Electrons and nucleus of an atom.

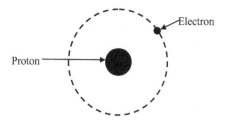

FIGURE 8.2   One proton and one electron = electrically neutral.

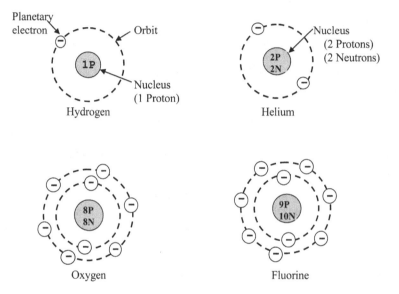

FIGURE 8.3   Atomic structure of elements.

element; an element retains its characteristics when subdivided into atoms) in question. Figure 8.3 shows a simplified drawing of several atoms of different materials based on the conception of electrons orbiting about the nucleus. For example, hydrogen has a nucleus consisting of one proton, around, which rotates one electron. The helium atom has a nucleus containing two protons and two neutrons with two electrons encircling the nucleus. Both of these elements are electrically neutral (or balanced) because each has an equal number of electrons and protons. Since the negative (−) charge of each electron is equal in magnitude to the positive (+) charge of each proton, the two opposite charges cancel.

A balanced (neutral or stable) atom has a certain amount of energy, which is equal to the sum of the energies of its electrons. Electrons, in turn, have different energies called *energy levels*. The energy level of an electron is proportional to its distance from the nucleus. Therefore, the energy levels of electrons in shells farther from the nucleus are higher than that of electrons in shells nearer the nucleus.

When an electric force is applied to a conducting medium, such as copper wire, electrons in the outer orbits of the copper atoms are forced out of orbit (i.e., liberating or freeing electrons) and are impelled along the wire. This electrical force, which forces electrons out of orbit, can be produced in a number of ways, such as: by moving a conductor through a magnetic field; by friction, as when a glass rod is rubbed with cloth (silk); or by chemical action, as in a battery.

When the electrons are forced from their orbits, they are called *free electrons*. Some of the electrons of certain metallic atoms are so loosely bound to the nucleus that they are relatively free to move from atom to atom. These free electrons constitute the flow of an electric current in electrical conductors.

Note: When an electric force is applied to a copper wire, free electrons are displaced from the copper atoms and move along the wire, producing electric current as shown in Figure 8.4.

If the internal energy of an atom is raised above its normal state, the atom is said to be *excited*. Excitation may be produced by causing the atoms to collide with particles that are impelled by an electric force as shown in Figure 8.4. In effect, what occurs is that energy is transferred from the electric source to the atom. The excess energy absorbed by an atom may become sufficient to cause loosely bound outer electrons (as shown in Figure 8.4) to leave the atom against the force that acts to hold them within.

Note: An atom that has lost or gained one or more electrons is said to be ionized. If the atom loses electrons it becomes positively charged and is referred to as a positive ion. Conversely, if the atom gains electrons, it becomes negatively charged and is referred to as a negative ion.

Conductors, Semiconductors, and Insulators

Electric current moves easily through some materials but with greater difficulty through others. Substances that permit the free movement of a large number of electrons are called *conductors*. The most widely used electrical conductor is copper because of its high conductivity (how good a conductor the material is) and cost-effectiveness.

Electrical energy is transferred through a copper or other metal conductor by means of the movement of free electrons that migrate from atom to atom inside the conductor (see Figure 8.4). Each electron moves a very short distance to the neighboring atom where it replaces one or more electrons by forcing them out of their orbits. The replaced electrons repeat the process in other nearby atoms until the movement is transmitted throughout the entire length of the conductor. A good conductor is said to have a low opposition, or **resistance**, to the electron (current) flow.

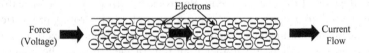

FIGURE 8.4   Electron flow in a copper wire.

Note: If lots of electrons flow through a material with only a small force (voltage) applied, we call that material a **conductor**.

Table 8.1 lists many of the metals commonly used as electric conductors. The best conductors appear at the top of the list, with the poorer ones shown last.

Note: The movement of each electron (e.g., in copper wire) takes a very small amount of time, almost instantly. This is an important point to keep in mind later in the book, when events in an electrical circuit seem to occur simultaneously.

While it is true that electron motion is known to exist to some extent in all matter, some substances, such as rubber, glass, and dry wood, have very few free electrons. In these materials, large amounts of energy must be expended in order to break the electrons loose from the influence of the nucleus. Substances containing very few free electrons are called *insulators*. Insulators are important in electrical work because they prevent the current from being diverted from the wires.

Note: If the voltage is large enough, even the best insulators will break down and allow their electrons to flow.

Table 8.2 lists some materials that we often use as insulators in electrical circuits. The list is in decreasing order of ability to withstand high voltages without conducting.

A material that is neither a good conductor nor a good insulator is called a *semiconductor*. Silicon and germanium are substances that fall into this category. Because of their peculiar crystalline structure, these materials may under certain conditions act as conductors; under other conditions, as insulators. As the temperature is raised, however, a limited number of electrons become available for conduction.

Static Electricity

Electricity at rest is often referred to as *static electricity*. More specifically, when two bodies of matter have unequal charges and are near one another, an electric force is exerted between them because of their unequal charges. However, because they are

---

TABLE 8.1
Electrical Conductors

Electrical Conductors

Silver
Copper
Gold
Aluminum
Zinc
Brass
Iron
Tin
Mercury

TABLE 8.2
Insulators

Common Insulators

Rubber
Mica
Wax or paraffin
Porcelain
Bakelite
Plastics
Glass
Fiberglass
Dry wood
Air

not in contact, their charges cannot equalize. The existence of such an electric force, where current cannot flow, is **static electricity**.

However, static, or electricity at rest, will flow if given the opportunity. An example of this phenomenon is often experienced when one walks across a dry carpet and then touches a doorknob—a slight shock is usually felt and a spark at the fingertips is likely noticed. In the workplace, static electricity is prevented from building up by properly bonding equipment to the ground or the earth.

Charged Bodies

To fully grasp a clear understanding of static electricity, it is necessary to know one of the fundamental laws of electricity and its significance.

The fundamental law of charged bodies states, *Like charges repel each other and unlike charges attract each other.*

A positive charge and negative charge, being opposite or unlike, tend to move toward each other—attracting each other. In contrast, like bodies tend to repel each other. Electrons repel each other because of their like negative charges, and protons repel each other because of their like positive charges. Figure 8.5 demonstrates the law of charged bodies.

It is important to point out another significant part of the fundamental law of charged bodies—that is, *force of attraction or repulsion existing between two magnetic poles decreases rapidly as the poles are separated from each other.* More specifically, the force of attraction or repulsion varies directly as the product of the separate pole strengths and inversely as the square of the distance separating the magnetic poles, provided the poles are small enough to be considered as points.

Let's look at an example.

If you increased the distance between two north poles from 2 ft to 4 ft, the force of repulsion between them is decreased to one-fourth of its original value. If either pole strength is doubled, the distance remaining the same, the force between the poles will be doubled.

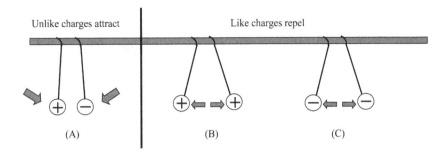

FIGURE 8.5 Reaction between two charged bodies. The opposite charge in (A) attracts. The like charges in (B) and (C) repel each other.

Coulomb's Law

Simply put, Coulomb's Law points out that the amount of attracting or repelling force which acts between two electrically charged bodies in free space depends on two things:

a. Their charges
b. The distance between them

Specifically, *Coulomb's Law* states, "Charged bodies attract or repel each other with a force that is directly proportional to the product of their charges and is inversely proportional to the square of the distance between them."

Note: The magnitude of the electric charge a body possesses is determined by the number of electrons compared with the number of protons within the body. The symbol for the magnitude of electric charge is Q, expressed in units of **coulombs (C)**. A charge of one positive coulomb means a body contains a charge of 6.25×10^{18}. A charge of one negative coulomb, $-Q$, means a body contains a charge of 6.25×10^{18} more electrons than protons.

Electrostatic Fields

The fundamental characteristic of an electric charge is its ability to exert force. The space between and around charged bodies in which their influence is felt is called an *electric field of force*. The electric field is always terminated on material objects and extends between positive and negative charges. This region of force can consist of air, glass, paper, or a vacuum. This region of force is referred to as an *electrostatic field*.

When two objects of opposite polarity are brought near each other, the electrostatic field is concentrated in the area between them. Lines that are referred to as electrostatic lines of force generally represent the field. These lines are imaginary and are used merely to represent the direction and strength of the field. To avoid confusion, the positive lines of force are always shown leaving charge, and for a negative charge, they are shown as entering. Figure 8.6 illustrates the use of lines to represent the field about charged bodies.

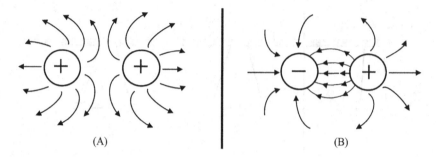

(A)   (B)

FIGURE 8.6 Electrostatic lines of force. (A) represents the repulsion of like-charged bodies and their associated fields. (B) represents the attraction between unlike-charged bodies and their associated fields.

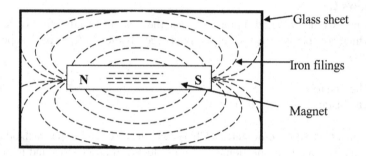

FIGURE 8.7 The magnetic field around a bar magnet. If the glass sheet is tapped gently, the filings will move into a definite pattern that describes the field of force around the magnet.

Note: A charged object will retain its charge temporarily if there is no immediate transfer of electrons to or from it. In this condition, the charge is said to be *at rest*. Remember, electricity at rest is called *static* electricity.

Magnetism

Most electrical equipment depends directly or indirectly upon magnetism. *Magnetism* is defined as a phenomenon associated with magnetic fields; that is, it has the power to attract such substances as iron, steel, nickel, or cobalt (metals that are known as magnetic materials). Correspondingly, a substance is said to be a magnet if it has the property of magnetism. For example, a piece of iron can be magnetized and is, thus, a magnet.

When magnetized, the piece of iron (note: we will assume a piece of flat bar 6 inches long × 1-inch-wide × 0.5 inches thick; a bar magnet—see Figure 8.7) will have two points opposite to each other, which most readily attract other pieces of iron. The points of maximum attraction (one on each end) are called the *magnetic poles* of the magnet: the north (N) pole and the south (S) pole. Just as like electric charges repel each other and opposite charges attract each other, like magnetic poles repel each other and unlike poles attract each other. Although invisible to the naked eye, its force can be shown to exist by sprinkling small iron filings on a glass covering a bar magnet as shown in Figure 8.7.

Figure 8.8 shows how the field looks without iron filings; it is shown as lines of force [known as *magnetic flux or flux lines*; the symbol for magnetic flux is the Greek lowercase letter φ (phi)] in the field, repelled away from the north pole of the magnet and attracted to its south pole.

Note: A *magnetic circuit* is a complete path through which magnetic lines of force may be established under the influence of a magnetizing force. Most magnetic circuits are composed largely of magnetic materials in order to contain the magnetic flux. These circuits are similar to the electric circuit (an important point), which is a complete path through which current is caused to flow under the influence of an electromotive force.

There are three types or groups of magnets:

a. *Natural Magnets*: found in the natural state in the form of a mineral (an iron compound) called magnetite.
b. *Permanent Magnets*: (artificial magnet) hardened steel or some alloy such as Alnico bars that have been permanently magnetized. The permanent magnet most people are familiar with is the horseshoe magnet (see Figure 8.9). Wind turbines use permanent magnets made from rare-earth materials.
c. *Electromagnets*: (artificial magnet) composed of soft-iron cores around which are wound coils of insulated wire. When an electric current flows through the coil, the core becomes magnetized. When the current ceases to flow, the core loses most of the magnetism.

Magnetic Materials

Natural magnets are no longer used (they have no practical value) in electrical circuitry because more powerful and more conveniently shaped permanent magnets can be produced artificially. Commercial magnets are made from special steels and alloys—magnetic materials.

Magnetic materials are those materials that are attracted or repelled by a magnet and that can be magnetized themselves. Iron, steel, and alloy bar are the most common magnetic materials. These materials can be magnetized by inserting the material (in bar form) into a coil of insulated wired and passing a heavy direct current through the coil. The same material may also be magnetized if it is stroked with a bar

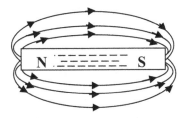

FIGURE 8.8 Magnetic field of force around a bar magnet, indicated by lines of force.

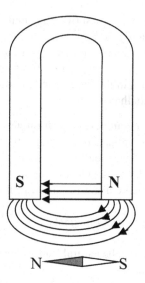

FIGURE 8.9   Horseshoe magnet

magnet. It will then have the same magnetic property that the magnet used to induce the magnetism has—namely, there will be two poles of attraction, one at either end. This process produces a permanent magnet by *induction*—that is, the magnetism is induced in the bar by the influence of the stroking magnet.

Note: Permanent magnets are those of hard magnetic materials (hard steel or alloys) that retain their magnetism when the magnetizing field is removed. A temporary magnet is one that has *no* ability to retain a magnetized state when the magnetizing field is removed.

Even though classified as permanent magnets, it is important to point out that hardened steel and certain alloys are relatively difficult to magnetize and are said to have a *low permeability* because the magnetic lines of force do not easily permeate or distribute themselves readily through the steel.

Note: *Permeability* refers to the ability of a magnetic material to concentrate magnetic flux. Any material that is easily magnetized has high permeability. A measure of permeability for different materials in comparison with air or vacuum is called *relative* permeability, symbolized by μ (or mu).

Once hard steel and other alloys are magnetized, however, they retain a large part of their magnetic strength and are called *permanent magnets*. Conversely, materials that are relatively easy to magnetize—such as soft iron and annealed silicon steel—are said to have a *high permeability*. Such materials retain only a small part of their magnetism after the magnetizing force is removed and are called *temporary magnets*.

The magnetism that remains in a temporary magnet after the magnetizing force is removed is called *residual magnetism*.

Early magnetic studies classified magnetic materials merely as being magnetic and nonmagnetic—that is, based on the strong magnetic properties of iron. However, because weak magnetic materials can be important in some applications, present studies classify materials into one of the following three groups: paramagnetic, diamagnetic, and ferromagnetic.

a. **Paramagnetic materials**: These include aluminum, platinum, manganese, and chromium—materials that become only slightly magnetized even though under the influence of a strong magnetic field. This slight magnetization is in the same direction as the magnetizing field. Relative permeability is slightly more than 1 (i.e., considered nonmagnetic materials).

b. **Diamagnetic materials**: These include bismuth, antimony, copper, zinc, mercury, gold, and silver—materials that can also be slightly magnetized when under the influence of a very strong field. Relative permeability is less than 1 (i.e., considered nonmagnetic materials).

c. **Ferromagnetic materials**: These include iron, steel, nickel, cobalt, and commercial alloys—materials that are the most important group for applications of electricity and electronics. Ferromagnetic materials are easy to magnetize and have high permeability, ranging from 50 to 3,000.

Magnetic Earth

The earth is a huge magnet, and surrounding earth is the magnetic field produced by the earth's magnetism. Most people would have no problem understanding or at least accepting this statement. However, if told that the earth's north magnetic pole is actually its south magnetic pole and that the south magnetic pole is actually the earth's north magnetic pole, they might not accept or understand this statement. However, in terms of a magnet, it is true.

As can be seen from Figure 8.10, the magnetic polarities of the earth are indicated. The geographic poles are also shown at each end of the axis of rotation of the earth. As shown in Figure 8.10, the magnetic axis does not coincide with the geographic axis, and therefore the magnetic and geographic poles are not at the same place on the surface of the earth.

Recall that magnetic lines of force are assumed to emanate from the north pole of a magnet and to enter the South Pole as closed loops. Because the earth is a magnet, lines of force emanate from its north magnetic pole and enter the south magnetic pole as closed loops. A compass needle aligns itself in such a way that the earth's lines of force enter at its south pole and leave at its north pole. Because the north pole of the needle is defined as the end that points in a northerly direction, it follows that the magnetic pole near the north geographic pole is in reality a south magnetic pole, and vice versa.

DIFFERENCE IN POTENTIAL

Because of the force of its electrostatic field, an electric charge has the ability to do the work of moving another charge by attraction or repulsion. The force that causes

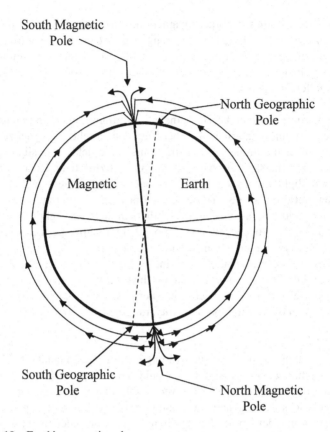

FIGURE 8.10 Earth's magnetic poles.

free electrons to move in a conductor as an electric current may be referred to as follows:

- Electromotive force (emf)
- Voltage
- Difference in potential

When a difference in potential exists between two charged bodies that are connected by a wire (conductor), electrons (current) will flow along the conductor. This flow is from the negatively charged body to the positively charged body until the two charges are equalized and the potential difference no longer exists.

Note: The basic unit of potential difference is the *volt* (V). The symbol for the potential difference is *V*, indicating the ability to do the work of forcing electrons (current flow) to move. Because the volt unit is used, the potential difference is called *voltage*.

The Water Analogy
In attempting, to train individuals in the concepts of basic electricity, especially in regard to difference of potential (voltage), current and resistance relationships in a

simple electrical circuit, it has been common practice to use what is referred to as the water analogy. We use the water analogy later to explain (in a simple straightforward fashion) voltage, current and resistance and their relationships in more detail, but for now we use the analogy to explain the basic concept of electricity: difference of potential, or voltage. Because a difference in potential causes current flow (against resistance), it is important that this concept be understood first before the concept of current flow and resistance are explained.

Consider the water tanks shown in Figure 8.11—two water tanks connected by a pipe and valve. At first, the valve is closed, and all the water is in Tank A. Thus, the water pressure across the valve is at maximum. When the valve is opened, the water flows through the pipe from A to B until the water level becomes the same in both tanks. The water then stops flowing in the pipe, because there is no longer a difference in water pressure (difference in potential) between the two tanks.

Just as the flow of water through the pipe in Figure 8.11 is directly proportional to the difference in water level in the two tanks, current flow through an electric circuit is directly proportional to the difference in potential across the circuit.

Important Point: a fundamental law of current electricity is that the current is directly proportional to the applied voltage; that is, if the voltage is increased, the current is increased. If the voltage is decreased, the current is decreased. Voltage does not move; instead, it pushes—it pushes current, electron flow, and electricity.

PRINCIPAL METHODS OF PRODUCING A VOLTAGE

There are many ways to produce electromotive force or voltage. Some of these methods are much more widely used than others. The following is a list of the six most common methods of producing electromotive force.

  a. **Friction**: voltage produced by rubbing two materials together
  b. **Pressure (Piezoelectricity)**: voltage produced by squeezing crystals of certain substances
  c. **Heat (Thermoelectricity)**: voltage produced by heating the joint (junction) where two unlike metals are joined
  d. **Light (photoelectricity)**: voltage produced by light striking photosensitive (light sensitive) substances

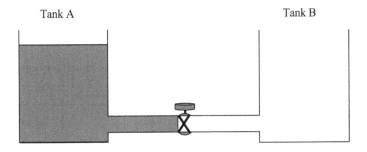

FIGURE 8.11   Water analogy of electric difference of potential

e. **Chemical Action**: voltage produced by chemical reaction in a battery cell
f. **Magnetism**: voltage produced in a conductor when the conductor moves through a magnetic field, or a magnetic field moves through the conductor in such a manner as to cut the magnetic lines of force of the field

In the study of basic electricity, we are most concerned with magnetism and chemistry as means to produce voltage. Friction has few practical applications, though we discussed it earlier in studying static electricity. Pressure, heat, and light do have useful applications, but we do not need to consider them in this text. Magnetism and chemistry, on the other hand, are the principal sources of voltage and are discussed at length in this text.

CURRENT

The movement or the flow of electrons is called *current*. To produce current, the electrons must be moved by a potential difference.

Note: The terms "current," "current flow," "electron flow," "electron current," etc. may be used to describe the same phenomenon.
    Electron flow, or current, in an electric circuit is from a region of less negative potential to a region of more positive potential.

Note: Current is represented by the letter I. The basic unit in which current is measured is the *ampere*, or *amp (A)*. One ampere of current is defined as the movement of one coulomb past any point of a conductor during one second of time.
    Recall that we used the water analogy to help us understand potential differences. We can also use the water analogy to help us understand current flow through a simple electric circuit.
    Consider Figure 8.12 that shows a water tank connected via a pipe to a pump with a discharge pipe. If the water tank contains an amount of water above the level of the pipe opening to the pump, the water exerts pressure (a difference in potential) against the pump. When sufficient water is available for pumping with the pump, water flows through the pipe against the resistance of the pump and pipe. The analogy should

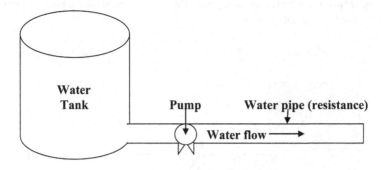

FIGURE 8.12   Water analogy: current flow

be clear—in an electric circuit, if a difference of potential exists, current will flow in the circuit.

Another simple way of looking at this analogy is to consider Figure 8.13 where the water tank has been replaced with a generator, the pipe with a conductor (wire), and water flow with the flow of electric current.

Again, the key point illustrated by Figures 8.12 and 8.13 is that to produce current, the electrons must be moved by a potential difference.

Electric current is generally classified into two general types:

- Direct current (d-c)
- Alternating current (a-c)

Direct current is current that moves through a conductor or circuit in one direction only. *Alternating current* periodically reverses direction.

RESISTANCE

Earlier it was pointed out that free electrons, or electric current, can move easily through a good conductor, such as copper, but that an insulator, such as glass, is an obstacle to current flow. In the water analogy shown in Figure 8.12 and the simple electric circuit shown in Figure 8.13, either the pipe or the conductor indicates resistance.

Every material offers some resistance, or opposition, to the flow of electric current through it. Good conductors, such as copper, silver, and aluminum, offer very little resistance. Poor conductors, or insulators, such as glass, wood, and paper, offer high resistance to current flow.

Note: The amount of current that flows in a given circuit depends on two factors: voltage and resistance.

Note: The letter R represents resistance. The basic unit in which resistance is measured is the *ohm* (Ω). One ohm is the resistance of a circuit element, or circuit, which permits a steady current of 1 ampere (1 coulomb per second) to flow when a steady electromotive force (emf) of 1 V is applied to the circuit. Manufactured circuit parts containing definite amounts of resistance are called *resistors*.

The size and type of material of the wires in an electric circuit are chosen to keep the electrical resistance as low as possible. In this way, current can flow easily through the conductors, just as water flows through the pipe between the tanks

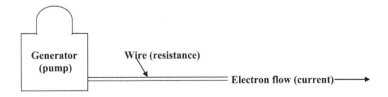

FIGURE 8.13 Simple electric circuit with current flow.

in Figure 8.11. If the water pressure remains constant, the flow of water in the pipe will depend on how far the valve is opened. The smaller the opening, the greater the opposition (resistance) to the flow, and the smaller will be the rate of flow in gallons per second.

In the electric circuit shown in Figure 8.13, the larger the diameter of the wire, the lower its electrical resistance (opposition) to the flow of current through it. In the water analogy, pipe friction opposes the flow of water between the tanks. This friction is similar to electrical resistance. The resistance of the pipe to the flow of water through it depends upon (1) the length of the pipe, (2) the diameter of the pipe, and (3) the nature of the inside walls (rough or smooth). Similarly, the electrical resistance of the conductors depends upon (1) the length of the wires, (2) the diameter of the wires, and (3) the material of the wires (copper, silver, etc.).

It is important to note that temperature also affects the resistance of electrical conductors to some extent. In most conductors (copper, aluminum, etc.) the resistance increases with temperature. Carbon is an exception. In carbon, the resistance decreases as temperature increases.

Important Note: Electricity is a study that is frequently explained in terms of opposites. The term that is exactly the opposite of resistance is *conductance*. Conductance (G) is the ability of a material to pass electrons. The unit of conductance is the *Mho*, which is ohm spelled backwards. The relationship that exists between resistance and conductance is the reciprocal. A reciprocal of a number is obtained by dividing the number into one. If the resistance of a material is known, dividing its value into one will give its conductance. Similarly, if the conductance is known, dividing its value into one will give its resistance.

BATTERY-SUPPLIED ELECTRICITY

Battery-supplied direct current electricity has many applications and is widely used in industrial operations. Applications include providing electrical energy in plant vehicles and emergency diesel generators, material handling equipment (forklifts), portable electric/electronic equipment, backup emergency power for light-packs, for hazard warning signal lights and flashlights, and as standby power supplies or uninterruptible power supplies (UPS) for computer systems. In some instances, they are used as the only source of power; while in others (as mentioned above), they are used as a secondary or standby power supply.

THE VOLTAIC CELL

The simplest cell (a device that transforms chemical energy into electrical energy) is known as a *voltaic* (or galvanic) cell (see Figure 8.14). It consists of a piece of carbon (C) and a piece of zinc (Zn) suspended in a jar that contains a solution of water (H_2O) and sulfuric acid (H_2SO_4).

Note: A simple cell consists of two strips, or *electrodes*, placed in a container that holds the *electrolyte*. A battery is formed when two or more cells are connected.

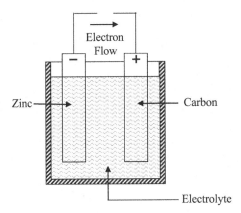

FIGURE 8.14 Simple voltaic cell.

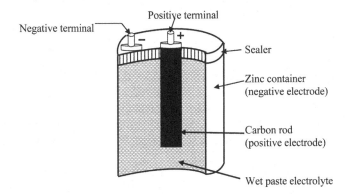

FIGURE 8.15 Dry cell (cross-sectional view).

The electrodes are the conductors by which the current leaves or returns to the electrolyte. In the simple cell described above, they are carbon and zinc strips placed in the electrolyte. Zinc contains an abundance of negatively charged atoms, while carbon has an abundance of positively charge atoms. When the plates of these materials are immersed in an electrolyte, chemical action between the two begins.

In the *dry cell* (see Figure 8.15), the electrodes are the carbon rod in the center and the zinc container in which the cell is assembled.

The electrolyte is the solution that acts upon the electrodes that are placed in it. The electrolyte may be a salt, an acid, or an alkaline solution. In the simple voltaic cell and in the automobile storage battery, the electrolyte is in a liquid form; while in the dry cell (see Figure 8.15), the electrolyte is a moist paste.

PRIMARY AND SECONDARY CELLS

Primary cells are normally those that cannot be recharged or returned to good condition after their voltage drops too low. Dry cells in flashlights and transistor radios are examples of primary cells. Some primary cells have been developed to the state where they can be recharged.

A *secondary cell* is one in which the electrodes and the electrolyte are altered by the chemical action that takes place when the cell delivers current. These cells are rechargeable. During recharging, the chemicals that provide electric energy are restored to their original condition. Recharging is accomplished by forcing an electric current through them in the opposite direction to that of discharge.

Connecting as shown in Figure 8.16 recharges a cell. Some battery chargers have a voltmeter and an ammeter that indicate the charging voltage and current.

The automobile storage battery is the most common example of the secondary cell.

BATTERY

As was stated previously, a *battery* consists of two or more cells placed in a common container. The cells are connected in series, in parallel, or in some combination of series and parallel, depending upon the amount of voltage and current required of the battery.

Battery Operation

The chemical reaction within a battery provides the voltage. This occurs when a conductor is connected externally to the electrodes of a cell, causing electrons to flow under the influence of a difference in potential across the electrodes from the zinc (negative) through the external conductor to the carbon (positive), returning within the solution to the zinc. After a short period, the zinc will begin to waste away because of the acid.

The voltage across the electrodes depends upon the materials from which the electrodes are made and the composition of the solution. The difference in potential between the carbon and zinc electrodes in a dilute solution of sulfuric acid and water is about 1.5 V.

The current that a primary cell may deliver depends upon the resistance of the entire circuit, including that of the cell itself. The internal resistance of the primary cell depends upon the size of the electrodes, the distance between them in the

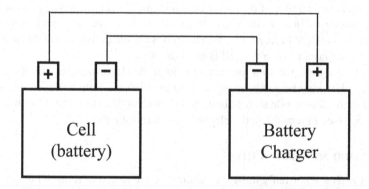

FIGURE 8.16   Hookup for charging a secondary cell with a battery charger.

solution, and the resistance of the solution. The larger the electrodes and the closer together they are in solution (without touching), the lower the internal resistance of the primary cell and the more current it is capable of supplying to the load.

Note: When current flows through a cell, the zinc gradually dissolves in the solution and the acid is neutralized.

Combining Cells

In many operations, battery-powered devices may require more electrical energy than one cell can provide. Various devices may require either a higher voltage or more current, and in some cases both. Under such conditions, it is necessary to combine, or interconnect, a sufficient number of cells to meet the higher requirements. Cells connected in series provide a higher voltage, while cells connected in parallel provide a higher current capacity. To provide adequate power when both voltage and current requirements are greater than the capacity of one cell, a combination series-parallel network of cells must be interconnected.

When cells are connected in **series** (see Figure 8.17), the total voltage across the battery of cells is equal to the sum of the voltage of each of the individual cells. In Figure 8.17, the four 1.5-V cells in series provide a total battery voltage of 6 V. When cells are placed in series, the positive terminal of one cell is connected to the negative terminal of the other cell. The positive electrode of the first cell and the negative electrode of the last cell then serve as the power takeoff terminals of the battery. The current flowing through such a battery of series cells is the same as from one cell because the same current flows through all the series cells.

To obtain a greater current, a battery has cells connected in *parallel* as shown in Figure 8.18. In this parallel connection, all the positive electrodes are connected to one line, and all negative electrodes are connected to the other. Any point on the positive side can serve as the positive terminal of the battery and any point on the negative side can be the negative terminal.

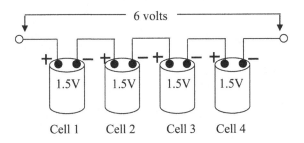

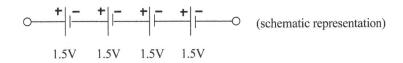

(schematic representation)

FIGURE 8.17 Cells in series.

The total voltage output of a battery of three parallel cells is the same as that for a single cell (Figure 8.18), but the available current is three times that of one cell; that is, the current capacity has been increased.

Identical cells in parallel all supply equal parts of the current to the load. For example, of three different parallel cells producing a load current of 210 ma, each cell contributes 70 ma.

Figure 8.19 depicts a schematic of a *series-parallel* battery network supplying power to a load requiring both a voltage and current greater than one cell can provide. To provide the required increased voltage, groups of three 1.5-V cells are connected in series. To provide the required increased amperage, four series groups are connected in parallel.

Types of Batteries

In the past 25 years, several different types of batteries have been developed. In this text, we briefly discuss five types: the dry cell, lead-acid battery, alkaline cell, nickel-cadmium, and mercury cell.

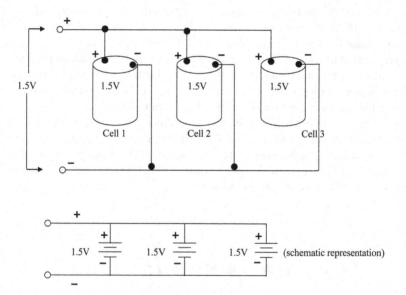

FIGURE 8.18 Cells in parallel.

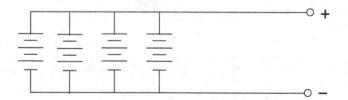

FIGURE 8.19 Series-parallel connected cells.

Dry Cell

The dry cell, or carbon-zinc cell, is so called because its electrolyte is not in a liquid state (however, the electrolyte is a moist paste). The dry cell battery is one of the oldest and most widely used commercial types of dry cell. The carbon, in the form of a rod that is placed in the center of the cell, is the positive terminal. The case of the cell is made of zinc, which is the negative terminal (see Figure 8.15). Between the carbon electrode and the zinc case is the electrolyte of a moist chemical paste-like mixture. The cell is sealed to prevent the liquid in the paste from evaporating. The voltage of a cell of this type is about 1.5 V.

Lead-Acid Battery

The *lead-acid battery* is a secondary cell—commonly termed a storage battery—that stores chemical energy until it is released as electrical energy.

Note: The lead-acid battery differs from the primary cell type battery mainly in that it may be recharged, whereas most primary cells are not normally recharged. In addition, the term "storage battery" is somewhat deceiving because this battery does not store electrical energy but is a source of chemical energy that produces electrical energy.

As the name implies, the lead-acid battery consists of a number of lead-acid cells immersed in a dilute solution of sulfuric acid. Each cell has two groups of lead plates; one set is the positive terminal and the other is the negative terminal. Active materials within the battery (lead plates and sulfuric acid electrolyte) react chemically to produce a flow of direct current whenever current consuming devices are connected to the battery terminal posts. This current is produced by a chemical reaction between the active material of the plates (electrodes) and the electrolyte (sulfuric acid).

This type of cell produces slightly more than 2 V. Most automobile batteries contain six cells connected in series so that the output voltage from the battery is slightly more than 12 V.

Besides being rechargeable, the main advantage of the lead-acid storage battery over the dry cell battery is that the storage battery can supply current for a much longer time than the average dry cell.

Safety Note: Whenever a lead-acid storage battery is charging, the chemical action produces dangerous hydrogen gas; thus, the charging operation should only take place in a well-ventilated area.

Alkaline Cell

The *alkaline cell* is a secondary cell that gets its name from its alkaline electrolyte—potassium hydroxide. Another type battery, sometimes called the "alkaline battery," has a negative electrode of zinc and a positive electrode of manganese dioxide. It generates 1.5 V.

Nickel-Cadmium Cell

The *nickel-cadmium cell*, or Ni-Cad cell, is the only dry battery that is a true storage battery with a reversible chemical reaction, allowing recharging many times. In the secondary nickel-cadmium dry cell, the electrolyte is potassium hydroxide, the negative electrode is nickel hydroxide, and the positive electrode is cadmium oxide. The operating voltage is 1.25 V.

Because of its rugged characteristics (stands up well to shock, vibration, and temperature changes) and availability in a variety of shapes and sizes, it is ideally suited for use in powering portable communication equipment.

Mercury Cell

The *mercury cell* was developed because of space exploration activities; the development of small transceivers and miniaturized equipment where a power source of miniaturized size was needed. In addition to reduced size, the mercury cell has a good shelf life and is very rugged; they also produce a constant output voltage under different load conditions.

There are two different types of mercury cells: one is a flat cell that is shaped like a button, while the other is a cylindrical cell that looks like a standard flashlight cell. The advantage of the button-type cell is that several of them can be stacked inside one container to form a battery. A cell produces 1.35 V.

BATTERY CHARACTERISTICS

Batteries are generally classified by their various characteristics. Parameters such as internal resistance, specific gravity, capacity, and shelf life are used to classify batteries by type.

Regarding *internal resistance*, it is important to keep in mind that a battery is a d-c voltage generator. As such, the battery has internal resistance. In a chemical cell, the resistance of the electrolyte between the electrodes is responsible for most of the cell's internal resistance. Because any current in the battery must flow through the internal resistance, this resistance is in series with the generated voltage. With no current, the voltage drop across the resistance is zero so that the full-generated voltage develops across the output terminals. This is the open-circuit voltage or no-load voltage. If a load resistance is connected across the battery, the load resistance is in series with internal resistance. When current flows in this circuit, the internal voltage drop decreases the terminal voltage of the battery.

The ratio of the weight of a certain volume of liquid to the weight of the same volume of water is called the *specific gravity* of the liquid. Pure sulfuric acid has a specific gravity of 1.835 since it weighs 1.835 times as much as water per unit volume. The specific gravity of a mixture of sulfuric acid and water varies with the strength of the solution from 1.000 to 1.830.

The specific gravity of the electrolyte solution in a lead-acid cell ranges from 1.210 to 1.300 for new, fully charged batteries. The higher the specific gravity, the less internal resistance of the cell and the higher the possible load current. As the cell discharges, the water formed dilutes the acid and the specific gravity gradually decreases to about 1.150, at which time the cell is considered to be fully discharged.

The specific gravity of the electrolyte is measured with a *hydrometer*, which has a compressible rubber bulb at the top, a glass barrel, and a rubber hose at the bottom of the barrel. In taking readings with a hydrometer, the decimal point is usually omitted. For example, a specific gravity of 1.260 is read simply as "twelve-sixty." A hydrometer reading of 1,210–1,300 indicates full charge; about 1,250 is half-charge, and 1,150–1,200 is complete discharge.

The *capacity* of a battery is measured in ampere-hours (Ah).

Note: The ampere-hour capacity is equal to the product of the current in amperes and the time in hours during which the battery is supplying this current. The ampere-hour capacity varies inversely with the discharge current. The size of a cell is determined generally by its ampere-hour capacity.

The capacity of a storage battery determines how long it will operate at a given discharge rate and depends upon many factors, the most important of these are as follows:

- The area of the plates in contact with the electrolyte
- The quantity and specific gravity of the electrolyte
- The type of separators
- The general condition of the battery (degree of sulfating, plates bucked, separators warped, sediment in bottom of cells, etc.)
- The final limiting voltage

The *shelf life* of a cell is that period of time during which the cell can be stored without losing more than approximately 10% of its original capacity. The loss of capacity of a stored cell is due primarily to the drying out of its electrolyte in a wet cell and to chemical actions, which change the materials within the cell. Keeping it in a cool, dry place can extend the shelf life.

SIMPLE ELECTRICAL CIRCUIT

An electric circuit includes: an *energy source* [source of electromotive force (emf) or voltage; that is, a battery or generator], a conductor (wire), a load, and a means of control (see Figure 8.20). The energy source could be a battery, as in Figure 8.20, or some other means of producing a voltage. The *load* that dissipates the energy could

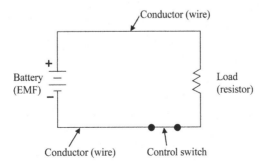

FIGURE 8.20 Simple closed circuit.

be a lamp, a resistor, or some other device (or devices) that does useful work, such as an electric toaster, a power drill, radio, or soldering iron. *Conductors* are wires that offer low resistance to current; they connect all the loads in the circuit to the voltage source. No electrical device dissipates energy unless current flows through it. Because conductors, or wires, are not perfect conductors, they heat up (dissipate energy), so they are actually part of the load. For simplicity, however, we usually think of the connecting wiring as having no resistance, since it would be tedious to assign a very low resistance value to the wires every time we wanted to solve a problem. *Control devices* might be switches, variable resistors, circuit breakers, fuses, or relays.

A complete pathway for current flow, or *closed circuit* (Figure 8.20), is an unbroken path for current from the emf, through a load, and back to the source. A circuit is called *open* (see Figure 8.21) if a break in the circuit (e.g., open switch) does not provide a complete path for current.

Important Point: Current flows from the negative (−) terminal of the battery, shown in Figures 8.20 and 8.21, through the load to the positive (+) battery terminal, and continues by going through the battery from the positive (+) terminal to the negative (−) terminal. As long as this pathway is unbroken, it is a closed circuit and the current will flow. However, if the path is broken at ANY point, it is an open circuit and no current flows.

To protect a circuit, a fuse is placed directly into the circuit (see Figure 8.22). A fuse will open the circuit whenever a dangerous large current starts to flow

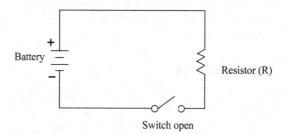

FIGURE 8.21   Open circuit.

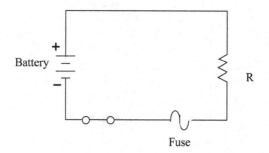

FIGURE 8.22   A simple fused circuit.

(i.e., a short circuit condition occurs, caused by an accidental connection between two points in a circuit which offers very little resistance). A fuse will permit currents smaller than the fuse value to flow but will melt and therefore break or open the circuit if a larger current flows.

SCHEMATIC REPRESENTATION

The simple circuits shown in Figures 8.20, 8.21, and 8.22 are displayed in *schematic* form. A *schematic diagram* (usually shortened to "schematic") is a simplified drawing that represents the *electrical*, not the physical, situation in a circuit. The symbols used in schematic diagrams are the electrician's "shorthand"; they make the diagrams easier to draw and easier to understand. Consider the symbol in Figure 8.23 used to represent a battery power supply. The symbol is rather simple, straightforward—but is also very important. For example, by convention, the shorter line in the symbol for a battery represents the *negative* terminal. It is important to remember this, because it is sometimes necessary to note the direction of the current flow, which is from negative to positive, when you examine the schematic. The battery symbol shown in Figure 8.23 has a single cell; so only one short and one long line are used. The number of lines used to represent a battery varies (and they are not necessarily equivalent to the number of cells), but they are always in pairs, with long and short lines alternating. In the circuit shown in Figure 8.22, the current would flow in a *counterclockwise* direction, that is, in the opposite direction that a clock's hands move. If the long and short lines of the battery symbol (symbol shown in Figure 8.23) were reversed, the current in the circuit shown in Figure 8.22 would flow *clockwise*, that is, in the direction of a clock's hands.

Note: In studies of electricity and electronics many circuits are analyzed which consist mainly of specially designed resistive components. As previously stated, these components are called resistors. Throughout the remaining analysis of the basic circuit, the resistive component will be a physical resistor. However, the resistive component could be any one of several electrical devices.

Keep in mind that in the simple circuits shown in the figures to this point we have only illustrated and discussed a few of the many symbols used in schematics to represent circuit components. (Other symbols will be introduced, as we need them.)

It is also important to keep in mind that a closed loop of wire (conductor) is not necessarily a circuit. A source of voltage must be included to make it an electric circuit. In any electric circuit where electrons move around a closed loop, current, voltage, and resistance are present. The physical pathway for current flow is actually the circuit. By knowing any two of the three quantities, such as voltage and current, the third (resistance) may be determined. This is done mathematically using Ohm's Law, which is the foundation on which electrical theory is based.

FIGURE 8.23   Schematic symbol for a battery.

OHM'S LAW

Simply put, *Ohm's Law* defines the relationship between current, voltage, and resistance in electric circuits. Ohm's Law can be expressed mathematically in three ways.

a. The *current* in a circuit is equal to the voltage applied to the circuit divided by the resistance of the circuit. Stated another way, the current in a circuit is DIRECTLY proportional to the applied voltage and INVERSELY proportional to the circuit resistance. Ohm's Law may be expressed as an equation:

$$I = \frac{E}{R} \tag{8.1}$$

where
    I = current in amperes (amps)
    E = voltage in volts
    R = resistance in ohms

b. The *resistance* of a circuit is equal to the voltage applied to the circuit divided by the current in the circuit:

$$R = \frac{E}{I} \tag{8.2}$$

c. The applied *voltage* (E) to a circuit is equal to the product of the current and the resistance of the circuit:

$$E = I \times R = IR \tag{8.3}$$

If any two of the quantities in Equations (8.1)–(8.3) are known, the third may be easily found. Let us look at an example.

Example 8.1

PROBLEM:

Figure 8.24 shows a circuit containing a resistance of 6 ohms and a source of voltage of 3 V. How much current flows in the circuit?
Given:

E = 3 V
R = 6 ohms
I = ?

SOLUTION:

$$I = \frac{E}{R}$$

$$I = \frac{3}{6}$$

I = 0.5 amperes

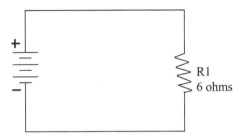

FIGURE 8.24 Determining current in a simple circuit.

To observe the effect of source voltage on circuit current, we use the circuit shown in Figure 8.24 but double the voltage to 6 V.

Notice that as the source of voltage doubles, the circuit current also doubles.

Example 8.2

PROBLEM:

Given:

E = 6 V
R = 6 ohms
I = ?

SOLUTION:

$$I = \frac{E}{R}$$

$$I = \frac{6}{6}$$

I = 1 amperes

Key Point: Circuit current is directly proportional to the applied voltage and will change by the same factor that the voltage changes.

To verify that current is inversely proportional to resistance, assume the resistor in Figure 8.24 to have a value of 12 ohms.

Example 8.3

PROBLEM:

Given:

E = 3 V
R = 12 ohms
I = ?

SOLUTION:

$$I = \frac{E}{R}$$

$$I = \frac{3}{12}$$

$$I = 0.25 \text{ amperes}$$

Comparing this current of 0.25 ampere for the 12-ohm resistor, to the 0.5-ampere of current obtained with the 6-ohm resistor, shows that doubling the resistance will reduce the current to one half the original value. *The point*: Circuit current is inversely proportional to the circuit resistance.

Recall that if you know any two quantities among E, I, and R, you can calculate the third. In many circuit applications, current is known and either the voltage or the resistance will be the unknown quantity. To solve a problem, in which current and resistance are known, the basic formula for Ohm's Law must be transposed to solve for E, for I, or for R.

However, the Ohm's Law equations can be memorized and practiced effectively by using an Ohm's Law circle (see Figure 8.25).

To find the equation for E, I, or R when two quantities are known, cover the unknown third quantity with your finger, ruler, piece of paper, etc. as shown in Figure 8.26.

Example 8.4

PROBLEM:

Find I when E = 120 V and R = 40 ohms (Ω).

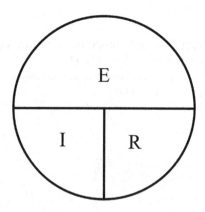

FIGURE 8.25   Ohm's Law circle.

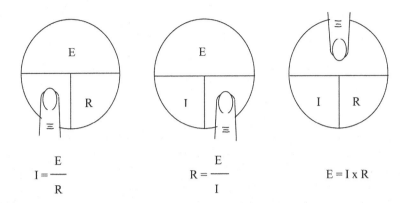

$$I = \frac{E}{R}$$

$$R = \frac{E}{I}$$

$$E = I \times R$$

FIGURE 8.26 Putting the Ohm's Law circle to work.

SOLUTION:

Place finger on I as shown in the figure below.
Use Equation (8.1) to find the unknown I.

$$I = \frac{E}{R}$$

$$\frac{120}{40} = 3A$$

$$I = 0.25 \text{ amperes}$$

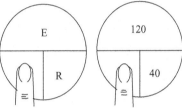

Example 8.5

PROBLEM:

Find R when E = 220 V and I = 10 A

SOLUTION:

Place finger on R as shown in the figure.
Use Equation (8.2) to find the unknown R.

$$R = \frac{E}{I}$$

$$= \frac{220}{10} = 22\Omega$$

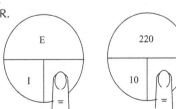

Example 8.6

PROBLEM:

Find E when I = 2.5 A and R = 25 Ω.

SOLUTION:

Place a finger on E as shown in the figure.

$E = IR = 2.5(25) = 62.5$ V

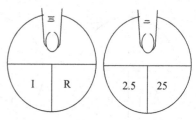

Note: In the previous examples we have demonstrated how the Ohm's Law circle can help solve simple voltage, current, and amperage problems. The beginning student is cautioned, however, not to rely wholly on the use of this circle when transposing simple formulas but rather to use it to supplement his/her knowledge of the algebraic method. Algebra is a basic tool in the solution of electrical problems and the importance of knowing how to use it should not be under-emphasized or bypassed after the operator has learned a shortcut method such as the one indicated in this circle.

Example 8.7

PROBLEM:

An electric light bulb draws 0.5 A when operating on a 120-V d-c circuit. What is the resistance of the bulb?

SOLUTION:

The first step in solving a circuit problem is to sketch a schematic diagram of the circuit itself, labeling each of the parts and showing the known values (see Figure 8.27).

Since I and E are known, we use Equation (8.2) to solve for R.

$$R = \frac{E}{I} = \frac{120}{0.5} = 240\,\Omega$$

ELECTRICAL POWER

Power, whether electrical or mechanical, pertains to the rate at which work is being done, so the power consumption at your site is related to current flow. A large electric motor or air dryer consumes more power (and draws more current) in a given

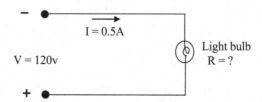

FIGURE 8.27 Simple circuit.

length of time than, for example, an indicating light on a motor controller. *Work* is done whenever a force causes motion. If a mechanical force is used to lift or move a weight, work is done. However, the force exerted WITHOUT causing motion, such as the force of a compressed spring acting between two fixed objects, does not constitute work.

Key Point: Power is the rate at which work is done.

ELECTRICAL POWER CALCULATIONS

The electric power P used in any part of a circuit is equal to the current I in that part multiplied by the V across that part of the circuit. In equation form,

$$P = EI \qquad (8.4)$$

where
   P = power, Watts (W)
   E = voltage, V
   I = current, A

If we know the current I and the resistance R but not the voltage V, we can find the power P by using Ohm's Law for voltage, so that substituting Equation (8.3)

$$E = IR$$

into (8.4) we have

$$P = IR \times I = I^2 R \qquad (8.5)$$

In the same manner, if we know the voltage V and the resistance R but not the current I, we can find the P by using Ohm's Law for current, so that substituting Equation (8.1)

$$I = \frac{E}{R}$$

into (8.4) we have

$$P = E\frac{E}{R} = \frac{E^2}{R} \qquad (8.6)$$

Key Point: If we know any two quantities, we can calculate the third.

Example 8.8

PROBLEM:

The current through a 200-Ω resistor to be used in a circuit is 0.25 A. Find the power rating of the resistor.

SOLUTION:

Since current (I) and resistance (R) are known, use Equation (8.5) to find P.

$$P = I^2R = (0.25)^2 (200) = 0.0625(200) = 12.5 \text{ W}$$

Important Point: The power rating of any resistor used in a circuit should be twice the wattage calculated by the power equation to prevent the resistor from burning out. Thus, the resistor used in Example 8.8 should have a power rating of 25 watts.

Example 8.9

PROBLEM:

How many kilowatts of power are delivered to a circuit by a 220-V generator that supplies 30 A to the circuit?

SOLUTION:

Since V and I are given, use Equation (8.4) to find P.

$$P = EI = 220(30) = 6600 \text{ W} = 6.6 \text{ Kw}$$

Example 8.10

PROBLEM:

If the voltage across a 30,000-Ω resistor is 450 V, what is the power dissipated in the resistor?

SOLUTION:

Since R and E are known, use Equation (8.6) to find P.

$$P = \frac{E^2}{R} = \frac{450^2}{30,000} = \frac{202,500}{30,000} = 6.75 \text{ W}$$

In this section, P was expressed in terms of alternate pairs of the other three basic quantities E, I, and R. In practice, you should be able to express any one of the three basic quantities, as well as P, in terms of any two of the others. Figure 8.28 is a summary of 12 basic formulas you should know. The four quantities E, I, R, and P are at the center of the figure.

Adjacent to each quantity are three segments. Note that in each segment, the basic quantity is expressed in terms of two other basic quantities, and no two segments are alike.

ELECTRICAL ENERGY

Energy (the mechanical definition) is defined as the ability to do work (energy and time are essentially the same and are expressed in identical units). Energy is

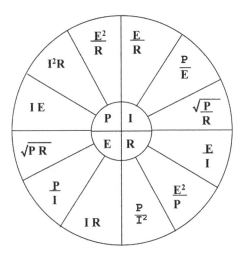

FIGURE 8.28   Ohm's Law circle – summary of basic formulas.

expended when work is done, because it takes energy to maintain a force when that force acts through a distance. The total energy expended to do a certain amount of work is equal to the working force multiplied by the distance through which the force moved to do the work.

In electricity, total energy expended is equal to the *rate* at which work is done, multiplied by the length of time the rate is measured. Essentially, energy W is equal to power P times time t.

The kilowatt-hour (kWh) is a unit commonly used for large amounts of electric energy or work. The amount of kilowatt-hours is calculated as the product of the power in kilowatts (kW) and the time in hours (h) during which the power is used.

$$kWh = kW \times h \tag{8.7}$$

Example 8.11

PROBLEM:

How much energy is delivered in 4 hours by a generator supplying 12 kW?

SOLUTION:

$kWh = kW \times h$

$\qquad = 12(4) = 48$

Energy delivered $= 48\,kWh$

SERIES D-C CIRCUIT CHARACTERISTICS

As previously mentioned, an electric circuit is made up of a voltage source, the necessary connecting conductors, and the effective load.

If the circuit is arranged so that the electrons have only ONE possible path, the circuit is called a *Series Circuit*. Therefore, a series circuit is defined as a circuit that contains only one path for current flow. Figure 8.29 shows a series circuit having several loads (resistors).

Key Point: A *series circuit* is a circuit in which there is only one path for current to flow along.

SERIES CIRCUIT RESISTANCE

Referring to Figure 8.30, the current in a series circuit, in completing its electrical path, must flow through each resistor inserted into the circuit. Thus, each additional resistor offers added resistance. In a series circuit, *the total circuit resistance (R_T) is equal to the sum of the individual resistances.*

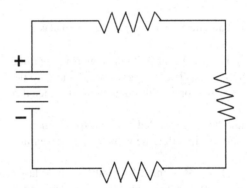

FIGURE 8.29   D-C series circuit.

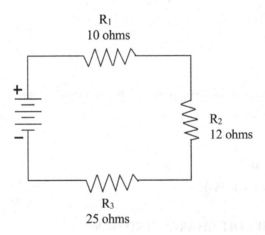

R_1
10 ohms

R_2
12 ohms

R_3
25 ohms

FIGURE 8.30   Solving for total resistance in a series circuit.

As an equation:

$$R_T = R_1 + R_2 + R_3 \ldots R_n \qquad (8.8)$$

where

R_T = total resistance,

R_1, R_2, R_3 = resistance in series,

R_n = any number of additional resistors in the equation

Example 8.12

PROBLEM:

Three resistors of 10 ohms, 12 ohms, and 25 ohms are connected in series across a battery whose emf is 110 V (Figure 8.30). What is the total resistance?

SOLUTION:

Given:

R_1 = 10 ohms
R_2 = 12 ohms
R_3 = 25 ohms
R_T = ?
$R_T = R_1 + R_2 + R_3$
$R_T = 10 + 12 + 25$
$R_T = 47$ ohms

Equation (8.8) can be transposed to solve for the value of an unknown resistance. For example, transposition can be used in some circuit applications where the total resistance is known but the value of a circuit resistor has to be determined.

Example 8.13

PROBLEM:

The total resistance of a circuit containing three resistors is 50 ohms (see Figure 8.31). Two of the circuit resistors are 12 ohms each. Calculate the value of the third resistor.

SOLUTION:

Given:

R_T = 50 ohms
R_1 = 12 ohms
R_2 = 12 ohms
R_3 = ?
$R_T = R_1 + R_2 + R_3$

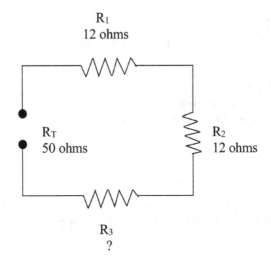

R_1
12 ohms

R_T
50 ohms

R_2
12 ohms

R_3
?

FIGURE 8.31   Calculating the value of one resistance in a series circuit.

Subtracting $(R_1 + R_2)$ from both sides of the equation

$R_3 = R_T - R_1 - R_2$

$R_3 = 50 - 12 - 12$

$R_3 = 50 - 24$

$R_3 = 26$ ohms

Key Point: When resistances are connected in series, the total resistance in the circuit is equal to the sum of the resistances of all the parts of the circuit.

Series Circuit Current

Because there is but one path for current in a series circuit, the same *current* (I) must flow through each part of the circuit. Thus, to determine the current throughout a series circuit, only the current through one of the parts need be known.

The fact that the same current flows through each part of a series circuit can be verified by inserting ammeters into the circuit at various points as shown in Figure 8.32. As indicated in Figure 8.32, each meter indicates the same value of current.

Key Point: In a series circuit, the same current flows in every part of the circuit. *Do not* add the currents in each part of the circuit to obtain I.

Series Circuit Voltage

The *voltage* drop across the resistor in the basic circuit is the total voltage across the circuit and is equal to the applied voltage. The total voltage across a series circuit is also equal to the applied voltage but consists of the sum of two or more individual voltage drops. This statement can be proven by an examination of the circuit shown in Figure 8.33.

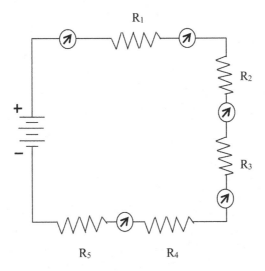

FIGURE 8.32 Current in a series circuit.

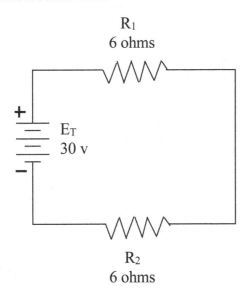

FIGURE 8.33 Calculating total resistance in a series circuit.

In this circuit a source potential (E_T) of 30 V is impressed across a series circuit consisting of two 6-ohm resistors. The total resistance of the circuit is equal to the sum of the two individual resistances, or 12 ohms. Using Ohm's Law, the circuit current may be calculated as follows:

$$I = \frac{E_T}{R_T}$$

$$I = \frac{30}{12}$$

$$I = 2.5\, amperes$$

Knowing the value of the resistors to be 6 ohms each, and the current through the resistors to be 2.5 amperes, the voltage drops across the resistors can be calculated. The voltage (E_1) across R_1 is therefore:

$$E_1 = IR_1$$

$$E_1 = 2.5\, amps \times 6\, ohms$$

$$E_1 = 15\, volts$$

Since R_2 is the same ohmic value as R_1 and carries the same current, the voltage drop across R_2 is also equal to 15 V. Adding these two 15 V drops together gives a total drop of 30 V exactly equal to the applied voltage. For a series circuit then,

$$E_T = E_1 + E_2 + E_3 \ldots E_n \tag{8.9}$$

where

E_T = total voltage, V
E_1 = voltage across resistance R_1, V
E_2 = voltage across resistance R_2, V
E_3 = voltage across resistance R_3, V

Example 8.14

PROBLEM:

A series circuit consists of three resistors having values of 10 ohms, 20 ohms, and 40 ohms, respectively. Find the applied voltage if the current through the 20-ohm resistor is 2.5 amperes.

SOLUTION:

To solve this problem, a circuit diagram is first drawn and labeled as shown in Figure 8.34.
    Given:

R_1 = 10 ohms
R_2 = 20 ohms
R_3 = 40 ohms
I = 2.5 amps

Since the circuit involved is a series circuit, the same 2.5 amperes of current flows through each resistor. Using Ohm's Law, the voltage drops across each of the three resistors can be calculated and are:

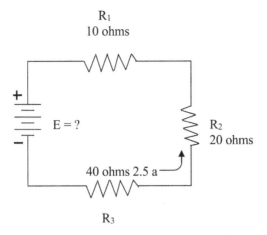

FIGURE 8.34   Solving for applied voltage in a series circuit.

$E_1 = 25$ V
$E_2 = 50$ V
$E_3 = 100$ V

Once the individual drops are known they can be added to find the total or applied voltage-using Equation (8.9):

$E_T = E_1 + E_2 + E_3$

$E_T = 25$ v $+ 50$ v $+ 100$ v

$E_T = 175$ volts

Key Point 1: The total voltage (E_T) across a series circuit is equal to the sum of the voltages across each resistance of the circuit.

Key Point 2: The voltage drops that occur in a series circuit are in direct proportions to the resistance across which they appear. This is the result of having the same current flow through each resistor. Thus, the larger the resistor, the larger will be the voltage drop across it.

Series Circuit Power

Each resistor in a series circuit consumes *power*. This power is dissipated in the form of heat. Because this power must come from the source, the total power must be equal in amount to the power consumed by the circuit resistances. In a series circuit, the total power is equal to the **sum** of the powers dissipated by the individual resistors. Total power (P_T) is thus equal to:

$$P_T = P_1 + P_2 + P_3 \dots P_n \qquad (8.10)$$

where

P_T = total power, W
P_1 = power used in first part, W
P_2 = power used in second part, W
P_3 = power used in third part, W
P_n = power used in nth part, W

Example 8.15

PROBLEM:

A series circuit consists of three resistors having values of 5 ohms, 15 ohms, and 20 ohms. Find the total power dissipation when 120 V is applied to the circuit (see Figure 8.35).

SOLUTION:

Given:

R_1 = 5 ohms
R_2 = 15 ohms
R_3 = 20 ohms
E = 120 V

The total resistance is found first.

$R_T = R_1 + R_2 + R_3$

$R_T = 5 + 15 + 20$

$R_T = 40$ Ohms

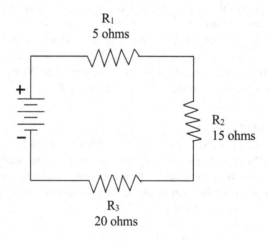

FIGURE 8.35   Solving for total power in a series circuit.

Using total resistance and the applied voltage, the circuit current is calculated.

$$I = \frac{E_T}{R_T}$$

$$I = \frac{120}{40}$$

$$I = 3 \text{ amps}$$

Using the power formula, the individual power dissipations can be calculated. For resistor R_1:

$$P_1 = I^2 R_1$$

$$P_1 = (3)^2 5$$

$$P_1 = 45 \text{ watts}$$

For R_2:

$$P_2 = I^2 R_2$$

$$P_2 = (3)^2 15$$

$$P_2 = 135 \text{ watts}$$

For R_3:

$$P_3 = I^2 R_3$$

$$P_3 = (3)^2 20$$

$$P_3 = 180 \text{ watts}$$

To obtain total power:

$$P_T = P_1 + P_2 + P_3$$

$$P_T = 45 + 135 + 180$$

$$P_T = 360 \text{ watts}$$

To check the answer, the total power delivered by the source can be calculated:

$$P = E \times I$$

$$P = 3\,a \times 120 \text{ v}$$

$$P = 360 \text{ watts}$$

Thus, the total power is equal to the sum of the individual power dissipations.

Key Point: We found that Ohm's Law can be used for total values in a series circuit as well as for individual parts of the circuit. Similarly, the formula for power may be used for total values.

$$P_T = IE_T \qquad\qquad (8.11)$$

SUMMARY OF THE RULES FOR SERIES D-C CIRCUITS

To this point, we have covered many of the important factors governing the operation of basic series circuits. In essence, what we have really done is to lay a strong foundation to build upon in preparation for more advanced circuit theory that follows. A summary of the important factors governing the operation of a series circuit is listed as follows:

a. The same current flows through each part of a series circuit.
b. The total resistance of a series circuit is equal to the sum of the individual resistances.
c. The total voltage across a series circuit is equal to the sum of the individual voltage drops.
d. The voltage drop across a resistor in a series circuit is proportional to the size of the resistor.
e. The total power dissipated in a series circuit is equal to the sum of the individual dissipations.

GENERAL SERIES CIRCUIT ANALYSIS

Now that we have discussed the "pieces" involved in putting together the puzzle for solving series circuit analysis, we move on to the next step in the process: solving series circuit analysis in total.

Example 8.16

PROBLEM:

Three resistors of 20 ohms, 20 ohms, and 30 ohms are connected across a battery supply rated at 100 V terminal voltage. Completely solve the circuit shown in Figure 8.36.

Note: In solving the circuit, the total resistance will be found first. Next, the circuit current will be calculated. Once the current is known the voltage drops and power dissipations can be calculated.

SOLUTION:

The total resistance is:

$$R_T = R_1 + R_2 + R_3$$
$$R_T = 20 \text{ ohms} + 20 \text{ ohms} + 30 \text{ ohms}$$
$$R_T = 70 \text{ ohms}$$

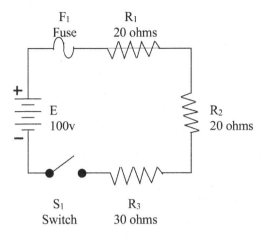

FIGURE 8.36 Solving for various values in a series circuit.

By Ohm's Law the current is:

$$I = \frac{E}{R_T}$$

$$I = \frac{100}{70}$$

$$I = 1.43 \text{ amps (rounded)}$$

The voltage (E_1) across R_1 is:

$$E_1 = IR_1$$
$$E_1 = 1.43 \text{ amps} \times 20 \text{ ohms}$$
$$E_1 = 28.6 \text{ volts}$$

The voltage (E_2) across R_2 is:

$$E_2 = IR_2$$
$$E_2 = 1.43 \text{ amps} \times 20 \text{ ohms}$$
$$E_2 = 28.6 \text{ volts}$$

The voltage (E_3) across R_3 is:

$$E_3 = IR_2$$
$$E_3 = 1.43 \text{ amps} \times 30 \text{ ohms}$$
$$E_3 = 42.9 \text{ volts}$$

The power dissipated by R_1 is:

$$P_1 = I \times E_1$$
$$P_1 = 1.43 \text{ amps} \times 28.6 \text{ volts}$$
$$P_1 = 40.9 \text{ Watts}$$

The power dissipated by R_2 is:

$$P_2 = I \times E_2$$
$$P_2 = 1.43 \text{ amps} \times 28.6 \text{ volts}$$
$$P_2 = 40.9 \text{ Watts}$$

The power dissipated by R_3 is:

$$P_3 = I \times E_3$$
$$P_3 = 1.43 \text{ amps} \times 42.9 \text{ volts}$$
$$P_3 = 61.3 \text{ Watts (rounded)}$$

The total power dissipated is:

$$P_T = E_T \times I$$
$$P_T = 100 \text{ volts} \times 1.43 \text{ amps}$$
$$P_T = 143 \text{ Watts}$$

Note: Keep in mind when applying Ohm's Law to a series circuit to consider whether the values used are component values or total values. When the information available enables the use of Ohm's Law to find total resistance, total voltage, and total current, total values must be inserted into the formula.

To find total resistance:

$$R_T = \frac{E_T}{I_T}$$

To find total voltage:

$$E_T = I_T \times R_T$$

To find total current:

$$I_T = \frac{E_T}{R_T}$$

KIRCHHOFF'S VOLTAGE LAW

Kirchhoff's voltage law states that the voltage applied to a closed circuit equals the sum of the voltage drops in that circuit. It should be obvious that this fact was used in the study of series circuits to this point. It was expressed as follows:

Voltage applied = sum of voltage drops

$$E_A = E_1 + E_2 + E_3$$

where E_A is the applied voltage and E_1, E_2, and E_3 are voltage drops.

Another way of stating Kirchhoff's Law is that the algebraic sum of the instantaneous emf's and voltage drops around any closed circuit is zero.

Through the use of Kirchhoff's Law, circuit problems can be solved which would be difficult and often impossible with only knowledge of Ohm's Law. When Kirchhoff's Law is properly applied, an equation can be set up for a closed loop and the unknown circuit values may be calculated.

POLARITY OF VOLTAGE DROPS

When there is a voltage drop across a resistance, one end must be more positive or more negative than the other end. The polarity of the voltage drop is determined by the direction of current flow. In the circuit shown in Figure 8.37, the current is seen to be flowing in a counterclockwise direction due to the arrangement of the battery source E. Notice that the end of resistor R_1 into which the current flows is marked *negative* (–). The end of R_1 at which the current leaves is marked **positive** (+). These polarity markings are used to show that the end of R_1 into which the current flows is at a higher negative potential than is the end of the resistor at which the current leaves. Point A is thus more negative than point B.

Point C, which is at the same potential as point B, is labeled negative. This is to indicate that point C, though positive with respect to point A, is more negative than point D. To say a point is positive (or negative), without stating what it is positive *in respect to*, has no meaning.

Kirchhoff's Voltage Law can be written as an equation as shown below:

$$E_a + E_b + E_c + \dots E_n = 0 \tag{8.12}$$

where E_a, E_b, etc. are the voltage drops and emf's around any closed-circuit loop.

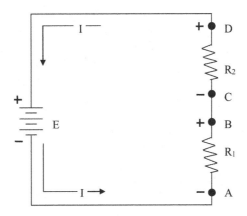

FIGURE 8.37 Polarity of voltage drops.

Example 8.17

PROBLEM:

Three resistors are connected across a 60-V source. What is the voltage across the third resistor if the voltage drops across the first two resistors are 10 V and 20 V?

SOLUTION:

First, draw a diagram like the one shown in Figure 8.38. Next, a direction of current is assumed as shown. Using this current, the polarity markings are placed at each end of each resistor and on the terminals of the source. Starting at point A, trace around the circuit in the direction of current flow recording the voltage and polarity of each component. Starting at point A these voltages would be as follows:

Basic formula:

$$E_a + E_b + E_c \ldots E_n = 0$$

From the circuit:

$$(+E_?) + (+E_2) + (+E_3) - (E_A) = 0$$

Substituting values from circuit:

$$E_? + 10 + 20 - 60 = 0$$

$$E_? - 30 = 0$$

$$E_? = 30 \text{ volts}$$

Thus, the unknown voltage $(E_?)$ is found to be 30 V.

Using the same idea as above, a problem can be solved in which the current is the unknown quantity.

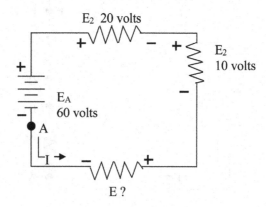

FIGURE 8.38   Determining unknown voltage in a series circuit.

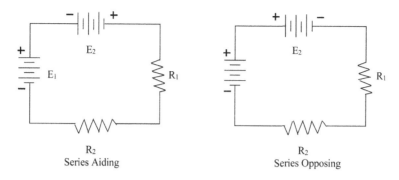

FIGURE 8.39 Aiding and opposing sources.

SERIES AIDING AND OPPOSING SOURCES

Sources of voltage that cause current to flow in the same direction are considered to be *series aiding,* and their voltages add. Sources of voltage that would tend to force current in opposite directions are said to be *series opposing,* and the effective source voltage is the difference between the opposing voltages. When two opposing sources are inserted into a circuit, current flow would be in a direction determined by the larger source. Examples of series aiding and opposing sources are shown in Figure 8.39.

KIRCHHOFF'S LAW AND MULTIPLE SOURCE SOLUTIONS

Kirchhoff's Law can be used to solve multiple source circuit problems. In applying this method, the exact same procedure is used for the multiple sources as was used for single source circuits. This is demonstrated by the following example.

Example 8.18

PROBLEM:

Find the amount of current in the circuit shown in Figure 8.40.

SOLUTION:

Start at point A.
Basic equation:

$$E_a + E_b + E_c + \ldots E_n = 0$$

From the circuit:

$$E_{b2} + E_1 - E_{b1} + E_{b3} + E_2 = 0$$

$$+40 + 40I - 140 + 20 + 20I = 0$$

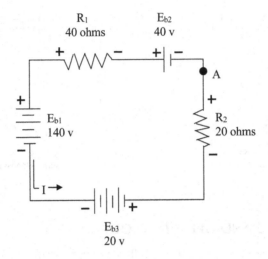

FIGURE 8.40 Solving for circuit current in a multiple source circuit.

Combining like terms:

$$+60I - 80 = 0$$

$$60I = 80$$

$$I = 1.33 \text{ amps}$$

GROUND

The term "ground" is used to denote a common electrical point of zero potential. The reference point of a circuit is always considered to be at zero potential; the earth (ground) is said to be at a zero potential.

The common ground for much electrical/electronic equipment is the metal chassis. The value of ground is noted when considering its contribution to economy, simplification of schematics, and ease of measurement. When completing each electrical circuit, common points of a circuit at zero potential are connected directly to the metal chassis thereby eliminating a large amount of connecting wire. An example of a grounded circuit is illustrated in Figure 8.41.

Note: Most voltage measurements used to check proper circuit operation in electronic equipment are taken with respect to ground. One-meter lead is attached to ground and the other meter lead is moved to various test points.

OPEN AND SHORT CIRCUITS

A circuit is *open* if a break in the circuit does not provide a complete path for current. Figure 8.42 shows an open circuit, because the fuse is blown.

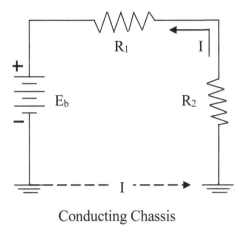

Conducting Chassis

FIGURE 8.41   Ground used as a conductor.

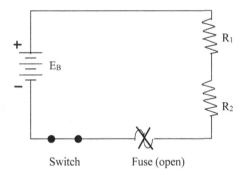

Switch   Fuse (open)

FIGURE 8.42   Open circuit – fuse blown.

To protect a circuit, a fuse is placed directly into the circuit. A fuse will open the circuit whenever a dangerously large current starts to flow. A fuse will permit currents smaller than the fuse value to flow but will melt and therefore break or open the circuit if a larger current flows. A dangerously large current will flow when a "short circuit" occurs. A short circuit is usually caused by an accidental connection between two points in a circuit that offers very little resistance that passes an abnormal amount of current. A short circuit often occurs because of improper wiring or broken insulation.

PARALLEL D-C CIRCUITS

The principles we applied to solving simple series circuit calculations for determining the reactions of such quantities as voltage, current, and resistance can be used in parallel and series-parallel circuits.

Parallel Circuit Characteristics

A *parallel circuit* is defined as one having two or more components connected across the same voltage source (see Figure 8.4). Recall that in a series circuit there is only one path for current flow. As additional loads (resistors, etc.) are added to the circuit, the total resistance increases and the total current decreases. This is *not the case* in a *parallel* circuit. In a parallel circuit, each load (or branch) is connected directly across the voltage source. In Figure 8.43, commencing at the voltage source (E_b) and tracing counterclockwise around the circuit, two complete and separate paths can be identified in which current can flow. One path is traced from the source through resistance R_1 and back to the source; the other, from the source through resistance R_2 and back to the source.

Voltage in Parallel Circuits

Recall that in a series circuit the source voltage divides proportionately across each resistor in the circuit. In a parallel circuit (see Figure 8.44), the same voltage is present across all the resistors of a parallel group. This voltage is equal to the applied voltage (E_b) and can be expressed in equation form as:

$$E_b = E_{R1} = E_{R2} = E_{Rn} \tag{8.13}$$

We can verify Equation (8.13) by taking voltage measurements across the resistors of a parallel circuit, as illustrated in Figure 8.44. Notice that each voltmeter indicates the same amount of voltage; that is, the voltage across each resistor is the same as the applied voltage.

Key Point: In a parallel circuit the voltage remains the same throughout the circuit.

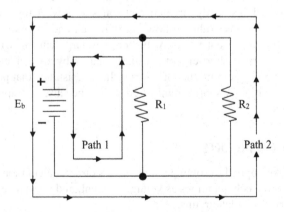

FIGURE 8.43 Basic parallel circuit.

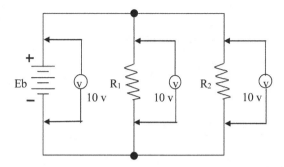

FIGURE 8.44 Voltage comparison in a parallel circuit.

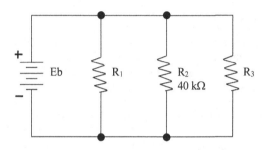

FIGURE 8.45 For Example 8.19.

Example 8.19

PROBLEM:

Assume that the current through a resistor of a parallel circuit is known to be 4.0 milliamperes (ma) and the value of the resistor is 40,000 ohms. Determine the potential (voltage) across the resistor. The circuit is shown in Figure 8.45.

SOLUTION:

Given:

$R_2 = 40 \text{ k}\Omega$
$I_{R2} = 4.0 \text{ ma}$
Find:
$E_{R2} = ?$
$E_b = ?$

Select proper equation

$$E = IR$$

Substitute known values:

$$E_{R2} = I_{R2} \times R_2$$

$$E_{R2} = 4.0 \text{ ma} \times 40,000 \text{ ohms}$$

[use power of tens]

$$E_{R2} = \left(4.0 \times 10^{-3}\right) \times \left(40 \times 10^{3}\right)$$

$$E_{R2} = 4.0 \times 40$$

Resultant:

$$E_{R2} = 160 \text{ v}$$

Therefore:

$$E_b = 160 \text{ v}$$

CURRENT IN PARALLEL CIRCUITS

In a series circuit, a single current flows. Its value is determined in part by the total resistance of the circuit. However, the source current in a parallel circuit divides among the available paths in relation to the value of the resistors in the circuit. Ohm's Law remains unchanged. For a given voltage, current varies inversely with resistance.

Important Point: Ohm's Law states: The current in a circuit is inversely proportional to the circuit resistance. This fact, important as a basic building block of electrical theory, obviously, is also important in the following explanation of current flow in parallel circuits.

The behavior of current in a parallel circuit is best illustrated by an example. The example we use is Figure 8.46. The resistors R_1, R_2, and R_3 are in parallel with each other and with the battery. Each parallel path is then a branch with its own individual current. When the total current I_T leaves the voltage source E, part I_1 of the current I_T will flow through R_1, part I_2 will flow through R_2, and the remainder I_3 through R_3. The branch current I_1, I_2, and I_3 can be different. However, if a voltmeter (used for measuring the voltage of a circuit) is connected across R_1, R_2, and R_3, the respective voltages E_1, E_2, and E_3 will be equal. Therefore,

$$E = E_1 = E_2 = E_3 \tag{8.14}$$

The total current I_T is equal to the sum of all branch currents.

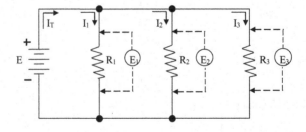

FIGURE 8.46 Parallel circuit.

$$I_T = I_1 = I_2 = I_3 \tag{8.15}$$

This formula applies for any number of parallel branches whether the resistances are equal or unequal.

By Ohm's Law, each branch current equals the applied voltage divided by the resistance between the two points where the voltage is applied. Hence for each branch we have the following equations:

$$\text{Branch 1: } I_1 = \frac{E_1}{R_1} = \frac{E}{R_1}$$

$$\text{Branch 2: } I_2 = \frac{E_2}{R_2} = \frac{E}{R_2}$$

$$\text{Branch 3: } I_3 = \frac{E_3}{R_3} = \frac{V}{R_3} \tag{8.16}$$

With the same applied voltage, any branch that has less resistance allows more current through it than a branch with higher resistance.

Example 8.20

PROBLEM:

Two resistors each drawing 2 A and a third resistor drawing 1 A are connected in parallel across a 100-V line (see Figure 8.47). What is the total current?

SOLUTION:

The formula for total current is

$$I_T = I_1 + I_2 + I_3$$

$$= 2 + 2 + 1$$

$$= 5 \text{ A}$$

The total current is 5 A.

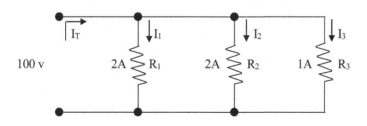

FIGURE 8.47 For Example 8.20.

Example 8.21

PROBLEM:

Two branches R_1 and R_2 across a 100-V power line draw a total line current of 20 A (Figure 8.48). Branch R_1 takes 10 A. What is the current I_2 in branch R_2?

SOLUTION:

Starting with Equation (8.15), transpose to find I_2 and then substitute given values.

$$I_T = I_1 + I_2$$
$$I_2 = I_T - I_1$$
$$= 20 - 10 = 10 \text{ A}$$

The current in branch R_2 is 10 A.

Example 8.22

PROBLEM:

A parallel circuit consists of two 15-ohm and one 12-ohm resistor across a 120-V line (see Figure 8.49). What current will flow in each branch of the circuit and what is the total current drawn by all the resistors?

SOLUTION:

There is 120-V potential across each resistor. Using Equation (8.16), apply Ohm's Law to each resistor.

$$I_1 = \frac{V}{R_1} = \frac{120}{15} = 8 \text{ A}$$

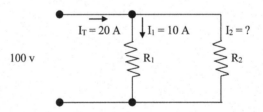

FIGURE 8.48   For Example 8.21.

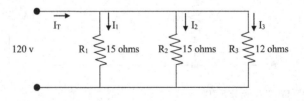

FIGURE 8.49   For Example 8.22.

$$I_2 = \frac{V}{R_2} = \frac{120}{15} = 8 \text{ A}$$

$$I_3 = \frac{V}{R_3} = \frac{120}{12} = 10 \text{ A}$$

Now find total current, using Equation (8.16).

$$I_T = I_1 + I_2 + I_3$$

$$= 8 + 8 + 10 = 26 \text{ A}$$

PARALLEL CIRCUITS AND KIRCHHOFF'S CURRENT LAW

The division of current in a parallel network follows a definite pattern. This pattern is described by *Kirchhoff's Current Law*, which is stated as follows:

The algebraic sum of the currents entering and leaving any junction of conductors is equal to zero. This can be stated mathematically as

$$I_a + I_b + \ldots + I_n = 0 \qquad (8.17)$$

where I_a, I_b + etc. are the currents entering and leaving the junction. Currents entering the junction are assumed positive, and currents leaving the junction are considered negative. When solving a problem using Equation (8.17), the currents must be placed into the equation with the proper polarity.

Example 8.23

PROBLEM:

Solve for the value of I_3 in Figure 8.50.

SOLUTION:

First, the currents are given proper signs.

$I_1 = + 10$ amps
$I_2 = - 3$ amps
$I_3 = ?$ amps
$I_4 = - 5$ amps

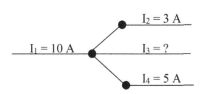

FIGURE 8.50 For Example 8.23.

Then, place these currents into Equation (8.17) with the proper signs as follows:

Basic equation:

$$I_a + I_b + \ldots + I_n = 0$$

Substitution:

$$I_1 + I_2 + I_3 + I_4 = 0$$

$$(+10) + (-3) + (I_3) + (-5) = 0$$

Combining like terms:

$$I_3 = 2 = 0$$

$$I_3 = -2 \text{ amps}$$

thus, I_3 has a value of 2 amps, and the negative sign shows it to be a current leaving the junction.

PARALLEL CIRCUIT RESISTANCE

Unlike series circuits, where total resistance (R_T) is the sum of the individual resistances, in a parallel circuit the total resistance is *not* the sum of the individual resistances.

In a parallel circuit, we can use Ohm's Law to find total resistance. We use the equation

$$R = \frac{E}{I}$$

or

$$R_T = \frac{E_S}{I_T}$$

R_T is the total resistance of all the parallel branches across the voltage source E_S, and I_T is the sum of all the branch currents.

Example 8.24

PROBLEM:

What is the total resistance of the circuit shown in Figure 8.51?

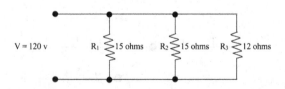

FIGURE 8.51   For Example 8.24.

Given: $E_S = 120v$
$I_T = 26A$

SOLUTION:

In Figure 8.51 the line voltage is 120 V, and the total line current is 26 A. Therefore,

$$R_T = \frac{E}{I_T} = \frac{120}{26} = 4.26 \text{ ohms}$$

Important Point: Notice that R_T is smaller than any of the three resistances in Figure 8.51. This fact may surprise you—it may seem strange that the total circuit resistance is *less* than that of the smallest resistor (R_3-12 ohms). However, if we refer back to the water analogy we have used previously, it makes sense. Consider water pressure and water pipes—and assume there is some way to keep the water pressure constant. A small pipe offers more resistance to the flow of water than a larger pipe; but if we add another pipe in parallel, one of even smaller diameter, the total resistance to water flow is decreased. In an electrical circuit, even a larger resistor in another parallel branch provides an additional path for current flow, so the total resistance is less. Remember, if we add one more branch to a parallel circuit, the total resistance decreases and the total current increases.

Back to Example 8.24 and Figure 8.51. What we essentially demonstrated in working this particular problem is that the total load connected to the 120-V line is the same as the single equivalent resistance of 4.62 Ω connected across the line. (It is probably more accurate to call this total resistance the "equivalent resistance," but by convention R_t, or total resistance, is used—but they are often used interchangeably, too.)

The "equivalent resistance" in the equivalent circuit is shown in Figure 8.52.

There are other methods used to determine the equivalent resistance of parallel circuits. The most appropriate method for a particular circuit depends on the number and value of the resistors. For example, consider the parallel circuit shown in Figure 8.53.

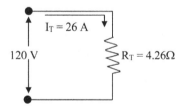

$I_T = 26$ A

120 V

$R_T = 4.26\Omega$

FIGURE 8.52   Equivalent circuit to that of Figure 8.51.

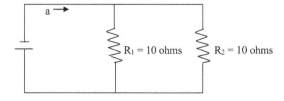

a

$R_1 = 10$ ohms

$R_2 = 10$ ohms

FIGURE 8.53   Two equal resistors connected in parallel.

For this circuit, the following simple equation is used:

$$R_{eq} = \frac{R}{N} \qquad\qquad (8.18)$$

where

R_{eq} = equivalent parallel resistance
R = ohmic value of one resistor
N = number of resistors

Thus,

$$R_{eq} = \frac{10 \text{ ohms}}{2}$$

$$R_{eq} = 5 \text{ ohms}$$

Note: Equation (8.18) is valid for any number of equal value parallel resistors.

Key Point: When two equal value resistors are connected in parallel, they present a total resistance equivalent to a single resistor of one-half the value of either of the original resistors.

Example 8.25

PROBLEM:

Five 50-ohm resistors are connected in parallel. What is the equivalent circuit resistance?

SOLUTION:

Using Equation (8.18)

$$R_{eq} = \frac{R}{N} = \frac{50}{5} = 10 \text{ ohms}$$

What about parallel circuits containing resistance of unequal value? How is equivalent resistance determined?
Example 8.26 demonstrates how this is accomplished.

Example 8.26

PROBLEM:

Refer to Figure 8.54.

SOLUTION:

Given:

$R_1 = 3\Omega$

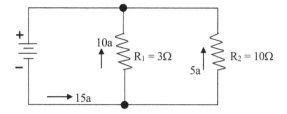

FIGURE 8.54   For Example 8.26.

$R_2 = 6\Omega$
$E_a = 30v$

Known:

$I_1 = 10$ amps
$I_2 = 5$ amps
$I_t = 15$ amps

Determine:

$R_{eq} = ?$

$R_{eq} = \dfrac{E_a}{I_t}$

$R_{eq} = \dfrac{30}{15} = 2$ ohms

Key Point: In Example 8.26 the equivalent resistance of two ohms is less than the value of either branch resistor. Remember, in parallel circuits the equivalent resistance will always be smaller than the resistance of any branch.

RECIPROCAL METHOD

When circuits are encountered in which resistors of unequal value are connected in parallel, the equivalent resistance may be computed by using the *reciprocal method*.

Note: A *reciprocal* is an inverted fraction; the reciprocal of the fraction 3/4, for example is 4/3. We consider a whole number to be a fraction with 1 as the denominator, so the reciprocal of a whole number is that number divided into 1. For example, the reciprocal of R_t is $1/R_t$.

The equivalent resistance in parallel is given by the formula

$$\frac{1}{R_T} = \frac{1}{R_1} + \frac{1}{R_2} + \frac{1}{R_3} \cdots \frac{1}{R_n} \tag{8.19}$$

where

R_T is the total resistance in parallel, and R_1, R_2, R_3, and R_n are the branch resistances.

Example 8.27

PROBLEM:

Find the total resistance of a 2-ohm, a 4-ohm, and an 8-ohm resistor in parallel (Figure 8.55).

SOLUTION:

Write the formula for three resistors in parallel.

$$\frac{1}{R_T} = \frac{1}{R_1} + \frac{1}{R_2} + \frac{1}{R_3}$$

Substituting the resistance values.

$$\frac{1}{R_T} = \frac{1}{2} + \frac{1}{4} + \frac{1}{8}$$

Add fractions.

$$\frac{1}{R_T} = \frac{4}{8} + \frac{2}{8} + \frac{1}{8} = \frac{7}{8}$$

Invert both sides of the equation to solve for R_T.

$$R_T = \frac{8}{7} = 1.14\,\Omega$$

Note: When resistances are connected in parallel, the total resistance is always **less** than the smallest resistance of any single branch.

PRODUCT OVER THE SUM METHOD

When any two unequal resistors are in parallel, it is often easier to calculate the total resistance by multiplying the two resistances and then dividing the product by the sum of the resistances.

$$R_T = \frac{R_1 \times R_2}{R_1 + R_2} \qquad (8.20)$$

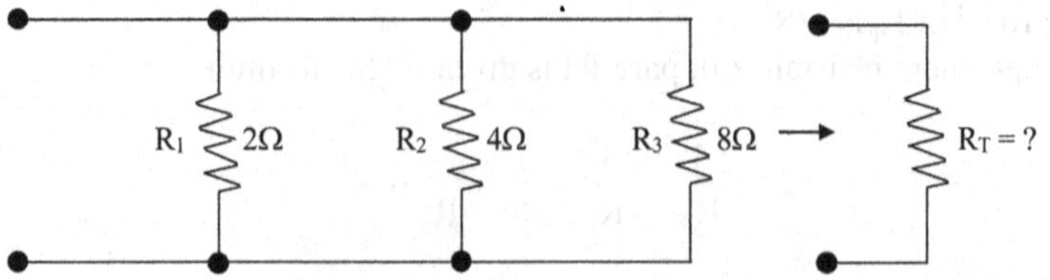

FIGURE 8.55   For Example 8.27.

where
R_T is the total resistance in parallel, and R_1 and R_2 are the two resistors in parallel.

Example 8.28

PROBLEM:

What is the equivalent resistance of a 20-ohm and a 30-ohm resistor connected in parallel?

SOLUTION:

Given: $R_1 = 20$
$R_2 = 30$

$$R_T = \frac{R_1 \times R_2}{R_1 + R_2}$$

$$R_T = \frac{20 \times 30}{20 + 30}$$

$R_T = 12$ ohms

REDUCTION TO AN EQUIVALENT CIRCUIT

In the study of basic electricity, it is often necessary to resolve a complex circuit into a simpler form. Any complex circuit consisting of resistances can be reduced to a basic equivalent circuit containing the source and total resistance. This process is called *reduction to an equivalent circuit*. An example of circuit reduction was demonstrated in Example 8.28 and is illustrated in Figure 8.56.

The circuit shown in Figure 8.56(A) is reduced to the simple circuit shown in (B).

POWER IN PARALLEL CIRCUITS

As in the series circuit, the total *power* consumed in a parallel circuit is equal to the sum of the power consumed in the individual resistors.

Note: Because power dissipation in resistors consists of a heat loss, power dissipations are additive regardless of how the resistors are connected in the circuit.

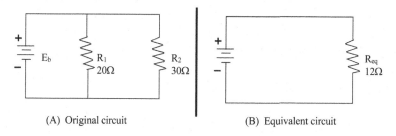

(A) Original circuit   (B) Equivalent circuit

FIGURE 8.56   Parallel circuit with equivalent circuit.

$$P_T = P_1 + P_2 + P_3 + \ldots + P_n \tag{8.21}$$

where

P_T is the total power, and P_1, P_2, P_3, and P_n are the branch powers.

Total power can also be calculated by the equation

$$P_T = EI_T \tag{8.22}$$

where

P_T is the total power, E is the voltage source across all parallel branches, and I_T is the total current.

The power dissipated in each branch is equal to EI and equal to V^2/R.

Note: In both parallel and series arrangements, the sum of the individual values of power dissipated in the circuit equals the total power generated by the source. The circuit arrangements cannot change the fact that all power in the circuit comes from the source.

RULES FOR SOLVING PARALLEL D-C CIRCUITS

Problems involving the determination of resistance, voltage, current, and power in a parallel circuit are solved as simply as in a series circuit. The procedure is the same— (1) draw a circuit diagram, (2) state the values given and the values to be found, (3) state the applicable equations, and (4) substitute the given values and solve for the unknown.

Along with following the problem-solving procedure above, it is also important to remember and apply the rules for solving parallel d-c circuits. These rules are:

  a. The same voltage exists across each branch of a parallel circuit and is equal to the source voltage.
  b. The current through a branch of a parallel network is inversely proportional to the amount of resistance of the branch.
  c. The total current of a parallel circuit is equal to the sum of the currents of the individual branches of the circuit.
  d. The total resistance of a parallel circuit is equal to the reciprocal of the sum of the reciprocals of the individual resistances of the circuit.
  e. The total power consumed in a parallel circuit is equal to the sum of the power consumption of the individual resistances.

SERIES-PARALLEL CIRCUITS

To this point, we have discussed series and parallel d-c circuits. However, operators will seldom encounter a circuit that consists solely of either type of circuit. Most circuits consist of both series and parallel elements. A circuit of this type is referred to as a *series-parallel circuit*, or as a combination circuit. The solution of a series-parallel (combination) circuit is simply a matter of application of the laws and rules discussed before this point.

SOLVING A SERIES-PARALLEL CIRCUIT

At least three resistors are required to form a series-parallel circuit: two parallel resistors connected in series with at least one other resistor. In a circuit of this type, the current I_T divides after it flows through R_1, and part flows through R_2 and part flows through R_3. Then, the current joins at the junction of the two resistors and flows back to the positive terminal of the voltage source (E) and through the voltage source to the positive terminal.

In solving for values in a series-parallel circuit (current, voltage, and resistance), follow the rules that apply to a series circuit for the series part of the circuit, and follow the rules that apply to a parallel circuit for the parallel part of the circuit. Solving series-parallel circuits is simplified if all parallel and series groups are first reduced to single equivalent resistances and the circuits redrawn in simplified form. Recall, the redrawn circuit is called an *equivalent circuit*.

Note: There are no general formulas for the solution of series-parallel circuits because there are so many different forms of these circuits.

Note: The total current in the series-parallel circuit depends on the effective resistance of the parallel portion *and* on the other resistances.

CONDUCTORS

Earlier we mentioned that electric current moves easily through some materials but with greater difficulty through others. Three good electrical conductors are copper, silver, and aluminum (generally, we can say that most metals are good conductors). Now copper is the material of choice used in electrical conductors. Under special conditions, certain gases are also used as conductors (e.g., neon gas, mercury vapor, and sodium vapor are used in various kinds of lamps).

The function of the wire conductor is to connect a source of the applied voltage to a load resistance with a minimum IR voltage drop in the conductor so that most of the applied voltage can produce current in the load resistance. Ideally, a conductor must have very low resistance (e.g., a typical value for a conductor—copper—is less than 1 Ω per 10 ft).

Because all electrical circuits utilize conductors of one type or another, in this section, we discuss the basic features and electrical characteristics of the most common types of conductors.

Moreover, because conductor splices and connections (and insulation of such connections) are also an essential part of any electric circuit, they are also discussed.

UNIT SIZE OF CONDUCTORS

A standard (or unit size) of a conductor has been established to compare the resistance and size of one conductor with another. The unit of linear measurement used is (in regard to the diameter of a piece of wire) the *mil* (0.001 of an inch). A convenient unit of wire length used is the foot. Thus, the standard unit of size in most cases is the *mil-foot* (i.e., a wire will have unit size if it has a diameter of 1 mil and a length of 1 ft). The resistance in ohms of a unit conductor or a given substance is called the *resistivity* (or specific resistance) of the substance.

As a further convenience, *gage* numbers are also used in comparing the diameter of wires. The B and S (Browne and Sharpe) gage was used in the past; now the most commonly used gage is the *American wire gage* (AWG).

SQUARE MIL

Figure 8.57 shows a square mil. The *square mil* is a convenient unit of cross-sectional area for square or rectangular conductors. As shown in Figure 8.57, a square mil is the area of a square, the sides of which are 1 mil. To obtain the cross-sectional area in square mils of a square conductor, square one side measured in mils. To obtain the cross-sectional area in square mils of a rectangular conductor, multiply the length of one side by that of the other, each length being expressed in mils.

Example 8.29

PROBLEM:

Find the cross-sectional area of a large rectangular conductor 5/8 in. thick and 5 in. wide.

SOLUTION:

The thickness may be expressed in mils as $0.625 \times 1,000 = 625$ mils and the width as $5 \times 1,000 = 5,000$ mils. The cross-sectional area is $625 \times 5,000$, or 3,125,000 square mils.

CIRCULAR MIL

The *circular mil* is the standard unit of wire cross-sectional area used in most wire tables. To avoid the use of decimals (because most wires used to conduct electricity may be only a small fraction of an inch), it is convenient to express these diameters

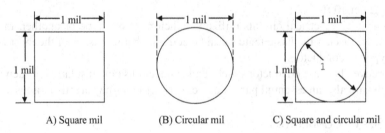

A) Square mil   (B) Circular mil   C) Square and circular mil

FIGURE 8.57   (A) Square mil; (B) Circular mil; (C) Comparison of circular to square mil.

in mils. For example, the diameter of a wire is expressed as 25 mils instead of 0.025 in. A circular mil is the area of a circle having a diameter of 1 mil, as shown in Figure 8.57 (B). The area in circular mils of a round conductor is obtained by squaring the diameter measured in mils. Thus, a wire having a diameter of 25 mils has an area of 25^2 or 625 circular mils. By way of comparison, the basic formula for the area of a circle is

$$A = \pi R^2 \tag{8.23}$$

and in this example the area in square inches is

$$A = \pi R^2 = 3.14(0.0125)^2 = 0.00049 \text{ sq. in.}$$

If D is the diameter of a wire in mils, the area in square mils can be determined using

$$A = \pi(D/2)^2 \tag{8.24}$$

which translates to

$$= 3.14/4 \; D^2$$
$$= 0.785 \; D^2 \text{ sq. mils}$$

Thus, a wire 1 mil in diameter has an area of

$$A = 0.785 \times 1^2 = 0.785 \text{ sq. mils,}$$

which is equivalent to 1 circular mil. The cross-sectional area of a wire in circular mils is therefore determined as

$$A = \frac{0.785 \; D^2}{0.785} = D^2 \text{ circular mils,}$$

where D is the diameter in mils. Therefore, the constant $\pi/4$ is eliminated from the calculation.

It should be noted that in comparing square and round conductors that the circular mil is a smaller unit of area than the square mil, and therefore there are more circular mils than square mils in any given area. The comparison is shown in Figure 8.57(C). The area of a circular mil is equal to 0.785 of a square mil.

Important Point: To determine the circular-mil area when the square-mil area is given, divide the area in square mils by 0.785. Conversely, to determine the square-mil area when the circular-mil area is given, multiply the area in circular mils by 0.785.

Example 8.30

PROBLEM:

A No. 12 wire has a diameter of 80.81 mils. What are (1) its area in circular mils and (2) its area in square mils?

SOLUTION:

(1) $A = D^2 = 80.81^2 = 6,530$ circular mils

(2) $A = 0.785 \times 6,530 = 5,126$ square mils

Example 8.31

PROBLEM:

A rectangular conductor is 1.5 in. wide and 0.25 in. thick. (1) What is its area in square mils? (2) What size of the round conductor in circular mils is necessary to carry the same current as the rectangular bar?

SOLUTION:

(1) $1.5" = 1.5 \times 1,000 = 1,500$ mils

$0.25" = 0.25 \times 1,000 = 250$ mils

$A = 1,500 \times 250 = 375,000$ sq. mils

(2) To carry the same current, the cross-sectional area of the rectangular bar and the cross-sectional area of the round conductor must be equal. There are more circular mils than square mils in this area, and therefore

$$A = \frac{375,000}{0.785} = 477,700 \text{ circular mils}$$

Note: Many electric cables are composed of stranded wires. The strands are usually single wires twisted together in sufficient numbers to make up the necessary cross-sectional area of the cable. The total area in circular mils is determined by multiplying the area of one strand in circular mils by the number of strands in the cable.

CIRCULAR-MIL-FOOT

As shown in Figure 8.58, a *circular-mil foot* is actually a unit of volume. More specifically, it is a unit conductor 1 ft in length and having a cross-sectional area of 1 circular mil. The circular-mil-foot is useful in making comparisons between wires that are made of different metals because it is considered a unit conductor. Because it is considered a unit conductor, the circular-mil-foot is useful in making comparisons

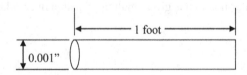

FIGURE 8.58 Circular—mil—foot.

between wires that are made of different metals. For example, a basis of comparison of the **resistivity** of various substances may be made by determining the resistance of a circular-mil-foot of each of the substances.

Note: It is sometimes more convenient to employ a different unit of volume when working with certain substances. Accordingly, unit volume may also be taken as the centimeter cube. The inch cube may also be used. The unit of volume employed is given in tables of specific resistances.

RESISTIVITY

All materials differ in their atomic structure and therefore in their ability to resist the flow of an electric current. The measure of the ability of a specific material to resist the flow of electricity is called its *resistivity*, or specific resistance—the resistance in ohms offered by unit volume (the circular-mil-foot) of a substance to the flow of electric current. Resistivity is the reciprocal of conductivity (i.e., the ease by which current flows in a conductor). A substance that has a high resistivity will have a low conductivity, and vice versa.

The resistance of a given length, for any conductor, depends upon the resistivity of the material, the length of the wire, and the cross-sectional area of the wire according to the equation

$$R = \rho \frac{L}{A} \tag{8.25}$$

where
   R = resistance of the conductor,
   L = length of the wire, ft
   A = cross-sectional area of the wire, CM
   ρ = specific resistance or resistivity, CM $\times$ Ω/ft

The factor ρ (Greek letter rho, pronounced "roe") permits different materials to be compared for resistance according to their nature without regard to different lengths or areas. Higher values of ρ mean more resistance.

Key Point: The resistivity of a substance is the resistance of a unit volume of that substance.

Many tables of resistivity are based on the resistance in ohms of a volume of the substance 1-ft long and 1 circular mil in cross-sectional area. The temperature at which the resistance measurement is made is also specified. If the kind of metal of which the conductor is made is known, the resistivity of the metal may be obtained from a table. The resistivity, or specific resistance, of some common substances is given in Table 8.3.

Note: Because silver, copper, gold, and aluminum have the lowest values of resistivity, they are the best conductors. Tungsten and iron have a much higher resistivity.

TABLE 8.3
Resistivity (Specific Resistance)

Substance	Specific Resistance @ 20 ° (CM ft Ω)
Silver	9.8
Copper (drawn)	10.37
Gold	14.7
Aluminum	17.02
Tungsten	33.2
Brass	42.1
Steel (soft)	95.8
Nichrome	660.0

Example 8.32

PROBLEM:

What is the resistance of 1,000 ft of copper wire having a cross-sectional area of 10,400 circular mils (No. 10 wire), the wire temperature being 20°C?

SOLUTION:

The resistivity (specific resistance), from Table 8.3, is 10.37. Substituting the known values in the preceding Equation (8.24), the resistance, R, is determined as

$$R = \rho \frac{L}{A} = 10.37 \times \frac{1,000}{10,400} = 1\, \text{ohm, approximately}$$

WIRE MEASUREMENT

Wires are manufactured in sizes numbered according to a table known as the American wire gage (AWG). Table 8.4 lists the standard wire sizes that correspond to the AWG. The gage numbers specify the size of the round wire in terms of its diameter and cross-sectional area. Note the following:

(a) As the gage numbers increase from 1 to 40, the diameter and circular area decrease. Higher gage numbers mean smaller wire sizes. Thus, No. 12 is a smaller wire than No. 4.
(b) The circular area doubles for every three-gage size. For example, No 12 wire has about twice the area of No. 15 wire.
(c) The higher the gage number and the smaller the wire, the greater the resistance of the wire for any given length. Therefore, for 1,000 ft of wire, No. 12 has a resistance of 1.62 Ω while No. 4 has 0.253 Ω.

FACTORS GOVERNING THE SELECTION OF WIRE SIZE

Several factors must be considered in selecting the size of wire to be used for transmitting and distributing electric power. These factors include allowable power loss

TABLE 8.4
Copper Wire Table

Gage #	Diameter	Circular mils	Ohms/1000 ft @ 25°C
1	289.0	83,700.0	0.126
2	258.0	66,400.0	0.159
3	229.0	52,600.0	0.201
4	204.0	41,700.0	0.253
5	182.0	33,100.0	0.319
6	162.0	26,300.0	0.403
7	144.0	20,800.0	0.508
8	128.0	16,500.0	0.641
9	114.0	13,100.0	0.808
10	102.0	10,400.0	1.02
11	91.0	8,230.0	1.28
12	81.0	6,530.0	1.62
13	72.0	5,180.0	2.04
14	64.0	4,110.0	2.58
15	57.0	3,260.0	3.25
16	51.0	2,580.0	4.09
17	45.0	2,050.0	5.16
18	40.0	1,620.0	6.51
19	36.0	1,290.0	8.21
20	32.0	1,020.0	10.4
21	28.5	810.0	13.1
22	25.3	642.0	16.5
23	22.6	509.0	20.8
24	20.1	404.0	26.4
25	17.9	320.0	33.0
26	15.9	254.0	41.6
27	14.2	202.0	52.5
28	12.6	160.0	66.2
29	11.3	127.0	83.4
30	10.0	101.0	105.0
31	8.9	79.7	133.0
32	8.0	63.2	167.0
33	7.1	50.1	211.0
34	6.3	39.8	266.0
35	5.6	31.5	335.0
36	5.0	25.0	423.0
37	4.5	19.8	533.0
38	4.0	15.7	673.0
39	3.5	12.5	848.0
40	3.1	9.9	1,070.0

in the line; the permissible voltage drop in the line; the current-carrying capacity of the line; and the ambient temperatures in which the wire is to be used.

a. **Allowable power loss (I^2R) in the line**. This loss represents electrical energy converted into heat. The use of large conductors will reduce the resistance and therefore the I^2R loss. However, large conductors are heavier and require more substantial supports; thus, they are more expensive initially than small ones.

b. **Permissible voltage drop (IR drop) in the line**. If the source maintains a constant voltage at the input to the line, any variation in the load on the line will cause a variation in line current and a consequent variation in the IR drop in the line. A wide variation in the IR drop in the line causes poor voltage regulation at the load.

c. **The current-carrying capacity of the line**. When current is drawn through the line, heat is generated. The temperature of the line will rise until the heat radiated, or otherwise dissipated, is equal to the heat generated by the passage of current through the line. If the conductor is insulated, the heat generated in the conductor is not so readily removed, as it would be if the conductor were not insulated.

d. **Conductors installed where ambient temperature is relatively high**. When installed in such surroundings, the heat generated by external sources constitutes an appreciable part of the total conductor heating. Due allowance must be made for the influence of external heating on the allowable conductor current and each case has its own specific limitations.

Copper vs. Other Metal Conductors

If it were not cost prohibitive, silver, the best conductor of electron flow (electricity), would be the conductor of choice in electrical systems. Instead, silver is used only in special circuits where a substance with high conductivity is required.

The two most generally used conductors are copper and aluminum. Each has characteristics that make its use advantageous under certain circumstances. Likewise, each has certain disadvantages or limitations.

Copper has a higher conductivity, and it is more ductile (can be drawn out into wire), has relatively high tensile strength, and can be easily soldered. It is more expensive and heavier than aluminum.

Aluminum has only about 60% of the conductivity of copper, but its lightness makes possible long spans, and its relatively large diameter for a given conductivity reduces corona (i.e., the discharge of electricity from the wire when it has a high potential). The discharge is greater when a smaller diameter wire is used than when a larger-diameter wire is used. However, aluminum conductors are not easily soldered, and aluminum's relatively large size for a given conductance does not permit the economical use of an insulation covering.

Note: Recent practice involves using copper wiring (instead of aluminum wiring) in house and some industrial applications. This is the case because aluminum connections are not as easily made as they are with copper. In addition, over the years, many fires have been started because of improperly connected aluminum wiring (i.e., poor connections = high resistance connections, resulting in excessive heat generation).

A comparison of some of the characteristics of copper and aluminum is given in Table 8.5.

TEMPERATURE COEFFICIENT

The resistance of pure metals—such as silver, copper, and aluminum—increases as the temperature increases. The *temperature coefficient* of resistance, α (Greek letter alpha), indicates how much the resistance changes for a change in temperature. A positive value for α means R increases with temperature; a negative α means R decreases; and a zero α means R is constant, not varying with changes in temperature. Typical values of α are listed in Table 8.6.

TABLE 8.5
Characteristics of Copper and Aluminum

Characteristics	Copper	Aluminum
Tensile strength (lb/in^2)	55,000	25,000
Tensile strength for same conductivity (lb)	55,000	40,000
Weight for same conductivity (lb)	100	48
Cross-section for same conductivity (CM)	100	160
Specific resistance (Ω/mil. ft)	10.6	17

TABLE 8.6
Properties of Conducting Materials (Approximate)

Material	Temperature coefficient, Ω/°C, ?
Aluminum	0.004
Carbon	−0.0003
Constantan	0 (average)
Copper	0.004
Gold	0.004
Iron	0.006
Nichrome	0.0002
Nickel	0.005
Silver	0.004
Tungsten	0.005

The amount of increase in the resistance of a 1-ohm sample of the copper conductor per degree rise in temperature (i.e., the temperature coefficient of resistance) is approximately 0.004. For pure metals, the temperature coefficient of resistance ranges between 0.004 and 0.006 ohm.

Thus, a copper wire having a resistance of 50 ohms at an initial temperature of 0°C will have an increase in resistance of 50 × 0.004, or 0.2 ohms (approximate) for the entire length of wire for each degree of temperature rise above 0°C. At 20°C, the increase in resistance is approximately 20 × 0.2, or 4 ohms. The total resistance at 20°C is 50 + 4, or 54 ohms.

Note: As shown in Table 8.6, carbon has a negative temperature coefficient. In general, α is negative for all semiconductors such as germanium and silicon. A negative value for α means less resistance at higher temperatures. Therefore, the resistance of semiconductor diodes and transistors can be reduced considerably when they become hot with a normal load current. Observe, also, that constantan has a value of zero for α (Table 8.6). Thus, it can be used for precision wire-wound resistors, which do not change resistance when the temperature increases.

CONDUCTOR INSULATION

Electric current must be contained; it must be channeled from the power source to a useful load—safely. To accomplish this, electric current must be forced to flow only where it is needed. Moreover, current-carrying conductors must not be allowed (generally) to come in contact with one another, their supporting hardware, or personnel working near them. To accomplish this, conductors are coated or wrapped with various materials. These materials have such a high resistance that they are, for all practical purposes, nonconductors. They are generally referred to as *insulators* or *insulating materials.*

There are a wide variety of insulated conductors available to meet the requirements of any job. However, only the necessary minimum of insulation is applied for any particular type of cable designed to do a specific job. This is the case because insulation is expensive and has a stiffening effect and is required to meet a great variety of physical and electrical conditions.

Two fundamental but distinctly different properties of insulation materials (e.g., rubber, glass, asbestos, and plastics) are insulation resistance and dielectric strength.

a. *Insulation resistance* is the resistance to current leakage through and over the surface of insulation materials.
b. *Dielectric strength* is the ability of the insulator to withstand potential differences and is usually expressed in terms of the voltage at which the insulation fails because of the electrostatic stress.

Various types of materials are used to provide insulation for electric conductors, including rubber, plastics, varnished cloth, paper, silk, cotton, and enamel.

CONDUCTORS SPLICES AND TERMINAL CONNECTIONS

When conductors join each other, or connect to a load, *splices* or *terminals* must be used. It is important that they be properly made, since any electric circuit is only as good as its weakest connection. The basic requirement of any splice or connection is that it be both mechanically and electrically as strong as the conductor or device with which it is used. High-quality workmanship and materials must be employed to ensure lasting electrical contact, physical strength, and insulation (if required).

Important Point: Conductor splices and connections are an essential part of any electric circuit.

Soldering Operations

Soldering operations are a vital part of electrical and/or electronics maintenance procedures. Soldering is a manual skill that must be learned by all personnel who work in the field of electricity. Obviously, practice is required to develop proficiency in the techniques of soldering.

In performing a soldering operation, both the solder and the material to be soldered (e.g., electric wire and/or terminal lugs) must be heated to a temperature, which allows the solder to flow. If either is heated inadequately, *cold* solder joints result (i.e., high resistance connections are created). Such joints do not provide either the physical strength or the electrical conductivity required. Moreover, in soldering operations it is necessary to select a solder that will flow at a temperature low enough to avoid damage to the part being soldered, or to any other part or material in the immediate vicinity.

Solderless Connections

Generally, terminal lugs and splicers, which do not require solder, are more widely used (because they are easier to mount correctly) than those which do require solder. Solderless connectors—made in a wide variety of sizes and shapes—are attached to their conductors by means of several different devices, but the principle of each is essentially the same. They are all crimped (squeezed) tightly onto their conductors. They afford adequate electrical contact, plus great mechanical strength.

Insulation Tape

The carpenter has his saw, the dentist his pliers, the plumber his wrench, and the electrician his insulating tape. Accordingly, one of the first things the newcomer maintenance operator learns (a newcomer who is also learning proper and safe techniques for performing electrical work) is the value of electrical insulation tape. Normally, the use of electrical insulating tape comes into play as the final step in completing a splice or joint, to place insulation over the bare wire at the connection point.

Typically, insulation tape used should be the same basic substance as the original insulation, usually a rubber-splicing compound. When using rubber (latex) tape as the splicing compound where the original insulation was rubber, it should be applied to the splice with a light tension so that each layer presses tightly against the one underneath it. In addition to the rubber tape application (which restores the insulation to its original form), restoring with friction tape is also often necessary.

In recent years, plastic electrical tape has come into wide use. It has certain advantages over rubber and friction tape. For example, it will withstand higher voltages for a given thickness. Single thin layers of certain commercially available plastic tape will stand several thousand volts without breaking down.

Important Point: Be advised that though the use of plastic electrical tape has become almost universal in industrial applications, it must be applied in more layers—because it is thinner than rubber or friction tape—to ensure an extra margin of safety.

ELECTROMAGNETISM

Earlier, we discussed the fundamental theories concerning simple magnets and magnetism. Those discussions dealt mainly with forms of magnetism that were not related directly to electricity—permanent magnets for instance. Further, only brief mention was made of those forms of magnetism having direct relation to electricity—producing electricity with magnetism for instance.

In medicine, anatomy and physiology are so closely related that the medical student cannot study one at length without involving the other. A similar relationship holds for the electrical field; that is, magnetism and basic electricity are so closely related that one cannot be studied at length without involving the other. This close fundamental relationship is continually borne out in subsequent sections of this chapter, such as in the study of generators, transformers, and motors. To be proficient in electricity, the operator must become familiar with such general relationships that exist between magnetism and electricity as follows:

a. Electric current flow will always produce some form of magnetism.
b. Magnetism is by far the most commonly used means for producing or using electricity.
c. The peculiar behavior of electricity under certain conditions is caused by magnetic influences.

Magnetic Field around a Single Conductor

In 1819, Hans Christian Oersted, a Danish scientist, discovered that a field of magnetic force exists around a single wire conductor carrying an electric current. In Figure 8.59, a wire is passed through a piece of cardboard and connected through a switch to a dry cell. With the switch open (no current flowing), if we sprinkle iron filings on the cardboard, then tap it gently, the filings will fall back haphazardly. Now, if we close the switch, current will begin to flow in the wire. If we tap the cardboard again, the magnetic effect of the current in the wire will cause the filings to fall back into a definite pattern of concentric circles with the wire as the center of the circles. Every section of the wire has this field of force around it in a plane perpendicular to the wire, as shown in Figure 8.60.

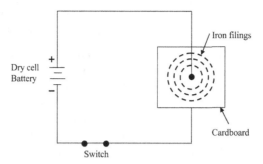

FIGURE 8.59 A circular patterns of magnetic force exists around a wire carrying an electric current.

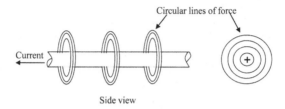

FIGURE 8.60 The circular fields of force around a wire carrying a current are in planes which are perpendicular to the wire.

FIGURE 8.61 The strength of the magnetic field around a wire carrying a current depends on the amount of current.

The ability of the magnetic field to attract bits of iron (as demonstrated in Figure 8.59) depends on the number of lines of force present. The strength of the magnetic field around a wire carrying a current depends on the current, because it is the current that produces the field. The greater the current, the greater the strength of the field. A large current will produce many lines of force extending far from the wire, while a small current will produce only a few lines close to the wire, as shown in Figure 8.61

POLARITY OF A SINGLE CONDUCTOR

The relation between the direction of the magnetic lines of force around a conductor and the direction of current flow along the conductor may be determined by means of the *left-hand rule for a conductor*. If the conductor is grasped in the left hand with the thumb extended in the direction of electron flow (– to +), the fingers will point in the direction of the magnetic lines of force. The north pole of a compass would point in this same direction if the compass was placed in the magnetic field.

Important Note: Arrows are generally used in electric diagrams to denote the direction of current flow along the length of the wire. Where cross sections of wire are shown, a special view of the arrow is used. A cross-sectional view of a conductor that is carrying current toward the observer is illustrated in Figure 8.62(A). The direction of the current is indicated by a dot, which represents the head of the arrow. A conductor that is carrying current away from the observer is illustrated in Figure 8.62(B). A cross, which represents the tail of the arrow, indicates the direction of the current.

FIELD AROUND TWO PARALLEL CONDUCTORS

When two parallel conductors carry current in the same direction, the magnetic fields tend to encircle both conductors, drawing them together with a force of attraction, as shown in Figure 8.63(A). Two parallel conductors carrying currents in opposite directions are shown in Figure 8.63(B). The field around one conductor is opposite in direction to the field around the other conductor. The resulting lines of force are crowded together in the space between the wires and tend to push the wires apart. Therefore, two parallel adjacent conductors carrying currents in the same direction attract each other and two parallel conductors carrying currents in opposite directions repel each other.

MAGNETIC FIELD OF A COIL

The magnetic field around a current-carrying wire exists at all points along its length. Bending the current-carrying wire into the form of a single loop has two results. First, the magnetic field consists of more dense concentric circles in a plane perpendicular to the wire, although the total number of lines is the same as for the straight conductor. Second, all the lines inside the loop are in the same direction. When this straight wire is wound around a core, it becomes a coil, and the magnetic

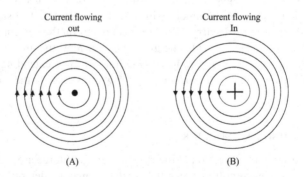

FIGURE 8.62 Magnetic field around a current-carrying conductor.

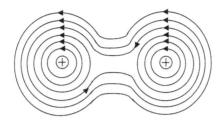

(A) Current flowing in the same direction

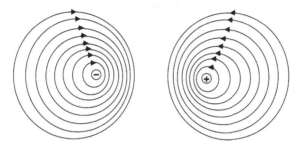

(B) Currents flowing in opposite directions

FIGURE 8.63 Magnetic field around two parallel conductors.

field assumes a different shape. When current is passed through the coiled conductor, the magnetic field of each turn of wire links with the fields of adjacent turns. The combined influence of all the turns produces a two-pole field similar to that of a simple bar magnet. One end of the coil will be a north pole and the other end will be a south pole.

POLARITY OF AN ELECTROMAGNETIC COIL

As mentioned, it was shown that the direction of the magnetic field around a straight conductor depends on the direction of current flow through that conductor. Thus, a reversal of current flow through a conductor causes a reversal in the direction of the magnetic field that is produced. It follows that a reversal of the current flow through a coil also causes a reversal of its two-pole field. This is true because that field is the product of the linkage between the individual turns of wire on the coil. Therefore, if the field of each turn is reversed, it follows that the total field (coils' field) is also reversed.

When the direction of electron flow through a coil is known, its polarity may be determined by the use of the *left-hand rule for coils*. This rule is stated as follows: Grasping the coil in the left hand, with the fingers "wrapped around" in the direction of electron flow, the thumb will point toward the north pole (see Figure 8.64).

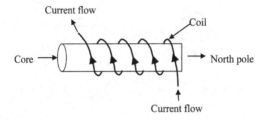

FIGURE 8.64 Current-carrying coil.

STRENGTH OF AN ELECTROMAGNETIC FIELD

The strength, or intensity, of the magnetic field of a coil depends on a number of factors.

- The *number of turns* of the conductor
- The *amount of current flow* through the coil
- The *ratio of the coil's length to its width*
- The *type of material in the core*

Magnetic Units

The law of current flow in the electric circuit is similar to the law for the establishing of flux in the magnetic circuit.

The *magnetic flux*, ϕ, (phi) is similar to current in the Ohm's Law formula and comprises the total number of lines of force existing in the magnetic circuit. The **Maxwell** is the unit of flux—that is, 1 line of force is equal to 1 Maxwell.

Note: The Maxwell is often referred to as simply a line of force, line of induction, or line.

The *strength* of a magnetic field in a coil of wire depends on how much current flows in the turns of the coil. The more the current, the stronger the magnetic field. In addition, the more the turns, the more concentrated are the lines of force. The *force* that produces the flux in the magnetic circuit (comparable to electromotive force in Ohm's Law) is known as *magnetomotive force*, or mmf. The practical unit of magnetomotive force is the **ampere-turn** (At). In equation form,

$$F = \text{ampere} - \text{turns} = NI \tag{8.26}$$

where
    F = magnetomotive force, At
    N = number of turns
    I = current, A

Example 8.33

PROBLEM:

Calculate the ampere-turns for a coil with 2,000 turns and a 5-Ma current.

SOLUTION:

Use Equation (8.26) and substitute N = 2,000 and I = 5 × 10⁻³ A.

$$NI = 2000\left(5 \times 10^{-3}\right) = 10 \text{ At}$$

The unit of *intensity* of magnetizing force per unit of length is designated as H and is sometimes expressed as Gilberts per centimeter of length. Expressed as an equation,

$$H = \frac{NI}{L} \qquad\qquad (8.27)$$

where
   H = magnetic field intensity, ampere-turns per meter (At/m)
   NI = ampere-turns, At
   L = length between poles of the coil, m

Note: Equation (8.26) is for a solenoid. *H* is the intensity of an air core. With an iron core, *H* is the intensity through the entire core and *L* is the length or distance between poles of the iron core.

PROPERTIES OF MAGNETIC MATERIALS

In this section, we discuss two important properties of magnetic materials: permeability and hysteresis.

Permeability

When the core of an electromagnet is made of annealed sheet steel, it produces a stronger magnet than if a cast iron core is used. This is the case because the magnetizing force of the coil more readily acts upon annealed sheet steel than is the hard cast iron. Simply put, soft sheet steel is said to have greater *permeability* because of the greater ease with which magnetic lines are established in it.

Recall that permeability is the relative ease with which a substance conducts magnetic lines of force. The permeability of air is arbitrarily set at 1. The permeability of other substances is the ratio of their ability to conduct magnetic lines compared to that of air. The permeability of nonmagnetic materials, such as aluminum, copper, wood, and brass is essentially unity, or the same as for air.

Note: The permeability of magnetic materials varies with the degree of magnetization, being smaller for high values of flux density.

Note: *Reluctance*, which is analogous to resistance, is the opposition to the production of flux in a material and is inversely proportional to permeability. Iron has high permeability and, therefore, low reluctance. Air has low permeability and hence high reluctance.

Hysteresis

When the current in a coil of wire reverses thousands of times per second, a considerable loss of energy can occur. This loss of energy is caused by *hysteresis*. Hysteresis means "a lagging behind"; that is, the magnetic flux in an iron core lags behind the increases or decreases of the magnetizing force.

The simplest method of illustrating the property of hysteresis is by graphical means such as the hysteresis loop shown in Figure 8.65.

The hysteresis loop (Fig. 8.65) is a series of curves that show the characteristics of a magnetic material. Opposite directions of the current result are in the opposite directions of $+H$ and $-H$ for field intensity. Similarly, opposite polarities are shown for flux density as $+B$ and $-B$. The current starts at the center 0 (zero) when the material is unmagnetized. Positive H values increase B to saturation at $+B_{max}$. Next H decreases to zero, but B drops to the value of B, because of hysteresis. The current that produced the original magnetization now is reversed so that H becomes negative. B drops to zero and continues to $-B_{max}$. As the $-H$ values decrease, B is reduced to $-B$, when H is zero. Now with a positive swing of current, H becomes positive, producing saturation at $+B_{max}$ again. The hysteresis loop is now completed. The curve does not return to zero at the center because of hysteresis.

ELECTROMAGNETS

An *electromagnet* is composed of a coil of wire wound around a core that is normally soft iron, because of its high permeability and low hysteresis. When direct current flows through the coil, the core will become magnetized with the same polarity that the coil would have without the core. If the current is reversed, the polarity of the coil and core are reversed.

The electromagnet is of great importance in electricity simply because the magnetism can be "turned on" or "turned off" at will. The starter solenoid (an electromagnet) in automobiles and powerboats is a good example. In an automobile or boat, an electromagnet is part of a relay that connects the battery to the induction

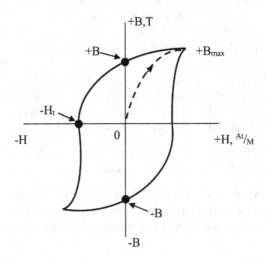

FIGURE 8.65 Hysteresis loop.

coil, which generates the very high voltage needed to start the engine. The starter solenoid isolates this high voltage from the ignition switch. When no current flows in the coil, it is an "air core," but when the coil is energized, a movable soft-iron core does two things. First, the magnetic flux is increased because the soft-iron core is more permeable than the air core. Second, the flux is more highly concentrated. All this concentration of magnetic lines of force in the soft-iron core results in a very good magnet when current flows in the coil. However, soft iron loses its magnetism quickly when the current is shut off. The effect of the soft iron is, of course, the same whether it is movable, as in some solenoids, or permanently installed in the coil. An electromagnet then consists of a coil and a core; it becomes a magnet when current flows through the coil.

The ability to control the action of magnetic force makes an electromagnet very useful in many circuit applications.

PERMANENT MAGNETS FOR WIND TURBINES

Before moving on to a discussion of alternating current (a-c), it is important to address permanent magnets that are currently used in many wind turbine generators. Permanent magnets are constructed from special alloys of ferromagnetic materials such as iron, nickel, and cobalt, several alloys of rare-earth metals, and minerals such as lodestone. Unlike electromagnets, permanent magnets produce a persistent magnetic field without the need for any external source of excitation such as electrical power.

Common practice today is to use rare-earth magnets in wind turbines, especially in offshore wind turbines, as they allow for above-average power density and diminished size and weight with peak efficiency at all speeds, offering a high annual production of energy with low lifetime expenditures. Most direct-drive turbines are equipped with permanent magnet generators that typically contain the rare-earth neodymium and smaller quantities of dysprosium. Although on a different extent or scale, the same is true for numerous gearbox designs. Note that for onshore wind turbine installations it is not necessary to utilize permanent magnet generators because the reduced size and weight is not a concern as it is with offshore installations.

DID YOU KNOW?

Although older and smaller wind turbines may generate DC voltage to provide charging current to batteries, it is more practical for larger grid-connected turbines to generate AC voltage. This eliminates the need to change the generated power to an AC voltage before it leaves the tower site.

A-C THEORY

Because voltage is induced in a conductor when lines of force are cut, the amount of the induced emf depends on the number of lines cut in a unit time. To induce an emf of 1 V, a conductor must cut 100,000,000 lines of force per second. To obtain this great number of "cuttings," the conductor is formed into a loop and rotated on an axis at great speed (see Figure 8.66). The two sides of the loop become individual conductors in series, each side of the loop cutting lines of force and inducing twice the voltage that a single conductor would induce. In commercial generators, the number of "cuttings" and the resulting emf are increased by: (1) increasing the number of lines of force by using more magnets or stronger electromagnets, (2) using more conductors or loops, and (3) rotating the loops faster.

How an a-c generator operates to produce an a-c voltage and current is a basic concept today, taught in elementary and middle school science classes. Of course, we accept technological advances as commonplace, today—we surf the internet, watch cable television, use our cell phones, and take space flight as a given—and consider producing the electricity that makes all these technologies possible as our right. These technologies are bottom shelf to us today—we have them available to us, so we simply use them. This point of view was surely not held initially—especially those who broke ground in developing technology and electricity.

In groundbreaking years of electric technology development, the geniuses of the science of electricity (including George Simon Ohm) performed their technological breakthroughs in faltering steps. We tend to forget that those first faltering steps of scientific achievement in the field of electricity were performed with crude, and for the most part, homemade apparatus.

Indeed, the innovators of electricity had to fabricate nearly all the laboratory equipment used in their experiments. At the time, the only convenient source of electrical energy available to these early scientists was the voltaic cell, invented some years earlier. Because of the fact that cells and batteries were the only sources of power available, some of the early electrical devices were designed to operate from *direct current.*

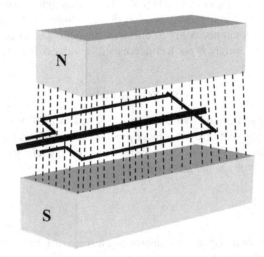

FIGURE 8.66 Loop rotating in a magnetic field produces an a-c voltage.

Thus, initially, direct current was used extensively. However, when the use of electricity became widespread, certain disadvantages in the use of direct current became apparent. In a direct current system, the supply voltage must be generated at the level required by the load. To operate a 240-V lamp, for example, the generator must deliver 240 V. A 120-V lamp could not be operated from this generator by any convenient means. A resistor could be placed in series with the 120-V lamp to drop the extra 120 V, but the resistor would waste an amount of power equal to that consumed by the lamp.

Another disadvantage of direct current systems is a large amount of power lost due to the resistance of the transmission wires used to carry current from the generating station to the consumer. This loss could be greatly reduced by operating the transmission line at very high voltage and low current. This is not a practical solution in a d-c system, however, since the load would also have to operate at high voltage. Because of the difficulties encountered with direct current, practically all modern power distribution systems use *alternating current (a-c)*, including wind turbine generated electrical power for distribution.

Unlike d-c voltage, a-c voltage can be stepped up or down by a device called a *transformer*. Transformers permit the transmission lines to be operated at high voltage and low current for maximum efficiency. Then at the consumer end the voltage is stepped down to whatever value the load requires by using a transformer. Due to its inherent advantages and versatility, alternating current has replaced direct current in all but a few commercial power distribution systems.

BASIC A-C GENERATOR

As shown in Figure 8.66, an a-c voltage and current can be produced when a conductor loop rotates through a magnetic field and cuts lines of force to generate an induced a-c voltage across its terminals. This describes the basic principle of operation of an alternating current generator or *alternator*. An alternator converts mechanical energy into electrical energy. It does this by utilizing the principle of *electromagnetic induction*. The basic components of an alternator are an armature, about which many turns of conductor are wound, which rotates in a magnetic field, and some means of delivering the resulting alternating current to an external circuit.

CYCLE

An a-c voltage is one that continually changes in magnitude and periodically reverses in polarity (see Figure 8.67). The zero axis is a horizontal line across the center. The vertical variations on the voltage wave show the changes in magnitude. The voltages above the horizontal axis have positive (+) polarity, while voltages below the horizontal axis have negative (–) polarity.

Figure 8.67 shows a suspended loop of wire (conductor or armature) being rotated (moved) in a counterclockwise direction through the magnetic field between the poles of a permanent magnet. For ease of explanation, the loop has been divided into a thick and thin half. Notice that in part (A), the thick half is moving along (parallel to) the lines of force. Consequently, it is cutting none of these lines. The same is true of the thin half, moving in the opposite direction. Because the conductors are not cutting any lines of force, no emf is induced. As the loop rotates toward the position shown in part (B), it cuts more and more lines of force per second because it is cutting more directly across the field (lines of force) as it approaches the position shown in (B). At position (B) the induced voltage is greatest because the conductor is cutting directly across the field.

As the loop continues to be rotated toward the position shown in part (C), it cuts fewer and fewer lines of force per second. The induced voltage decreases from its peak value. Eventually, the loop is once again moving in a plane parallel to the magnetic field, and no voltage (zero voltage) is induced. The loop has now been rotated through half a circle (one alternation, or 180°). The sine curve shown in the lower part of Figure 8.67 shows the induced voltage at every instant of rotation of the loop. Notice that this curve contains 360°, or two alternations. Two alternations represent one complete circle of rotation.

Important Point: Two complete alternations in a period is called a *cycle*.

In Figure 8.67, if the loop is rotated at a steady rate, and if the strength of the magnetic field is uniform, the number of cycles per second (cps), or *hertz*, and the voltage will remain at fixed values. Continuous rotation will produce a series of sine-wave voltage cycles, or, in other words, an a-c voltage. In this way, mechanical energy is converted into electrical energy.

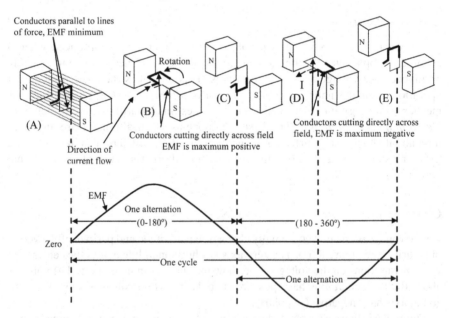

FIGURE 8.67 Basic A-C sine wave and alternating current generator.

FREQUENCY, PERIOD, AND WAVELENGTH

The *frequency* of an alternating voltage or current is the number of complete cycles occurring in each second of time. It is indicated by the symbol f and is expressed in hertz (Hz). One cycle per second equals one hertz. Thus, 60 cycles per second (cps) equals 60 Hz. A frequency of 2 Hz [Figure 8.68(A)] is twice the frequency of 1 Hz [Figure 8.68(B)].

The amount of time for the completion of 1 cycle is the *period*. It is indicated by the symbol T for time and is expressed in seconds (s). Frequency and period are reciprocals of each other.

$$f = \frac{1}{T} \tag{8.28}$$

$$T = \frac{1}{f} \tag{8.29}$$

Important Point: The higher the frequency, the shorter the period.

The angle of 360° represents the time for 1 cycle, or the period T. Therefore, we can show the horizontal axis of the sine wave in units of either electrical degrees or seconds (see Figure 8.69).

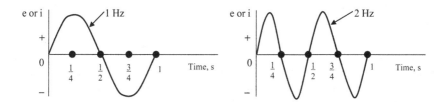

FIGURE 8.68 Comparison of frequencies.

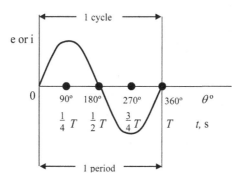

FIGURE 8.69 Relationship between electrical degrees and time.

The *wavelength* is the length of one complete wave or cycle. It depends upon the frequency of the periodic variation and its velocity of transmission. It is indicated by the symbol λ (Greek lowercase lambda). Expressed as a formula:

$$\lambda = \frac{\text{velocity}}{\text{frequency}} \tag{8.30}$$

CHARACTERISTIC VALUES OF A-C VOLTAGE AND CURRENT

Because an a-c sine wave voltage or current has many instantaneous values through-out the cycle, it is convenient to specify magnitudes for comparing one wave with another. The peak, average, or root-mean-square (rms) value can be specified (see Figure 8.70). These values apply to current or voltage.

Peak Amplitude

One of the most frequently measured characteristics of a sine wave is its amplitude. Unlike d-c measurement, the amount of alternating current or voltage present in a circuit can be measured in various ways. In one method of measurement, the maxi-mum amplitude of either the positive or the negative alternation is measured. The value of current or voltage obtained is called the *peak voltage* or the *peak current*. To measure the peak value of current or voltage, an oscilloscope must be used. The peak value is illustrated in Figure 8.70.

Peak-to-Peak Amplitude

A second method of indicating the amplitude of a sine wave consists of determining the total voltage or current between the positive and negative peaks. This value of current or voltage is called the *peak-to-peak value* (see Figure 8.70). Because both alternations of a pure sine wave are identical, the peak-to-peak value is twice the peak value. Peak-to-peak voltage is usually measured with an oscilloscope, although some voltmeters have a special scale calibrated in peak-to-peak volts.

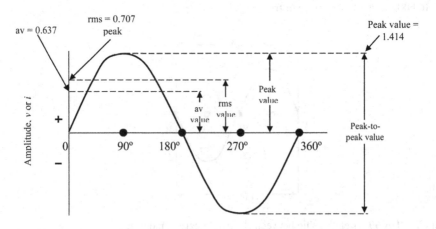

FIGURE 8.70 Amplitude values for a-c sine wave.

Instantaneous Amplitude

The *instantaneous value* of a sine wave of voltage for any angle of rotation is expressed by the formula:

$$e = E_m \times \sin \theta \qquad (8.31)$$

where
   e = the instantaneous voltage
   E_m = the maximum or peak voltage
   $\sin\theta$ = the sine of angle at which e is desired.

Similarly, the equation for the instantaneous value of a sine wave of current would be:

$$i = I_m \times \sin \theta \qquad (8.32)$$

where
   i = the instantaneous current
   I_m = the maximum or peak current
   $\sin\theta$ = the sine of the angle at which i is desired

Note: The instantaneous value of voltage constantly changes as the armature of an alternator moves through a complete rotation. Because current varies directly with voltage, according to Ohm's Law, the instantaneous changes in current also result in a sine wave whose positive and negative peaks and intermediate values can be plotted exactly as we plotted the voltage sine wave. However, instantaneous values are not useful in solving most a-c problems, so an *effective* value is used.

Effective or RMS Value

The *effective value* of an a-c voltage or current of a sine waveform is defined in terms of an equivalent heating effect of a direct current. The heating effect is independent of the direction of current flow.

Important Point: Since all instantaneous values of induced voltage are somewhere between zero and E_M (maximum, or peak voltage), the effective value of a sine wave voltage or current must be greater than zero and less than E_M (the maximum, or peak voltage).

The alternating current of the sine waveform having a maximum value of 14.14 amps produces the same amount of heat in a circuit having a resistance of one ohm as a direct current of 10 amps. Because this is true, we can work out a constant value for converting any peak value to a corresponding effective value. X represents this constant in the simple equation below. Solve for X to three decimal places.

$$14.14\,X = 10$$

$$X = 0.707$$

The effective value is also called the *root-mean-square (rms)* value because it is the square root of the average of the squared values between zero and maximum. The effective value of an a-c current is stated in terms of an equivalent d-c current. The phenomenon used as the standard comparison is the heating effect of the current.

Important Point: Anytime an a-c voltage or current is stated without any qualifications, it is assumed an effective value.

In many instances, it is necessary to convert from effective to peak or vice versa using a standard equation. Figure 8.70 shows that the peak value of a sine wave is 1.414 times the effective value; therefore, the equation we use is:

$$E_m = E \times 1.414 \tag{8.33}$$

where
   E_m = maximum or peak voltage
   E = effective or RMS voltage

and

$$I_m = I \times 1.414 \tag{8.34}$$

where
   I_m = maximum or peak current
   I = effective or RMS current

Upon occasion, it is necessary to convert a peak value of current or voltage to an effective value. This is accomplished by using the following equations:

$$E_m = E \times 0.707 \tag{8.35}$$

where
   E = effective voltage
   E_m = the maximum or peak voltage

$$I = I_m \times 0.707 \tag{8.36}$$

where
   I = the effective current
   I_m = the maximum or peak current

AVERAGE VALUE

Because the positive alternation is identical to the negative alternation, the *average value* of a complete cycle of a sine wave is zero. In certain types of circuits, however, it is necessary to compute the average value of one alternation. Figure 8.70 shows that the average value of a sine wave is 0.637 × peak value and therefore:

$$\text{Average Value} = 0.637 \times \text{peak value} \tag{8.37}$$

or

$$E_{avg} = E_m \times 0.637$$

where
   E_{avg} = the average voltage of one alternation
   E_m = the maximum or peak voltage

similarly

$$I_{avg} = I_m \times 0.637 \tag{8.38}$$

where
   I_{avg} = the average current in one alternation
   I_m = the maximum or peak current

Table 8.7 lists the various values of sine wave amplitude used to multiply in the conversion of a-c sine wave voltage and current.

RESISTANCE IN A-C CIRCUITS

If a sine wave of voltage is applied to a resistance, the resulting current will also be a sine wave. This follows Ohm's Law that states that the current is directly proportional to the applied voltage. Figure 8.71 shows a sine wave of voltage and the resulting sine wave of current superimposed on the same time axis. Notice that as the voltage increases in a positive direction, the current increases along with it. When the voltage reverses direction, the current reverses direction. At all times, the voltage and current pass through the same relative parts of their respective cycles at the same time. When two waves, such as those shown in Figure 8.71, are precisely in step with one another, they are said to be *in phase*. To be in phase, the two waves must go through their maximum and minimum points at the same time and in the same direction.

TABLE 8.7
A-C Sine Wave Conversion Table

Multiply the value	By	To get the value
Peak	2	Peak-to-peak
Peak-to-peak	0.5	Peak
Peak	0.637	Average
Average	1.637	Peak
Peak	0.707	RMS (effective)
RMS (effective)	1.414	Peak
Average	1.110	RMS (effective)
RMS (effective)	0.901	Average

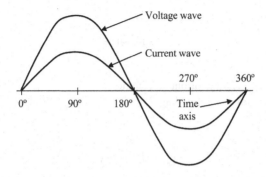

FIGURE 8.71 Voltage and current waves in phase.

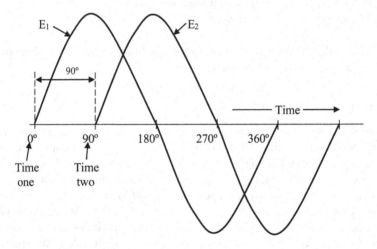

FIGURE 8.72 Voltage waves 90° out of phase.

In some circuits, several sine waves can be in phase with each other. Thus, it is possible to have two or more voltage drops in phase with each other and in phase with the circuit current.

Note: It is important to remember that Ohm's Law for d-c circuits is applicable to ac circuits *with resistance only*.

Voltage waves are not always in phase. For example, Figure 8.72 shows a voltage wave E_1 considered to start at 0° (time 1). As voltage wave, E_1 reaches its positive peak, a second voltage wave E_2 starts to rise (time 2). Because these waves do not go through their maximum and minimum points at the same instant of time, a *phase difference* exists between the two waves. The two waves are said to be out of phase. For the two waves in Figure 8.72, this phase difference is 90°.

Phase Relationships

In the preceding section, we discussed the important concepts of *in phase and phase difference*. Another important phase concept is the phase angle. The *phase angle* between two waveforms of the same frequency is the angular difference at a given instant of time. As an example, the phase angle between waves B and A (see Figure 8.73) is 90°. Take the instant of time at 90°. The horizontal axis is shown in angular units of time. Wave B starts at maximum value and reduces to zero value at 90°, while Wave A starts at zero and increases to a maximum value at 90°. Wave B reaches its maximum value 90° ahead of wave A, so wave B *leads* wave A by 90° (and wave A *lags* wave B by 90°). This 90° phase angle between Waves B and A is maintained throughout the complete cycle and all successive cycles. At any instant of time, wave B has the value that wave A will have 90° later. Wave B is a cosine wave because it is displaced 90° from wave A, which is a sine wave.

Important Point: The amount by which one wave leads or lags another is measured in degrees.

To compare phase angles or phases of alternating voltages or currents, it is more convenient to use vector diagrams corresponding to the voltage and current waveforms. A *vector* is a straight line used to denote the magnitude and direction of a given quantity. The length of the line drawn to scale denotes magnitude, and the direction is indicated by the arrow at one end of the line, together with the angle that the vector makes with a horizontal reference vector.

Note: In electricity, since different directions really represent *time* expressed as a phase relationship, an electrical vector is called a *phasor*. In an a-c circuit containing only resistance, the voltage and current occur at the *same time*, or are in phase. To indicate this condition by means of phasors, all that is necessary is to draw the phasors for the voltage and current in the same direction. The length of the phasor indicates the value of each.

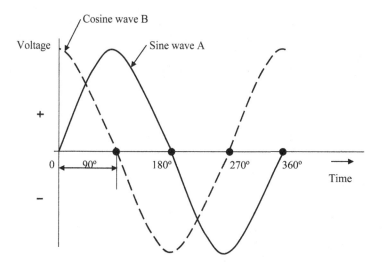

FIGURE 8.73   Wave B leads wave A by a phase angle of 90°.

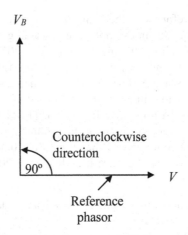

FIGURE 8.74 Phasor diagram.

A vector, or phasor, diagram is shown in Figure 8.74 where vector V_B is vertical to show the phase angle of 90° with respect to vector V_A, which is the reference. Because lead angles are shown in the counterclockwise direction from the reference vector, V_B leads V_A by 90°.

INDUCTANCE

To this point, we have learned the following key points about magnetic fields:

- A field of force exists around a wire carrying a current.
- This field has the form of concentric circles around the wire, in planes perpendicular to the wire, and with the wire at the center of the circles.
- The strength of the field depends on the current. Large currents produce large fields; small currents produce small fields.
- When lines of force cut across a conductor, a voltage is induced in the conductor.

Moreover, to this point we have studied circuits that have been *resistive* (i.e., resistors presented the only opposition to current flow). Two other phenomena—inductance and capacitance—exist in d-c circuits to some extent, but they are major players in a-c circuits. Both inductance and capacitance present a kind of opposition to current flow that is called *reactance*.

Inductance is the characteristic of an electrical circuit that makes itself evident by opposing the starting, stopping, or changing of current flow. A simple analogy can be used to explain inductance. We are all familiar with how difficult it is to push a heavy load (a cart full of heavy materials, etc.). It takes more work to start the load moving than it does to keep it moving. This is because the load possesses the property of *inertia*. Inertia is the characteristic of mass that opposes a *change* in velocity. Therefore, inertia can hinder us in some ways and help us in others. Inductance exhibits the same effect on current in an electric circuit as inertia does on the velocity of a mechanical object. The effects of inductance are sometimes desirable—sometimes undesirable.

Important Point: Simply put, *inductance* is the characteristic of an electrical conductor that opposes a *change* in current flow.

This means that because inductance is the property of an electric circuit that opposes any *change* in the current through that circuit, if the current increases, a self-induced voltage opposes this change and delays the increase. On the other hand, if the current decreases, a self-induced voltage tends to aid (or prolong) the current flow, delaying the decrease. Thus, current can neither increase nor decrease as fast in an inductive circuit as it can in a purely resistive circuit.

In a-c circuits, this effect becomes very important because it affects the *phase* relationships between voltage and current. In a-c circuits voltages (or currents) can be out of phase if they are induced in separate armatures of an alternator. When inductance is a factor in a circuit, the voltage and current generated by the *same* armature are out of phase. The important point to remember about a purely inductive a-c circuit is that the current through the circuit will lag the applied voltage by 90° angle. Thus, an inductor will not consume any active power; instead, it only consumes reactive power.

The phase angle between current and voltage is called the *power factor angle* because it is the basis for determining the power factor. Note that the power factor angle may be either positive or negative, corresponding to whether the current sine wave is leading the voltage sine wave or vice versa. If the power factor angle is positive, the power factor is angle is said to be leading; if it is negative, the power factor is lagging.

Note: Power factor correction capacitors are frequently used to improve the power factor of the wind turbine generator when viewed from the wind farm distribution system and utility. The correction capacitors are usually located at the base of the wind turbine tower or in the nearby control house or station. The idea is to place the correction capacitors as close to the generator as is feasible.

The unit for measuring inductance, L, is the *Henry* (named for the American physicist, Joseph Henry), abbreviated h and normally written in lower case, henry. Figure 8.75 shows the schematic symbol for an inductor. An inductor has an inductance of 1 henry if an emf of 1 V is induced in the inductor when the current through the inductor is changing at the rate of 1 ampere per second.

FIGURE 8.75   Schematic symbol for an inductor.

The relation between the induced voltage, inductance, and rate of change of current with respect to time is stated mathematically as

$$E = L \frac{\Delta I}{\Delta t} \tag{8.39}$$

where
    E = the induced emf in volts
    L = the inductance in henrys
    ΔI = is the change in amperes occurring in Δt seconds

Note: The symbol Δ (Delta) means "a change in …".

The henry is a large unit of inductance and is used with relatively large inductors. The unit employed with small inductors is the millihenry (mh). For still smaller inductors the unit of inductance is the microhenry (μh).

SELF-INDUCTANCE

As previously, explained, current flow in a conductor always produces a magnetic field surrounding, or linking with, the conductor. When the current changes, the magnetic field changes, and an emf is induced in the conductor. This emf is called a *self-induced emf* because it is induced in the conductor carrying the current.

    Note: Even a perfectly straight length of conductor has some inductance.

The direction of the induced emf has a definite relation to the direction in which the field that induces the emf varies. When the current in a circuit is increasing, the flux linking with the circuit is increasing. This flux cuts across the conductor and induces an emf in the conductor in such a direction to oppose the increase in current and flux. This emf is sometimes referred to as *counterelectromotive force* (cemf). The two terms are used synonymously throughout this manual. Likewise, when the current is decreasing, an emf is induced in the opposite direction and opposes the decrease in current.

Important Point: The effects just described are summarized by *Lenz's Law*, which states that the induced emf in any circuit is always in a direction as opposed to the effect that produced it.

Shaping a conductor so that the electromagnetic field around each portion of the conductor cuts across some other portion of the same conductor increases the inductance. This is shown in its simplest form in Figure 8.76(A). A loop of the conductor is looped so that two portions of the conductor lie adjacent and parallel to one another. These portions are labeled Conductor 1 and Conductor 2. When the switch is closed, electron flow through the conductor establishes a typical concentric field around *all* portions of the conductor. The field is shown in a single plane (for simplicity) that is perpendicular to both conductors. Although the field originates simultaneously in both conductors it is considered as originating in Conductor 1 and its effect on Conductor 2 will be noted. With increasing current, the field expands outward, cutting across a portion of Conductor 2. The dashed arrow shows the resultant induced

(A)

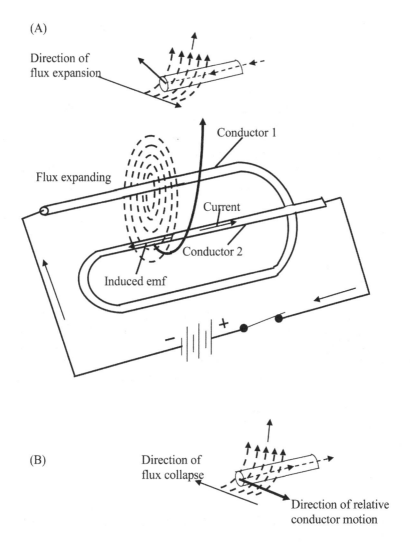

Direction of
flux expansion

Conductor 1

Flux expanding

Current

Conductor 2

Induced emf

+

−

(B)   Direction of
flux collapse

Direction of relative
conductor motion

FIGURE 8.76 Self-inductance.

emf in Conductor 2. Note that it is in *opposition* to the battery current and voltage, according to Lenz's Law.

In Figure 8.76(B), the same section of Conductor 2 is shown, but with the switch opened and the flux collapsing.

Important Point: From Figure 8.76, the important point to note is that the voltage of self-induction opposes both *changes* in current. It delays the initial buildup of current by opposing the battery voltage and delays the breakdown of current by exerting an induced voltage in the same direction that the battery voltage acted.

Four major factors affect the self-inductance of a conductor or circuit.

1. Number of turns: Inductance depends on the number of wire turns. Wind more turns to increase inductance. Take turns off to decrease the inductance. Figure 8.77 compares the inductance of two coils made with different numbers of turns.

2. Spacing between turns: Inductance depends on the spacing between turns, or the inductor's length. Figure 8.78 shows two inductors with the same number of turns. The first inductor's turns have a wide spacing. The second inductor's turns are close together. The second coil, though shorter, has a larger inductance value because of its close spacing between turns.

3. Coil diameter: Coil diameter, or cross-sectional area, is highlighted in Figure 8.79. The larger-diameter inductor has more inductance. Both coils shown have the same number of turns, and the spacing between turns is the same. The first inductor has a small diameter and the second one has a larger diameter. The second inductor has more inductance than the first one.

4. Type of core material: *Permeability*, as pointed out earlier, is a measure of how easily a magnetic field goes through a material. Permeability also tells us how much stronger the magnetic field will be with the material inside the coil.

    Figure 8.80 shows three identical coils. One has an air core; one has a powdered-iron core in the center and the other has a soft iron core. This figure illustrates the effects of core material on inductance. The inductance of a coil is affected by the magnitude of current when the core is a magnetic material. When the core is air, the inductance is independent of the current.

Key Point: The inductance of a coil increases very rapidly as the number of turns is increased. It also increases as the coil is made shorter, the cross-sectional area is made larger, or the permeability of the core is increased.

MUTUAL INDUCTANCE

When the current in a conductor or coil changes, the varying flux can cut across any other conductor or coil located nearby, thus inducing voltages in both. A varying

FIGURE 8.77 (A) Few turns, low inductance; (B) more turns, higher inductance.

FIGURE 8.78 (A) Wide spacing between turns, low inductance; (B) close spacing between turns, higher inductance.

FIGURE 8.79   (A) Small diameter, low inductance; (B) larger diameter, higher inductance.

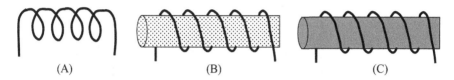

FIGURE 8.80   (A) Air core, low inductance; (B) powdered iron core, higher inductance; (C) soft iron core, highest inductance.

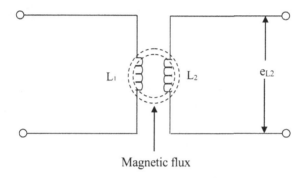

Magnetic flux

FIGURE 8.81   Mutual inductance between L_1 and l_2.

current in L_1, therefore, induces voltage across L_1 and across L_2 (see Figures 8.81 and 8.82 for the schematic symbol for two coils with mutual inductance). When the induced voltage e_{L2} produces current in L_2, its varying magnetic field induces voltage in L_1. Hence, the two coils L_1 and L_2 have *mutual inductance* because current change in one coil can induce voltage in the other. The unit of mutual inductance is the henry, and the symbol is L_M. Two coils have L_M of 1H when a current change of 1 A/s in one coil induces 1 E in the other coil.

The factors affecting the mutual inductance of two adjacent coils is dependent upon

• Physical dimensions of the two coils
• Number of turns in each coil
• Distance between the two coils
• Relative positions of the axes of the two coils
• The permeability of the cores

Important Point: The amount of mutual inductance depends on the relative position of the two coils. If the coils are separated a considerable distance, the amount

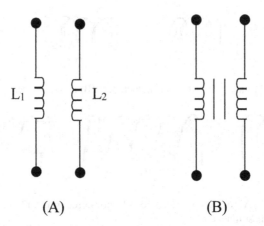

FIGURE 8.82 (A) Schematic symbol for two coils (air core) with mutual inductance; (B) two coils (iron core) with mutual inductance.

of flux common to both coils is small and the mutual inductance is low. Conversely, if the coils are close together, nearly all the flow of one coil links the turns of the other mutual inductance is high. The mutual inductance can be increased greatly by mounting the coils on a common iron core.

CALCULATION OF TOTAL INDUCTANCE

Note: In the study of advanced electrical theory, it is necessary to know the effect of mutual inductance in solving for total inductance in both series and parallel circuits. However, for our purposes, in this manual we do not attempt to make these calculations. Instead, we discuss the basic total inductance calculations that the maintenance operator should be familiar with.

If inductors in series are located far enough apart, or well shielded to make the effects of mutual inductance negligible, the total inductance is calculated in the same manner as for resistances in series; we merely add them:

$$L_t = L_1 + L_2 + L_3 \ldots (\text{etc.}) \qquad (8.40)$$

Example 8.34

PROBLEM:

If a series circuit contains three inductors whose values are 40 μh, 50 μh, and 20 μh, what is the total inductance?

SOLUTION:

$$L_t = 40\,\mu h + 50\,\mu h + 20\,\mu$$

$$= 110\,\mu h$$

In a parallel circuit containing inductors (without mutual inductance), the total inductance is calculated in the same manner as for resistances in parallel:

$$\frac{1}{L_t} = \frac{1}{L_1} + \frac{1}{L_2} + \frac{1}{L_3} + \dots (\text{etc.}) \tag{8.41}$$

Example 8.35

PROBLEM:

A circuit contains three totally shielded inductors in parallel. The values of the three inductances are: 4 mh, 5 mh, and 10 mh. What is the total inductance?

SOLUTION:

$$\frac{1}{L_t} = \frac{1}{4} + \frac{1}{5} + \frac{1}{10}$$

$$= 0.25 + 0.2 + 0.1$$

$$= 0.55$$

$$L_t = \frac{1}{0.55}$$

$$= 1.8 \text{mh}$$

CAPACITANCE

The ability of an electrical component or circuit to collect and store energy in the form of an electrical charge is known as capacitance. The energy storing devices are called *capacitors*; they are available in many shapes and sizes. Capacitors consist usually of two plates of thin metal sandwiched between a *dielectric*; this is an insulator made of ceramic, film, glass, air, or other materials and is symbolized as -| |-. The dielectric boosts a capacitor's charging capacity. Sometimes called condensers they are commonly used in the aviation, automotive, and marine industries. To a degree capacitors are like batteries, they both store energy but the battery releases the energy slowly while capacitors discharge quickly. Note that when an electrical circuit containing capacitors is turned off, the capacitors retain their energy charge, though usually they leak a slight amount of energy.

The ratio of the electric charge on each conductor of the capacitor to the potential difference (voltage) between them is known as *capacitance*. The capacitance value of a capacitor is measure in *farads* (F). One farad is a very large quantity of capacitance; thus, most commonly used household type capacitors are rated at only a fraction of a farad, often in microfarads (μf—a thousandth of a farad) or smaller in picofarads (a trillionth, pF). Supercapacitors (sometimes used to store energy produced by wind turbines) can store very large electrical charges of thousands of farads.

The large capacitors used for the storage of energy produced by wind turbines are made so that capacitance is increased by positioning a capacitor's plates closer together or the plates are increased in size so that there is more surface area. Also, a capacitor's capacitance can be increased by selecting a dielectric designed to increase the capacitance. It is important to point out that in electrical circuits, capacitors are frequently used to block direct current (dc) while permitting alternating (ac) current to flow.

REACTANCE

Reactance, symbolized as X and like resistance measured in ohms, is the opposition to alternating current flow by inductance or capacitance. Greater reactance gives a smaller current for the same applied voltage. Reactance is similar to resistance in some respects but differs in that reactance does not lead to dissipation of electrical energy as heat but stores energy for a quarter cycle and then returned to the circuit. It is important to remember that resistance continuously loses energy.

There are other differences between reactance and resistance. For example, reactance changes the phase so that the current through the element is shifted by a quarter of a cycle relative to the voltage applied across the element. Moreover, in purely reactive elements power is not dissipated but instead is stored. Also, reactance can be negative so that they essentially cancel each other out. Lastly, capacitors and inductors have a frequency-dependent reactance, but resistors maintain the same resistance for all frequencies.

Capacitive reactance, symbolized as Xc, is the opposition to the change of voltage across an element and is inversely proportional to frequency and the capacitance C. Capacitive reactance is found using Equation (8.42).

$$Xc = 1/2\pi fC \tag{8.42}$$

where
  Xc = capacitive reactance
  f = frequency
  C = capacitance

Inductive reactance, symbolized as X_L, is the opposition to the change of current through an element and is proportional to the frequency f and the inductance L as shown in Equation (8.43).

$$X_L = 2\pi fL \tag{8.43}$$

where
  X_L = inductive reactance
  f = frequency
  L = inductance

POWER FACTOR

We know that power consumed in a d-c circuit is given by the multiplication of voltage (V) and current (I), meaning Power = VI = Watts. But for an a-c circuit this formula does not work; it is incorrect. In an a-c circuit a third parameter, called *power factor* (PF), is included and the correct formula becomes Power = VI × (pf)—where pf is the power factor of the load. Thus, and the point is that pf is an important parameter that has significance *only* for a-c circuits and expresses energy efficiency. Moreover, the power factor is an important parameter for the calculation of active and reactive power in electrical circuits. Usually expressed as a percentage, the lower the percentage, the less efficient the power usage is.

So, exactly what is power factor? Power factor is the ratio of true power (TP; aka working power), measured in kilowatts (kW), to apparent power (AP), measured in kilowatt amperes (kVA). Apparent power is known as demand; that is, it is the measure of the amount of power used to run machinery and equipment during a certain period and is found by multiplying (kVA = V × A). The results are expressed as kVA units. To determine PF, we use PF = TP/AP. In mathematical terms, pf is defined as the cosine of the angle between the voltage phasor and current phasor in an a-c circuit.

REACTIVE POWER

One of the major technical challenges for wind farms or plants is fluctuating power output. Power fluctuations are caused by variations in wind speed inducing the generator to produce different power levels at varying times. Not in the too distant past it was common practice for the generated power to be fed directly to the grid without any kind energy storage. This was common practice regardless of power demands during the peak and off-peak hours. In a large wind farm, power fluctuations can lead to voltage variation of the interconnection point of the grid. Experience has demonstrated that using energy storage devices can smooth power fluctuations; subsequently, it will improve the distribution of power from the wind farm and also maintain stability control of the farm whenever subjected to voltage flicker. Voltage flicker in a wind turbine occurs due to wind speed turbulence and can be apparent during normal operations and during connect and disconnection operations, while the repeated connections and disconnection of capacitor banks also be the cause of fast flicker.

Another technical problem is the reactive power regulation in a large wind farm. Because of the power changes at different wind speeds, the amount of reactive power produced or absorbed by the wind farm and the grid changes. At the same time, the number and size of wind farms contributing to energy production are growing, meaning that the reactive produced can't be overlooked. As the scale of wind power continues to expand, the influence of wind power on energy quality borne by the power grid is not economic and all the more so unaffordable. The bottom line is that it is necessary to make appropriate reactive power compensation within wind farms when they are constructed.

PRACTICAL ELECTRICAL APPLICATIONS

As mentioned, wind turbine maintenance operators normally have little difficulty recognizing electrical equipment within the wind turbine and wind farm site. This is the case, of course, because there are few places within the turbine and site (that is, in the majority of sites) that an operator can look or go where electricity is not performing some important function. Simply stated, whether the "important function" be powering lighting, heating, air conditioning, pump motors, control systems, communications equipment, or computerized systems, it would be difficult for the modern wind turbine operator to imagine turbine and/or site operations without the use of electrical power. And keep in mind that the site is producing electricity. When you get right down to it the production of electricity to feed the site and grid is what a wind turbine farm is all about.

To this point, we have concentrated (in brief fashion, of course) on the fundamentals of electricity and electric circuits—very basic fundamentals. This was the goal. Moreover, along with satisfying the goal, also understand that having a basic knowledge of electrical theory is a great accomplishment. However, knowledge of basic theory (of any type) that is not put to practical use is analogous to understanding the operation of an internal combustion engine without ever having the opportunity to work on one.

In short, for the wind turbine maintenance operator, understanding basic electrical fundamentals assists in the successful advancement to be "qualified" to operate the site, maintain operation and monitoring of the turbine, and conduct repair and preventive maintenance on wind turbine and its associated machinery—much of which is electrical equipment. To this end, pertinent information on electrical applications most important to the wind turbine and wind farm operator in his/her daily task of operating turbines and other site electrical equipment as it is intended to be operated has been presented.

ELECTRICAL POWER GENERATION

Generators can be designed to supply small amounts of power, or they can be designed to supply many thousands of kilowatts of power. In addition, generators may be designed to supply either direct current or alternating current.

D-C GENERATORS

A *d-c generator* is a rotating machine that converts mechanical energy into electrical energy. This conversion is accomplished by rotating an armature, which carries conductors, in a magnetic field, thus inducing an emf in the conductors. As stated previously, in order for an emf to be induced in the conductors, a relative motion must always exist between the conductors and the magnetic field in such a manner that conductors cut through the field. In most d-c generators, the armature is the rotating member, and the field is the stationary member. A mechanical force is applied to the shaft of the rotating member to cause the relative motion. Thus, when mechanical energy is put into the machine in the form of a mechanical force or twist on the shaft, causing the shaft to turn at a certain speed, electrical energy in the form of voltage and current is delivered to the external load circuit.

Important Point: Mechanical power must be applied to the shaft constantly so long as the generator is supplying electrical energy to the external load circuit.

To gain a basic understanding of the operation of a d-c generator, consider the following explanation.

A simple d-c generator consists of an armature coil with a single turn of wire [see Figure 8.83(A) and (B)]. (**Note**: The armature coils used in large d-c machines are

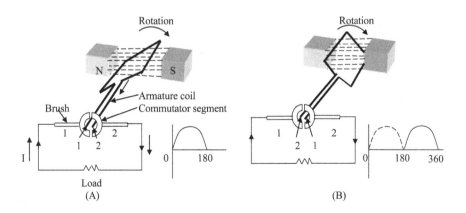

FIGURE 8.83   Basic operation of a d-c generator.

usually wound in their final shape before being put on the armature. The sides of the preformed coil are placed in the slots of the laminated armature core.) This armature coil cuts across the magnetic field to produce voltage. If a complete path is present, current will move through the circuit in the direction shown by the arrows [see Figure 8.83(A)]. In this position of the coil, commutator segment 1 is in contact with brush 1, while commutator segment 2 is in contact with brush 2. As the armature rotates a half turn in a clockwise direction, the contacts between the commutator segments and the brushes are reversed [see Figure 8.83(B)]. At this moment, segment 1 is in contact with brush 2 and segment 2 is in contact with brush 1. Because of this commutator action, that side of the armature coil that is in contact with either of the brushes is always cutting across the magnetic field in the same direction. Thus, brushes 1 and 2 have constant polarity, and a *pulsating d-c current* is delivered to the external load circuit.

Note: In d-c generators, voltage induced in individual conductors is a-c. It is converted to d-c (rectified) by the commutator that rotates in contact with carbon brushes so that the current generated is in one direction or direct current.

There are several different types of d-c generators. They take their names from the type of field excitation used (i.e., they are classified according to the manner in which the field windings are connected to the armature circuit). For example, when the generator's field is excited (or supplied) from a separate d-c source (such as a battery) other than its own armature, it is called a *separately excited d-c generator* (see Figure 8.84).

A *shunt generator* (self-excited) has its field windings connected in series with a rheostat, across the armature in shunt with the load, as shown in Figure 8.85. The shunt generator is widely used in industry.

A *series generator* (self-excited) has its field windings connected in series with the armature and load, as shown in Figure 8.86. Series generators are seldom used.

Compound generators (self-excited) contain both series and shunt field windings, as shown in Figure 8.87. Compound generators are widely used in industry.

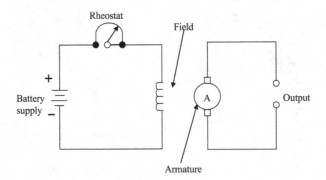

FIGURE 8.84   Separately excited d-c generator.

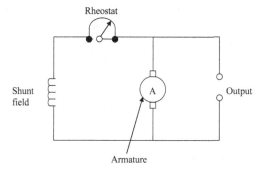

FIGURE 8.85   d-c shunt generator.

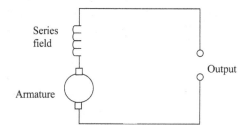

FIGURE 8.86   d-c series generator.

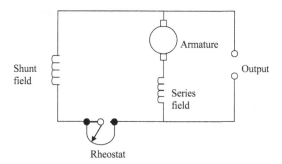

FIGURE 8.87   d-c compound generator.

Note: As central generating stations increased in size along with number and power distribution distances, d-c generating systems, because of the high-power losses in long d-c transmission lines, were replaced by a-c generating systems to reduce power transmission costs.

A-C GENERATORS

Most electric power utilized today is generated by *alternating current generators* (also called *alternators*). They are made in many different sizes, depending on their intended use. Regardless of size, however, all generators operate on the same basic principle—a magnetic field cutting through conductors, or conductors passing through a magnetic field. They are (1) a group of conductors in which the output voltage is generated, and (2) a second group of conductors through which direct current is passed to obtain an electromagnetic field of fixed polarity. The conductors in which the electromagnetic field originates are always referred to as the field windings.

In addition to the armature and field, there must also be motion between the two. To provide this, a-c generators are built in two major assemblies, the *stator* and the *rotor*. The rotor rotates inside the stator.

The revolving-field a-c generator (see Figure 8.88) is the most widely used type. In this type of generator, direct current from a separate source is passed through windings on the rotor by means of sliprings and brushes. [**Note:** Sliprings and brushes are adequate for the d-c field supply because the power level in the field is much smaller than in the armature circuit.] This maintains a rotating electromagnetic field of fixed polarity. The rotating magnetic field, following the rotor, extends outward and cuts through the armature windings embedded in the surrounding stator. As the rotor turns, a-c voltages are induced in the windings since magnetic fields of first one polarity and then the other cut through them. Since the output power is taken from stationary windings, the output may be connected through fixed output terminals T1 and T2 in Figure 8.88. This is advantageous, in that there are no sliding contacts and the whole output circuit is continuously insulated.

Important Point: In a-c generators, frequency and electromagnetic wave cycles per second depend on how fast the rotor turns and the number of electromagnetic field poles. Voltage generated depends on the rotor speed, the number of coils in the armature, and strength of the magnetic field.

Motors

A considerable amount of the electrical power fed to a typical wind turbine is consumed by electric motors. One thing is certain: there are a numerous variety of tasks that electric motors perform in wind turbine operations.

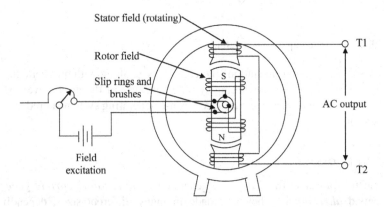

FIGURE 8.88 Essential parts of a rotating field a-c generator.

An *electric motor* is a machine used to change electrical energy to mechanical energy to do the work. [**Note**: Recall that a generator does just the opposite; that is, a generator changes mechanical energy to electrical energy.]

Previously, we pointed out that when a current passes through a wire, a magnetic field is produced around the wire. If this magnetic field passes through a stationary magnetic field, the fields either repel or attract, depending on their relative polarity. If both are positive or negative, they repel. If they are opposite polarity, they attract.

Applying this basic information to motor design, an electromagnetic coil, the armature, rotates on a shaft. The armature and shaft assembly are called the rotor. The rotor is assembled between the poles of a permanent magnet and each end of the rotor coil (armature) is connected to a commutator also mounted on the shaft. A commutator is composed of copper segments insulated from the shaft and from each other by an insulting material. As like poles of the electromagnet in the rotating armature pass the stationary permanent magnet poles, they are repelled, continuing the motion. As the opposite poles near each other, they attract, continuing the motion.

D-C Motors

The construction of a d-c motor is essentially the same as that of a d-c generator. However, it is important to remember that the d-c generator converts mechanical energy into the electrical energy back into mechanical energy. A d-c generator may be made to function as a motor by applying a suitable source of d-c voltage across the normal output electrical terminals.

There are various types of d-c motors, depending on the way the field coils are connected. Each has characteristics that are advantageous under given load conditions.

Shunt motors (see Figure 8.89) have the field coils connected in parallel with the armature circuit. This type of motor, with constant potential applied, develops variable torque at an essentially constant speed, even under changing load conditions. Such loads are found in machine-shop equipment such as lathes, shapes, drills, milling machines, and so forth.

Series motors (see Figure 8.90) have the field coils connected in series with the armature circuit. This type of motor, with constant potential applied, develops variable torque but its speed varies widely under changing load conditions. That is, the

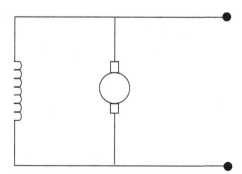

FIGURE 8.89   d-c shunt motor.

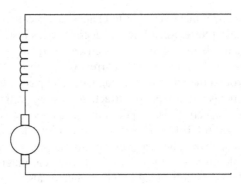

FIGURE 8.90   d-c series motor.

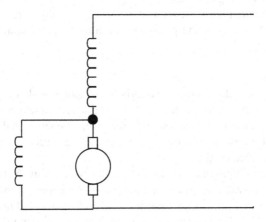

FIGURE 8.91   d-c compound motor.

speed is low under heavy loads, but becomes excessively high under light loads. Series motors are commonly used to drive electric hoists, winches, cranes, and certain types of vehicles (e.g., electric trucks). In addition, series motors are used extensively to start internal combustion engines.

Compound motors (see Figure 8.91) have one set of field coils in parallel with the armature circuit, and another set of field coils in series with the armature circuit. This type of motor is a compromise between shunt and series motors. It develops an increased starting torque over that of the shunt motor and has less variation in speed than the series motor.

The speed of a d-c motor is variable. It is increased or decreased by a rheostat connected in series with the field or in parallel with the rotor. Interchanging either the rotor or field winding connections reverses direction.

A-C Motors

A-c voltage can be easily transformed from low voltages to high voltages or vice versa and can be moved over a much greater distance without too much loss in efficiency. Most of the power generating systems today, therefore, produce alternating current. Thus, it logically follows that a great majority of the electrical motors utilized today are designed to operate on alternating current. However, there are other advantages in the use of a-c motors besides the wide availability of a-c power. In general, a-c motors are less expensive than d-c motors. Most types of a-c motors do not employ brushes and commutators. This eliminates many problems of maintenance and wear and eliminates dangerous sparking.

A-c motors are manufactured in many different sizes, shapes, and ratings, for use on an even greater number of jobs. They are designed for use with either polyphase or single-phase power systems.

This section cannot possibly cover all aspects of the subject of a-c motors. Consequently, it will deal mainly with the operating principles of the two most common types—the induction and synchronous motor.

Induction Motors

The induction motor is the most commonly used type of a-c motor because of its simple, rugged construction and good operating characteristics. It consists of two parts: the *stator* (stationary part) and the *rotor* (rotating part).

The most important type of polyphase induction motor is the three-phase motor.

Important Note: A three-phase (3-θ) system is a combination of three single-phase (1-θ) systems. In a 3-θ balanced system, the power comes from an a-c generator that produces three separate but equal voltages, each of which is out of phase with the other voltages by 120°. Although 1-θ circuits are widely used in electrical systems, most generation and distribution of a-c current is 3-θ.

The driving torque of both d-c and a-c motors is derived from the reaction of current-carrying conductors in a magnetic field. In the d-c motor, the magnetic field is stationary and the armature, with its current-carrying conductors, rotates. The current is supplied to the armature through a commutator and brushes.

In *induction motors*, the rotor currents are supplied by electromagnet induction. The stator windings, connected to the a-c supply, contain two or more out-of-time-phase currents, which produce corresponding mmf's. These mmf's establish a rotating magnetic field across the air gap. This magnetic field rotates continuously at a constant speed regardless of the load on the motor. The stator winding corresponds to the armature winding of a d-c motor or to the primary winding of a transformer. The rotor is not connected electrically to the power supply.

The induction motor derives its name from the fact that mutual induction (or transformer action) takes place between the stator and the rotor under operating conditions. The magnetic revolving field produced by the stator cuts across the rotor conductors, inducing a voltage in the conductors. This induced voltage causes rotor current to flow. Hence, motor torque is developed by the interaction of the rotor current and the magnetic revolving field.

Synchronous Motors

Like induction motors, *synchronous motors* have stator windings that produce a rotating magnetic field. However, unlike the induction motor, the synchronous motor requires a separate source of d-c from the field. It also requires special starting components. These include a salient-pole field with starting grid winding. The rotor of the conventional type of synchronous motor is essentially the same as that of the salient-pole a-c generator. The stator windings of induction and synchronous motors are essentially the same.

In operation, the synchronous motor rotor locks into step with the rotating magnetic field and rotates at the same speed. If the rotor is pulled out of step with the rotating stator field, no torque is developed and the motor stops. Since a synchronous motor develops torque only when running at synchronous speed, it is not self-starting and hence needs some device to bring the rotor to synchronous speed. For example, a synchronous motor may be started rotating with a d-c motor on a common shaft. After the motor is brought to synchronous speed, a-c current is applied to the stator windings. The d-c starting motor now acts as a d-c generator, which supplies d-c field excitation for the rotor. The load then can be coupled to the motor.

Single-Phase Motors

Single-phase (1-θ) motors are so called because their field windings are connected directly to a single-phase source. These motors are used extensively in fractional horsepower sizes in commercial and domestic applications. The advantages of using single-phase motors in small are that they are less expensive to manufacture than other types, and they eliminate the need for 3-phase a-c lines. Single-phase motors are used in fans, refrigerators, portable drills, grinders, and so forth.

A single-phase induction motor with only one stator winding and a cage rotor is like a 3-phase induction motor with a cage rotor except that the single-phase motor has no magnetic revolving field at the start and hence no starting torque. However, if the rotor is brought up to speed by external means, the induced currents in the rotor will cooperate with the stator currents to produce a revolving field, which causes the rotor to continue to run in the direction, which it was started.

Several methods are used to provide the single-phase induction motor with starting torque. These methods identify the motor as split-phase, capacitor, shaded-pole, and repulsion-start induction motor.

Another class of single-phase motors is the a-c series (universal) type. Only the more commonly used types of single-phase motors are described. These include the (1) split-phase motor, (2) capacitor motor, (3) shaded-pole motor, (4) repulsion-start motor, and (5) a-c series motor.

1.   Split-Phase Motors

The *split-phase motor* (see Figure 8.92) has a stator composed of slotted lamination that contains a starting winding and a running winding.

Note: If two stator windings of unequal impedance are spaced 90 electrical degrees apart but connected in parallel to a single-phase source, the field produced will appear to rotate. This is the principle of phase splitting.

The starting winding has fewer turns and smaller wire than the running winding hence has higher resistance and less reactance. The main winding occupies the lower half of the slots, and the starting winding occupies the upper half. When the same voltage is applied to both windings, the current in the main winding lags behind the

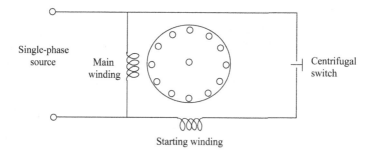

FIGURE 8.92   Split-phase motor.

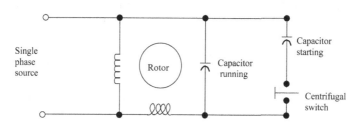

FIGURE 8.93   Capacitor motor.

current in the starting winding. The angle θ between the main and starting windings is enough phase difference to provide a weak rotating magnetic field to produce a starting torque. When the motor reaches a predetermined speed, usually 75% of synchronous speed, a centrifugal switch mounted on the motor shaft opens, thereby disconnecting the starting winding.

Because it has a low starting torque, fractional horsepower split-phase motors are used in a variety of equipment such as washers, oil burners, ventilating fans, and woodworking machines. Interchanging the starting winding leads can reverse the direction of rotation of the split-phase motor.

2.   Capacitor Motors

The *capacitor motor* is a modified form of split-phase motor, having a capacitor in series with the starting winding. The capacitor motor operates with an auxiliary winding and series capacitor permanently connected to the line (see Figure 8.93). The capacitance in series may be of one value for starting and another value for running. As the motor approaches synchronous speed, the centrifugal switch disconnects one section of the capacitor.

If the starting winding is cut out after the motor has increased in speed, the motor is called a *capacitor-start motor*. If the starting winding and capacitor are designed to be left in the circuit continuously, the motor is called capacitor-run motor. Capacitor motors are used to drive grinders, drill presses, refrigerator compressors, and other loads that require relatively high starting torque. Interchanging the starting winding leads may reverse the direction of rotation of the capacitor motor.

3. Shaded-Pole Motor

A *shaded-pole* motor employs a salient-pole stator and a cage rotor. The projecting poles on the stator resemble those of d-c machines except that the entire magnetic circuit is laminated, and a portion of each pole is split to accommodate a short-circuited coil called a *shading coil* (see Figure 8.94). The coil is usually a single band or strap of copper. The effect of the coil is to produce a small sweeping motion of the field flux from one side of the pole piece to the other as the field pulsates. This slight shift in the magnetic field produces a small starting torque. Thus, shaded-pole motors are self-starting. This motor is generally manufactured in very small sizes, up to 1/20 horsepower, for driving small fans, small appliances, and clocks.

In operation, during that part of the cycle when the main pole flux is increasing, the shading coil is cut by the flux, and the resulting induced emf and current in the shading coil tend to prevent the flux from rising readily through it. Thus, the greater portion of the flux rises in that portion of the pole that is not near the shading coil. When the flux reaches its maximum value, the rate of change of flux is zero, and the voltage and current in the shading coil are zero. At this time, the flux is distributed more uniformly over the entire pole face. Then as the main flux decreases toward zero, the induced voltage and current in the shading coil reverse their polarity, and the resulting mmf tends to prevent the flux from collapsing through the iron in the region of the shading coil. The result is that the main flux first rises in the unshaded portion of the pole and later in the shaded portion. This action is equivalent to a sweeping movement of the field across the pole face in the direction of the shaded pole. This moving field cuts the rotor conductors and the force exerted on them causes the rotor to turn in the direction of the sweeping field. The shaded-pole method of starting is used in very small motors, up to about 1/25 hp, for driving small fans, small appliances, and clocks.

4. Repulsion-Start Motor

Like a d-c motor, the *repulsion-start motor* has a form-wound rotor with commutator and brushes. The stator is laminated and contains a distributed single-phase winding. In its simplest form, the stator resembles that of the single-phase motor.

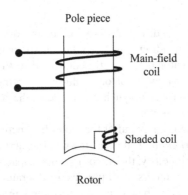

FIGURE 8.94   Shaded pole.

In addition, the motor has a centrifugal device, which removes the brushes from the commutator and places a short-circuiting ring around the commutator. This action occurs at about 75% of synchronous speed. Thereafter, the motor operates with the characteristics of the single-phase induction motor. This type of motor is made in sizes ranging from 1/2 to 15 hp and is used in applications requiring a high starting torque.

5. Series Motor

The *a-c series motor* will operate on either a-c or d-c circuits. When an ordinary d-c series motor is connected to an a-c supply, the current drawn by the motor is low due to the high series-field impedance. The result is low running torque. To reduce the field reactance to a minimum, a-c series motors are built with as few turns as possible. Armature reaction is overcome by using *compensating windings* (see Figure 8.95) in the pole pieces.

As with d-c series motors, in an a-c series motor, the speed increases to a high value with a decrease in load. The torque is high for high armature currents so that the motor has a good starting torque. A-c series motors operate more efficiently at low frequencies.

Fractional horsepower a-c series motors are called *universal motors.* They do not have compensating windings. They are used extensively to operate fans and portable tools, such as drills, grinders, and saws.

TRANSFORMERS

A *Transformer* is an electric control device (with no moving parts) that raises or lowers voltage or current in an electric distribution system. The basic transformer consists of two coils electrically insulated from each other and wound upon a common core (see Figure 8.96). Magnetic coupling is used to transfer electric energy from one coil to another. The coil, which receives energy from an a-c source, is called the *primary.* The coil that delivers energy to an a-c load is called the *secondary.* The core of transformers used at low frequencies is generally made of magnetic material, usually laminated sheet steel. Cores of transformers used at higher frequencies are made of powdered iron and ceramics, or nonmagnetic materials. Some coils are simply wound on nonmagnetic hollow forms such as cardboard or plastic so that the core material is actually air.

In operation, an alternating current will flow when an a-c voltage is applied to the primary coil of a transformer. This current produces a field of force that changes

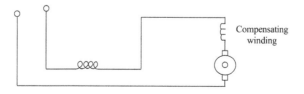

FIGURE 8.95   a-c series motor.

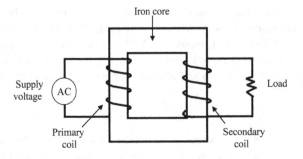

FIGURE 8.96   Basic transformer.

as the current changes. The changing magnetic field is carried by the magnetic core to the secondary coil, where it cuts across the turns of that coil. In this way, an a-c voltage in one coil is transferred to another coil, even though there is no electrical connection between them. The primary voltage and the number of turns on the primary determine the number of lines of force available in the primary—each turn producing a given number of lines. If there are many turns on the secondary, each line of force will cut many turns of wire and induce a high voltage. If the secondary contains only a few turns, there will be few cuttings and low induced voltage. The secondary voltage, then, depends on the number of secondary turns as compared with the number of primary turns. If the secondary has twice as many turns as the primary, the secondary voltage will be twice as large as the primary voltage. If the secondary has half as many turns as the primary, the secondary voltage will be one-half as large as the primary voltage.

Important Point: The voltage on the coils of a transformer is directly proportional to the number of turns on the coils.

A voltage ratio of 1:4 (read as "1 to 4") means that for each volt on the primary, there are 4 V on the secondary. This is called a *step-up* transformer. A step-up transformer receives a low voltage on the primary and delivers a high voltage from the secondary. A voltage ratio of 4:1 (read as "4 to 1") means that for 4 V on the primary, there is only 1 V on the secondary. This is called a *step-down* transformer. A step-down transformer receives a high voltage on the primary and delivers a low voltage from the secondary.

POWER DISTRIBUTION SYSTEM PROTECTION

Interruptions are very rare in a power distribution system that has been properly designed. Still, protective devices are necessary because of the load diversity. Most installations are quite complex. In addition, externally caused variations might overload them or endanger personnel.

Figure 8.97 shows the general relationship between protective devices and different components of a complete system. Each part of the circuit has its own protective device or devices that protect not only the load but also the wiring and control devices themselves. These disconnect and protective devices are described in the following sections.

1. FUSES

The passage of an electric current produces heat. The larger the current, the more the heat is produced. In order to prevent large currents from accidentally flowing through expensive apparatus and burning it up, a *fuse* is placed directly into the circuit, as in Figure 8.97, so as to form a part of the circuit through which all the current must flow.

Key Point: A fuse is a thin strip of easily melted material. It protects a circuit from large currents by melting quickly, thereby breaking the circuit.

The fuse will permit currents smaller than the fuse value to flow but will melt and therefore break the circuit if a larger, dangerous current ever appears. For instance, a dangerously large current will flow when a "short circuit" occurs. A short circuit is usually caused by an accidental connection between two points in a circuit, which offer very little resistance to the flow of electrons. If the resistance is small, there will

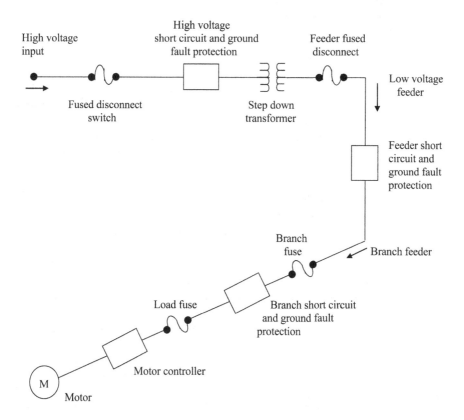

FIGURE 8.97   Motor power distribution system.

be nothing to stop the flow of the current, and the current will increase enormously. The resulting heat generated might cause a fire. However, if the circuit is protected by a fuse, the heat caused by the short-circuit current will melt the fuse wire, thus breaking the circuit and reducing the current to zero.

The number of amps of current that can flow through them before they melt and break the circuit rates fuses. Thus, we have 10-, 15-, 20-, 30-A, etc. fuses. We must be careful that any fuse inserted in a circuit be rated low enough to melt, or "blow," before the apparatus is damaged. For example, in a plant building wired to carry a current of 10 A it is best to use a fuse not larger than 10 A so that a current larger than 10 A could never flow.

Some equipment, such as an electric motor, requires more current during starting than for normal running. Thus, fast-time or medium-time fuse rating that will give running protection might blow during the initial period when a high starting current is required. *Delayed action* fuses are used to handle these situations.

2. CIRCUIT BREAKERS

Circuit breakers are protective devices that open automatically at a preset ampere rating to interrupt an overload or short circuit. Unlike fuses, they do not require replacement when they are activated. They are simply reset to restore power after the overload has been cleared.

Key Point: A circuit breaker is designed to break the circuit and stop the current flow when the current exceeds a predetermined value.

Circuit breakers are made in both plug-in and bolt-on designs. Plug-in breakers are used in load centers. Bolt-ons are used in panelboards and exclusively for high interrupting current applications.

Circuit breakers are rated according to current and voltage, as well as short circuit interrupting current. A single handle opens or closes contacts between two or more conductors. Breakers are single-pole but can be ganged single-pole units—forming double-or-triple-pole devices opened with a single handle.

Several types of circuit breakers are commonly used. They may be thermal, magnetic, or a combination of the two. Thermal breakers are tripped when the temperature rises because of heat created by the overcurrent condition. Bimetallic strips provide the time delay for overload protection. Magnetic breakers operate on the principle that a sudden current rise creates enough magnetic field to turn an armature, tripping the breaker and opening the circuit. Magnetic breakers provide the instantaneous action needed for short-circuit protection. Thermal-magnetic breakers combine features of both types of breakers.

Magnetic breakers are also used in circumstances where the ambient temperature might adversely affect the action of a thermal breaker.

An important feature of the circuit breaker is its arc chutes, which enable the breaker to extinguish very hot arcs harmlessly. Some circuit breakers must be reset by hand, while others reset themselves automatically. When the circuit breaker is reset, if the overload condition still exists, the circuit breaker will trip again to prevent damage to the circuit.

3. CONTROL DEVICES

Control devices are those electrical accessories (switches and relays), which govern the power delivered to any electrical load.

In its simplest form, the control applies voltage to or removes it from a single load. In more complex control systems, the initial switch may set into action other control devices (relays) that govern motor speeds, servomechanisms, temperatures, and numerous other equipment. In fact, all electrical systems and equipment are controlled in some manner by one or more controls. A controller is a device or group of devices, which serves to govern, in some predetermined manner, the device to which it is connected.

In large electrical systems, it is necessary to have a variety of controls for the operation of the equipment. These controls range from simple pushbuttons to heavy-duty contractors that are designed to control the operation of large motors. The push-button is manually operated while a contactor is electrically operated.

ENERGY STORAGE

Storing electrical energy for future use, standby-emergency supply, and in some cases for stabilizing electrical distribution systems is an ongoing technical area of intense study, research, investigation, experimentation, and testing—and a whole bunch of wishful thinking to achieve a "Eureka moment," which is possible, but not yet found.

Anyway, while we wait for the inevitable discovery of the way to store electrical energy for instantaneous use by ignoring a few important laws of physics in any and all situations for the moment, the here, and the now, we can store energy produced by wind farms electromagnetically (e.g., induction), electrochemically (e.g., batteries, supercapacitors, and ultracapacitors), kinetically (energy stored in an object because its motion), or as potential energy (energy of position) (Ribeiro et al., 2001). Note that the energy storage techniques or technology we use today usually includes a power conversion unit of some type to convert energy from one form to another. These energy storage converters are installed to control the output power to the energy storage device and are typically fashioned together using three-phase bidirectional converters (Ribeiro et al., 2001). The power converter acts as an inverter to transmit from DC to AC power, and it acts as a rectifier to transmit from AC to DC power. In the majority of cases, at present, battery storage is used.

When batteries are used to store wind-generated excess energy, the standard lead-acid type battery are generally the storage device of choice it is well suited for trickle charging and possesses a high electrical output charging efficiency.

Wind turbines can also use excess power to compress air that is commonly stored in large above-ground tanks or in underground caverns. When needed the compressed air can be used through direct expansion into a compressed air motor. It is also possible to inject compressed air into an internal combustion turbine, where it is burnt with fuel to provide mechanical energy which is the motive force that powers a generator.

Another wind turbine energy storage technique sometimes used is the utilization of hydrogen fuel cells. The cells use power generated by wind power to electrolyze water and storing the resulting hydrogen and then converting it back to electricity via a fuel cell system when needed.

One storage technique that is on the drawing board, so to speak, but is yet to be employed using wind power is to pump water up to an elevated storage reservoir and then used to power, much like a hydroelectric dam, a water turbine which in turn functions as the motive force for an electric generator.

THE BOTTOM LINE

While in the past wind power was exclusively used to pump water, grind grains, and power sawmills, today the primary function of wind power is to generate electricity. Consider that in the United States electricity generator from wind power increased from about 6 billion kilowatt-hours (kWh) in 2,000 to more than 330 billion kWh in 2020. Moreover, in 2020, wind turbines were the source of about 8.4% of the total US utility-scale electricity generation. The utility scale includes facilities with at least one megawatt (1,000 watts) of electricity generation capacity (EIA, 2021).

So, the real bottom line is that future use of wind-powered electricity printed boldly in electric lamps lit throughout the globe—so long as the wind continues to blow.

REFERENCES

EIA. 2021. *Electricity Generation from the Wind*. Washington, DC: US Energy Information Administration.

Ribeiro, P.F., Johnson, B.K., and Crow, M.L., et al., 2001. Energy storage systems for advanced power applications. *Proceedings of IEEE* 89(12):1744–1756.

9 Where Eagles Don't Dare

The wind goeth toward the south, and turneth about unto the north; it whirleth about continually, and the wind returneth again.

—Ecclesiastes 1:6

GOOD, BAD, AND WORSE

Good: As long as Earth exists, the wind will always exist. The energy in the winds that blow across the United States each year could produce more than 16 billion GJ of electricity—more than 1.5 times the electricity consumed in the United States in 2000.

Bad: Turbines are expensive. Wind doesn't blow all the time, so they have to be part of a larger plan. Turbines make noise. Turbine blades kill birds.

Ugly: Some look upon giant wind turbine blades cutting through the air as grotesque scars on the landscape, visible polluters.

The bottom line: Do not expect Don Quixote, mounted in armor on his old nag, Rocinante, with or without Sancho Panza, to charge those windmills. Instead, expect—you can count on it, bet on it, and rely on it—that the charge to build those windmills will be done by the rest of us, to satisfy our growing, inexorable need for renewable energy. What other choice do we have?

ENVIRONMENTAL IMPACT OF WIND TURBINES

The potential impacts of wind energy during site development/evaluation, site construction, and site operations and maintenance are discussed in the following sections.

WIND ENERGY SITE EVALUATION IMPACTS

Site evaluation phase activities, such as monitoring and testing, are temporary and are conducted at a smaller scale than those at the construction and operation phases. Potential impacts of these activities are presented below by type of affected resource. The impacts described are for typical site evaluation and exploration activities, such as ground clearing (removal of vegetative cover), vehicular and pedestrian traffic, borings for geotechnical surveys and guy-wire installation, and position of equipment, such as meteorological towers. If excavation of road construction is necessary during this phase, potential impacts would be similar in character to those for the construction phase but generally of smaller magnitude.

DOI: 10.1201/9781003288947-11

Air Quality (Including Global Climate Change and Carbon Footprint)

Impacts on air quality during monitoring and testing activities would be limited to temporary and local generation of vehicle emissions and fugitive dust. These impacts are unlikely to cause an exceedance of air quality standards or impact climate change.

Cultural Resources

Surface disturbance is minimal during the site evaluation phase, and cultural resources buried below the surface are unlikely to be affected. The cultural material present on the surface could be disturbed by vehicular traffic, ground clearing, and pedestrian activity (including the collection of artifacts). Monitoring and testing activities could affect areas of interest to Native Americans depending on their physical placement and/or level of visual intrusion. Surveys conducted during this phase to evaluate the presence and/or significance of cultural resources in the area would assist developers in designing the project to avoid or minimize impacts to these resources.

Ecological Resources

Impacts on vegetation, wildlife habitat, and aquatic habitat would be minimal during site monitoring and testing because of the limited nature of activities. The introduction and spread of invasive vegetation could occur as a result of vehicular traffic. Surveys conducted during this phase to evaluate the presence and/or significance of ecological resources in the area would assist developers in designing the project to avoid or minimize impacts to these resources.

Water Resources (Surface Water and Groundwater)

There likely would be minimal impact on water resources, local water quality, water flows, and surface water/groundwaters interactions. Very little water would likely be used during the site evaluation phase. Any water needed could be trucked in from off-site.

Land Use

Monitoring and testing activities would likely result in temporary and localized impacts to land use. These activities could create a temporary disturbance to wildlife and cattle in the immediate vicinity of the monitoring/testing site while workers are present. However, monitoring equipment is unlikely to change land use patterns over a longer period of time. Although a buffer area may be established around equipment to protect the public, wildlife, and equipment, access to the area for continued recreational use would not be affected. However, there may be visual impacts from the presence of equipment and access roads, potentially impacting the recreational experience. Monitoring and testing activities are unlikely to affect mining activities, military operations, or aviation.

Soils and Geologic Resources (Including Seismicity/Geo Hazards)

Surface disturbance and use of geologic materials are minimal during the site evaluation phase, and soils and geologic resources are unlikely to be affected. These activities would also be unlikely to activate geological hazards or increase soil erosion.

Borings for soil testing and geotechnical surveys provide useful site-specific data on these resources. Surface effects from pedestrian and vehicular traffic could occur in areas that contain special (e.g., cryptobiotic) soils.

Paleontological Resources

Surface disturbance is minimal during the site evaluation phase, and paleontological resources buried below the surface are unlikely to be affected. Fossil material present on the surface could be disturbed by vehicular traffic, ground clearing, and pedestrian activity (including the collection of fossils). Surveys conducted during this phase to evaluate the presence and/or significance of paleontological resources in the area would assist in designing the project to avoid or minimize impacts to these resources.

Transportation

No impacts on transportation are anticipated during the site evaluation phase. Transportation activities would be temporary and intermittent, and limited to low volumes of heavy- and medium-duty pickup trucks and personal vehicles.

Visual Resources

Monitoring and testing activities would have temporary and minor visual effects caused by the presence of equipment.

Socioeconomics

Site evaluation and exploration activities are temporary and limited and would not result in socioeconomic impacts on employment, local services, or property values.

Environmental Justice

Site evaluation and exploration activities are limited and would not result in significant high and adverse impacts in any resource area; therefore, environmental justice impacts are not expected during this phase.

Hazardous Materials and Waste Management

Impacts from the use, storage, and disposal of hazardous materials and waste would be minimal to nonexistent if appropriate management practices are followed.

Acoustics (Noise)

Activities associated with site monitoring and testing would generate low levels of temporary and intermittent noise. However, it is important to point out that wind turbine noise due to construction, operation, and maintenance activities is a major environmental issue and complaint. Because noise associated with wind turbines is a major environmental impact issue, an in-depth discussion about noise is provided below. Keep in mind that the noise generated and discussed refers to that generated during all phases of wind turbine operation.

Noise is commonly defined as any unwanted sound. Based on the author's observation of the operation of wind farms in Washington, Oregon, California, Indiana,

West Virginia, North Dakota, South Dakota, New Mexico, and Wyoming, turbine-generated noise can be characterized as ranging from the swooshing sound of rotating rotor blades to a deep, bass-like hum produced by a single operating wind turbine. Using a calibrated sound pressure level (SPL) decibel (dB) measuring device, I determined that the wind turbine-generated noise monitored and measured varied depending on the size of the turbines, their location, and distance from the turbine and/or wind farm.

Noise literally surrounds us every day and is with us just about everywhere we go. However, the noise we are concerned with here is that produced by wind turbines. Excessive amounts of noise in the wind farm environment (and outside it) cause many problems for people, including increased stress levels, interference with communication, disrupted concentration, and most importantly, varying degrees of hearing loss. Exposure to high noise levels also adversely affects the quality of life and increases accident rates.

One of the major problems with attempting to protect the inhabitant's hearing acuity is the tendency of many people to ignore the dangers of noise. Because hearing loss, like cancer, is insidious (you do not know you have cancer until you have it or that you have lost hearing until you have lost it), it's easy to ignore. It sort of sneaks up on you slowly and is not apparent (in many cases) until after the damage is done. Alarmingly, hearing loss from noise exposure has been well documented since the 18th century, yet since the advent of the industrial revolution, the number of exposed people has greatly increased.

WIND ENERGY CONSTRUCTION IMPACTS

Typical activities during the wind energy facility construction phase include ground clearing (removal of vegetative cover), grading, excavation, blasting, trenching, vehicular and pedestrian traffic, and drilling. Activities conducted in locations other than the facility site include excavation/blasting for construction materials such as sands and gravels and access road construction.

Air Quality (Including Global Climate Change and Carbon Footprint)

Emissions resulting from construction activities include vehicle emissions, diesel emissions from large construction equipment and generators; volatile organic compound (VOC) releases from storage and transfer of vehicle/equipment fuels and small amounts of carbon monoxide and nitrogen oxides; and particulates from blasting activities and fugitive dust. Fugitive dust would be caused by:

- Disturbing and moving soils (clearing, grading, excavation, trenching backfilling, dumping, and truck and equipment traffic)
- Mixing concrete and associated storage piles
- Drilling and pile driving

A construction permit is needed from the state or local air agency to control or mitigate these emissions; therefore, these emissions would not likely cause an exceedance of air quality standards or have an impact on climate change.

Cultural Resources

Direct impacts to cultural resources could occur from construction activities, and indirect impacts might be caused by soil erosion and increased accessibility to possible site locations. Potential impacts include:

- Complete destruction of the resource if present in areas undergoing surface disturbance or excavation
- Degradation or destruction of near-surface cultural resources on- and offsite resulting from topographic or hydrological pattern changes, or from soil movement (removal, erosion, sedimentation) (Note: The accumulation of sediment could protect some localities by increasing the amount of protective cover.)
- Unauthorized removal of artifacts or vandalism at the site could occur as a result of an increase in human access to previously inaccessible areas if significant cultural resources are present
- Visual impacts resulting from vegetation clearing, increases in dust, and the presence of large-scale equipment, machinery, and vehicles (if the resources have an associated landscape component that contributes to their significance, such as a sacred landscape or historic trail)

Ecological Resources

Ecological resources that could be affected include vegetation, fish, and wildlife, as well as their habitats. Adverse ecological effects could occur during construction from

- Erosion and runoff
- Fugitive dust
- Noise
- Introduction and spread of invasive vegetation
- Modification, fragmentation, and reduction of habitat
- Mortality of biota (i.e., death of plants and animals)
- Exposure to contaminants
- Interference with behavioral activities

Site clearing and grading, along with the construction of access roads, towers, and support facilities, could reduce, fragment, or dramatically alter existing habitat in the disturbed portions of the project area. Ecological resources would be most affected during construction by the disturbance of habitat in areas near turbines, support facilities, and access roads. Wildlife in surrounding habitats might also be affected if the construction activity (and associated noise) disturbs normal behaviors, such as feeding and reproduction.

Water Resources (Surface Water and Groundwater)

Water Use

Water would be used for dust control when clearing vegetation and grading, and for road traffic; for making concrete for foundations of towers, substations, and other buildings; and for consumptive use by the construction crew. Water could be trucked in from off-site or obtained from local groundwater wells or nearby surface water bodies, depending on availability.

Water Quality

Water quality could be affected by:

- Activities that cause soil erosion
- Weathering of newly exposed soils that could cause leaching and oxidation, thereby releasing chemicals into the water
- Discharge of waste or sanitary water
- Pesticide applications

Flow Alteration

Surface and groundwater flow systems could be affected by withdrawals made for water use, wastewater and stormwater discharges, and the diversions of surface water flow for access road construction of stormwater control systems. Excavation activities and the extraction of geological materials could affect surface and groundwater flow. The interaction between surface water and groundwater could also be affected if the surface water and groundwater were hydrologically connected, potentially resulting in unwanted dewatering or recharging of water resources.

Land Use

Impacts to land use could occur during construction if there were conflicts with existing land use plans and community goals; conflicts with existing recreational, educational, religious, scientific, or other use areas; or conversion of the existing commercial and use for the area (e.g., mineral extraction).

During construction, impacts to most land uses would be temporary, such as removal of livestock from grazing areas during blasting or heavy equipment operations, or temporary effects to the character of a recreation area because of construction noise, dust, and visual intrusions. Long-term land use impacts would occur if existing land uses were not compatible with wind energy development, such as remote recreational experiences; however, those uses could potentially be resumed if the land is reclaimed to predevelopment conditions.

When wind farm construction spreads, local opposition to the mass towers (some over 400 ft tall) impacts land owners' opinions, land values, and state regulations. However, this local opposition is relative in the sense that opinions are based on the receipt of or nonreceipt of monetary rewards for use of the land. That is, the opposition is based on whether the wind farms in a local area constructed on the personal property accrue a monetary reward or usage fee for property owners for allowing the presence of turbines on their land. This view is typically different for residents who do not reap an economical benefit from the presence of wind turbines in their backyards.

Impacts to aviation could be possible if the project is located within 20,000 ft (6,100 m) or less of an existing public or military airport or if proposed construction involves objects greater than 200 ft (61 m) in height. The Federal Aviation Administration (FAA) must be notified if either of these two conditions occurs, and the FAA would be responsible for determining if the project would adversely affect commercial, military, or personal air navigation safety. Similarly, impacts to military operations could occur if a project were located near a military facility if that facility conducts low-altitude military testing and training activities.

Soils and Geologic Resources (Including Seismicity/Geo Hazards)

Sands, gravels, and quarry stones would be excavated for the construction of access roads; for concrete for buildings, substations, transformer pads, foundations, and other ancillary structures; and for improving ground surface for lay-down areas and crane-staging areas.

Possible geological hazards such as landslides could be activated by excavation and blasting for raw materials, increasing slopes during site grading and construction of access roads, altering natural drainage patterns, and toe-cutting bases of slopes. Altering drainage patterns could also accelerate erosion and create slope instability.

Surface disturbance, heavy equipment traffic, and changes to surface runoff patterns could cause soil erosion and impacts on special soils (e.g., cryptobiotic soils). Impacts of soil erosion could include soil nutrient loss and reduced water quality in nearby surface water bodies.

Paleontological Resources

Impacts on paleontological resources could occur directly from the construction activities or indirectly from soil erosion and increased accessibility to fossil locations. Potential impacts include:

- Complete destruction of the resource if present in areas undergoing surface disturbance or excavation.
- Degradation or destruction of near-surface fossil resources on- and off-site due to changes in topography, changes in hydrological patterns, and soil movement (removal, erosion, sedimentation). (Note: The accumulation of sediment could serve to protect some locations by increasing the amount of protective cover.)
- Unauthorized removal of fossil resources or vandalism to the site could occur as a result of increased human access to previously inaccessible areas if significant paleontological resources are present.

Transportation

Short-term increases in the use of local roadways would occur during the construction period. Heavy equipment likely would remain at the site. Shipments of materials are unlikely to affect primary or secondary road networks significantly, but this would depend on the location of the project site relative to the material source. Oversized loads could cause temporary transportation disruptions and require some modifications to roads or bridges (such as fortifying bridges to accommodate the size or weight). Shipment weight might also affect the design of access roads for grade determinations and turning clearance requirements.

Visual Resources

Although many of us consider wind turbines to be visually acceptable and, in some cases, even pleasing to look at, wind turbines disturb the visual area of other people by creating negative changes in the natural environment. The test on whether a wind turbine or wind farm is a visual pollutant is to ask the question? How many of us would seriously like one or more wind turbines a few hundred feet or meters from our homes? It is important to remember that wind turbines can be anywhere from a few meters to a hundred meters high. Having a wind turbine tower over one's home is the last thing many residents want. Having said this, the possible sources of visual impacts during construction include the following:

- Road development (e.g., new roads or expansion of existing roads) and parking areas could introduce strong visual contrasts in the landscape, depending on the route relative to surface contours, and the width, length, and surface treatment of the roads.
- Conspicuous and frequent small-vehicle traffic for worker access and frequent large-equipment (e.g., trucks, graders, excavators, and cranes) traffic for road construction, site preparation, and turbine installation could produce visible activity and dust in dry soils. Suspension and visibility of dust would be influenced by vehicle speeds and road surface materials.
- Site development could be intermittent, staged, or phased, giving the appearance that work starts and stops. Depending on the length of time required for development, the project site could appear to be "under construction" for an extended period. This could give rise to perceptions of lost benefit and productivity, like those alleged for the equipment. Timing and duration concerns may result.
- There would be a temporary presence of large cranes or other large machines to assemble towers, nacelles, and rotors. This equipment would also produce emissions while operational and could create visible exhaust plumes. Support facilities and fencing associated with the construction work world also be visible.
- Ground disturbance and vegetation removal could result in visual impacts that produce a contrast of color, form, texture, and line. Excavation for turbine foundations and ancillary structures, trenching to bury electrical distribution system, grading and surfacing roads, cleaning and leveling staging areas, and stockpiling soil and spoils (if not removed) would (1) damage or remove vegetation, (2) expose bare soil, and (3) suspend dust. Soil scars and exposed slope faces would result from excavation, leveling, and equipment movement. Invasive species could colonize disturbed and stockpile soils and compacted areas.

SIDEBAR 9.1 VISUAL CONSERVATION:
BANANA VS. NIMBY OR BOTH
Renewable Energy Paradox

Ask an environmentalist if renewable energy is a good idea … it is a no-brainer they say… hell yes, they reply. If you tell them that you will build a wind farm or a series of electrical power towers in their backyard, they scream, NO WAY, JOSE!

As they are with fire- or waterfall-gazing, a few drivers are also mesmerized whenever they drive Interstate-40 through Oklahoma and Texas and/or Interstate-15 through Tehachapi Pass, California, on clear days and are greeted by a host of hundreds of whirling, swooshing, slashing wind turbines. Many of these folks do not need a road-to-Damascus change of view to accept these massive turbines. Moreover, they are not necessarily advocates for renewable energy; instead, they are simply captivated by the human-made machines as they stand tall against a backdrop of plains, desert, hills, and mountains and blue- or cloud-filled nothingness.

As mentioned above, the preceding view of wind farms is held by a few people here and there. However, it is safe to say, without qualification, that many other people have a different view of whirling, swooshing wind turbines scattered helter-skelter within the US landscape. Even though very few of these people dispute the environmental benefits of wind energy (or solar and other renewable energy producers), many feel that the construction of wind farms (anywhere) ruins or spoils the otherwise pristine landscape.

Not only does the possible or potential construction of wind farms (or solar or other renewable energy sources) elicit the Not In My Back Yard (NIMBY) phenomenon but even more so, in some cases, for the associated electric power transmission towers and lines. Generally, NIMBY opponents acknowledge the need for wind turbines and transmission lines, while arguing that they "just don't want them nearby to them." Again, most people understand the need for them, yet, hardly anyone wants to live within sight of them—because they look "ugly" or for personal safety concerns.

The opposition generated by the NIMBY phenomenon is one thing, but the opposition generated by the Build Absolutely Nothing Anywhere Near Anything (BANANA) phenomenon in which the protest, unlike NIMBY, is done with questioning the overall necessity of the development is another thing. The opponents are often environmentalists, in which case the argument is generally we don't need new wind turbine farms and associated transmission lines and power stations. Simply, BANANA phenomenon activists argue against any more of "whatever" (emphasis added) is being planned. With regard to the construction of new wind turbines and associated equipment, their argument is that instead we need to use power more wisely, not generate more.

Whether identified as the NIMBY or BANANA phenomenon, what we are talking about is complaints about visual pollution. Visual pollution is an aesthetic issue and refers to the impacts of pollution impairing one's ability to enjoy a view or vista. Visual pollution disturbs the visual areas people view by creating what they perceive as negative changes in the natural environment; that is, they intrude on, ruin, spoil, mar the natural landscape and, by many, are best described as an eyesore. The most common negative changes or forms of visual pollution are buildings, automobiles, trash dumps, space debris, telephone and electric towers, and wires and electrical substations (see Figures 9.1–9.3).

One thing is certain; you can't hide wind turbines from view. They can't be hidden; nor can they be camouflaged like cell phone towers and satellite dishes (Saito, 2014; Komanoff, 2003; Righter, 2002; Pasqualetti, 2002). Because of their visibility (and for other reasons), siting and establishing wind energy production farms and transmission infrastructure present unique challenges.

With regard to transmission siting, the remote location of much of the utility-scale wind power capacity requires the construction of new high-voltage transmission lines to transport electricity to population centers. Because transmission lines can cross private, public (state and federal), and tribal lands, the process of planning, permitting, and building new lines is highly visible and

FIGURE 9.1 High-voltage power lines.

FIGURE 9.2 Electrical substation.

implicates many diverse interests—and it can be costly, time-consuming, and controversial.

Another major consideration for the installation of transmission lines from remote locations (and any other location) is right of way (ROW) (see Figure 9.3). The right of way for a transmission corridor includes land set aside for the transmission line and associated facilities needed to facilitate maintenance and avoid the risk of fires and other accidents. It provides a safety margin between the high-voltage lines and surrounding structures and vegetation. Some vegetation clearing may be needed for safety and/or access reasons. A ROW generally consists of native vegetation or plants selected for favorable growth patterns (slow growth and low mature heights). However, in some cases, access roads constitute a portion of the ROW and provide more convenient access for repair and inspection vehicles. Vegetation clearing and/ or recontouring of land may be required for access road construction. The width of a ROW varies depending on the voltage rating on the line from 50 ft to approximately 175 ft or more for 500 kilovolt (kV) lines.

Before approval for new transmission is granted, the regulatory authority must determine that the project is necessary. Non-transmission alternatives must often be considered, including energy conservation, energy efficiency, distributed generation, and fully using unused capacity on existing transmission lines. When new transmission lines are deemed necessary, developers

FIGURE 9.3   Power line right away.

and utilities must find the best routes to the greatest concentrations of renewable energy and build with the least possible impact on the environment. Transmission lines from tall transmission towers carry high-voltage energy (115–500 kV) over long distances to a substation. Both transmission and distribution lines (from substations) carry enough energy to harm or kill both people and birds. Kill birds? Yes, actually, contrary to popular belief, some birds are electrocuted by electrical power lines. Small birds don't usually get electrocuted because they fail to complete a circuit either by touching a grounded wire or structure, or another energized wire, so electricity stays in the line. Larger birds, however—such as the California Condor, which has a wingspan of up to 9.5 ft—are more likely to touch a power line and ground wire, another energized wire, or a pole at the same time, giving electricity a path to the ground. In both situations, the birds are electrocuted and killed, a fuse is blown, power fails, and everyone is impacted.

Birds also fly into power lines. It is generally believed that birds collide with power lines because the lines are invisible to them or they do not see the lines before it is too late to avoid them. Birds' limited ability to judge distance makes power lines especially difficult to see, even as they are flying closer to them. Large birds are especially vulnerable because they are not always quick enough to change their direction before it is too late. Poor weather conditions, such as fog, rain, or snow, as well as darkness, may make the lines even more difficult to see.

When birds collide with power lines, they are either killed outright by the impact or injured by contact with electrical lines, resulting in crippling, which is likely fatal. Electrocutions can also start wildfires and cause power outages. An estimated 5–15% of all power outages can be attributed to bird collisions with power lines (USFWS, 2005).

In addition to increasing wildlife mortality due to collisions, electrocutions, and also by serving as perches for predators, transmission lines can fragment and interfere with wildlife habitats and corridors (WGA, 2009). Again, there are also concerns about the visual impacts of transmission lines. Moreover, many people feel that living or working near transmissions lines is hazardous to health. Burying transmission lines can help avoid many environmental and aesthetic issues. However, burying liens may also have negative impacts on soil, vegetation, and other resources (Molburg et al., 2007), and underground lines are typically four times as expensive as overhead lines. Also, although high-voltage direct-current (DC) lines can be buried, there is a limit on the maximum voltage and length of alternating-current (AC) lines that can be buried.

In all, constructing a major new transmission infrastructure can require — seven to ten years from planning to operation: one year for final engineering, one to two years for construction, and the rest of the time for planning and permitting. Substantial time and controversy are added to the process when environmental and related concerns are addressed at the end instead of at the beginning. The specific environmental impact concerns are addressed in the following sections.

Socioeconomics

Direct impacts would include that the creation of new jobs for workers (approximately two workers per megawatt) at wind energy development projects, and the associated income and taxes paid. Indirect impacts would occur as a result of the new economic development and would include new jobs at businesses that support the expanded workforce or provide project materials, and associated income taxes. Wind energy development activities could also potentially affect property value, either positively from increased employment effects or image of "clean energy," or negatively from proximity to the wind farm and any associated or perceived environmental effect (noise, visual, etc.).

Adverse impacts could occur if a large in-migrant workforce, culturally different from the local indigenous group, is brought in during construction. This influx of migrant workers could strain the existing community infrastructure and social services.

Environmental Justice

If significant impacts occurred in any resource areas, and the impact disproportionately affected minority or low-income populations, then there could be environmental justice concerns. Potential issues during construction are noise, dust, and visual impacts from the construction site and possible impacts associated with the construction of new access roads.

Hazardous Materials and Waste Management

Solid and industrial waste would be generated during construction activities. The solid waste would likely be nonhazardous and consist mostly of containers, packing material, and wastes from equipment assembly and construction crews. Industrial wastes would include minor amounts of paints, coatings, and spent solvents. Hazardous materials stored on-site for vehicle and equipment maintenance would include petroleum fluids (lubricating oils, hydraulic fluid, fuels), coolants, and battery electrolytes. Oils, transmission fluids, and dielectric fluids would be brought to the site to fill turbine components and other large electrical devices. Also, compressed gases would be used for welding, cutting, brazing, etc. These materials would be transported off-site for disposal, but impacts could result if the wastes were not properly handled and were released into the environment.

WIND ENERGY OPERATION IMPACT

Typical activities during the wind energy facility operations phase include turbine operation, power generation, and associate maintenance activities that would require vehicular access and heavy equipment operation when large components are being replaced. Potential impacts from these activities are presented below, by the type of affected resource.

Air Quality (Including Global Climate Change and Carbon Footprint)

There are no direct air emissions from operating a wind turbine. Minor volatile organic compound (VOC) emissions are possible during routine maintenance activities of applying lubricants, cooling fluids, and greases. Minor amounts of carbon monoxide and nitrogen oxides would be produced during the periodic operation of diesel emergency generators as part of preventive maintenance. Vehicular traffic would continue to produce small amounts of fugitive dust and tailpipe emissions during the operations phase. These emissions would not likely exceed air quality standards or have any impact on climate change.

Cultural Resources

Impacts during the operations phase would be limited to unauthorized collection of artifacts and visual impacts. The threat of unauthorized collection would be present once access roads are constructed in the site evaluation or construction phase, making remote lands accessible to the public. Visual impacts resulting from the presence of large wind turbines and associated facilities and transmission lines could affect some cultural resources, such as sacred landscapes or historic trails.

Ecological Resources

During operation, adverse ecological effects could occur from (1) disturbance of wildlife by turbine noise and human activity; (2) site maintenance (e.g., mowing); (3)

exposure of biota to contaminants; and (4) mortality of birds and bats from colliding with the turbines and meteorological towers.

During the operation of a wind facility, plant and animal habitats could still be affected by habitat fragmentation due to the presence of turbines, support facilities, and access roads. In addition, the presence of an energy development project and its associated access roads may increase human use of surrounding areas, which could, in turn, impact ecological resources in the surrounding areas through:

1. Introduction and spread of invasive vegetation
2. Fragmentation of habitat
3. Disturbance of biota
4. Increased potential for fire

As discussed in detail later, the presence of a wind energy project (and its associated infrastructure) could also interfere with migratory and other behaviors of some wildlife.

Water Resources (Surface Water and Groundwater)

Impacts to water use and quality and flow systems during the operation phase would be limited to possible degradation of water quality resulting from vehicular traffic and pesticide application if conducted improperly.

Land Use

Impacts on land use would be minimal, as many activities would be minimal, as many activities can continue to occur among the operating turbines, such as agriculture and grazing. It might be possible to collocate other forms of energy development, provided the necessary facilities could be installed without interfering with the operation and maintenance of the wind farm. Collocation of other forms of energy development could include directionally drilled oil and gas wells, underground mining, and geothermal or solar energy development. Recreation activities [e.g., off-highway vehicle (OHV) use and hunting] are also possible, but activities centered on solitude and scenic beauty could be affected. Military operations and aviation could be affected by radar interference associated with the operating turbines, and low-altitude activities could be affected by the presence of turbines over 200 feet high.

Soils and Geologic Resources (Including Seismicity/Geo Hazards)

Following construction, disturbed portions of the site would be revegetated and the soil and geologic conditions would stabilize. Impacts during the operations phase would be limited largely to soil erosion impacts caused by vehicular traffic for operator maintenance.

Paleontological Resources

Impacts during the operations phase would be limited to unauthorized collection of fossils. This threat is present once the access roads are constituted in the site evaluation or construction phases, making remote land accessible to the public.

Transportation

No noticeable impacts to transportation are likely during operations. Low volumes of heavy- and medium-duty pickup trucks and personal vehicles are expected for routine maintenance and monitoring. Infrequent, but routine shipments of component replacements during maintenance procedures are likely over the period of operation.

Visual Resources

Wind energy development projects would be highly visible in rural or natural landscapes, many of which have few other comparable structures. The artificial appearance of wind turbines may have visually incongruous "industrial" associations for some, particularly in a predominantly natural landscape; however, other viewers may find wind turbines visually pleasing and consider them a positive visual impact. Visual evidence of wind turbines cannot easily be avoided, reduced, or concealed, owing to their size and exposed location; therefore, effective mitigation is often limited.

Additional issues of concern are shadow flicker (strobe-like effects from flickering shadows cast by the moving rotors), blade glint from the sun reflecting off moving blades, visual contrasts from support facilities, and light pollution from the lighting on facilities and towers (which are required safety features).

Additional visual impacts from vehicular traffic would occur during maintenance, and as towers, nacelles, and rotors are upgraded or replaced. When replacing turbines and other facility components, the opportunity and pressures to break the uniformity of spacing between turbines and uniformity of size, shape, and color among facility components could increase visual contrast and visual "clutter."

Infrequent outages, disassembly, and repair of equipment may occur, producing the appearance of idle or missing rotors, "headless" towers (when nacelles are removed), and lowered towers, negative visual perceptions of "lost benefits" (e.g., loss of wind power) and "bone yards" (for storage) may result.

Socioeconomics

Direct impacts would include the creation of approximately one new job for 3 MW of installed capacity for operations and maintenance workers at wind energy development projects, and the associated income and taxes paid. Indirect impacts would occur from new economic development and include new jobs at businesses that support the expanded workforce or provide project materials, and associated income and taxes. Wind energy development activities could also potentially affect property values, either positively from increased employment effects or image of "clean energy" or negatively from proximity to the wind farm and any associated or perceived environment effects (noise, visual, etc.).

Environmental Justice

Possible environmental justice impacts during operation include the alteration of scenic quality in areas of traditional or cultural significance to minority or low-income populations. Noise impacts and health and safety impacts are also possible sources of disproportionate effects.

Hazardous Materials and Waste Management

Industrial and sanitary wastes are generated during routine operations (e.g., lubricating oils, hydraulic fluids, coolants, solvents, cleaning agents, sanitary wastewaters). These wastes are typically put in containers, characterized and labeled, possibly stored briefly, and transported by a licensed hauler to an appropriate permitted off-site disposal facility as a standard practice. Impacts could result if these wastes were not properly handled and released into the environment. Releases could also occur if individual turbine components or electrical equipment were to fail.

IMPACTS ON WILDLIFE

> He clasps the crag with crooked hands;
> Close to the sun in lonely lands,
> Ringed with the azure world, he stands.
> The wrinkled sea beneath him crawls;
> He watches from his mountain walls,
> And like a thunderbolt he falls.

> **—Alfred Lord Tennyson,** *The Eagle*

Alfred Lord Tennyson, in his classic poem, wants us to see the eagle as both a swift predator and a powerful bird which is nonetheless susceptible to defeat by other forces (most likely by humans). Human-made and installed wind turbines are responsible for bird deaths. This has been a less documented impact of wind turbines and is mainly argued by wildlife groups. The United States Fish and Wildlife Service (USFWS, 2014) points out noise standards, for example, for wind turbines developed by countries such as Sweden and New Zealand and some specific site-level standards implemented in the United States focus primarily on sleep disturbance and annoyance to humans. However, noise standards do not generally exist for wildlife, except in a few instances, where federally listed species may be impacted. Findings from recent research clearly indicate the need to better address noise–wildlife issues. As such, noise impacts to wildlife should clearly be included as a factor in wind turbine siting, construction, and operation. Later in this section, a detailed description of eagle conservation guidance with regard to land-based wind energy is presented. For now, it is important to point out some of the key issues which include (1) how wind facilities affect background noise levels; (2) how and what fragmentation, including acoustical fragmentation, occurs especially to species sensitive to habitat fragmentation; (3) comparison of turbine noise levels at lower valley sites—where it may be quieter—to turbines placed on ridgelines above rolling terrain where significant topographic sound shadowing can occur having the potential to significantly elevate sound levels above ambient conditions; and (4) correction and accounting of a 15 dB underestimate from daytime wind turbine noise readings used to estimate nighttime turbine noise levels (van den Berg, 2004; Barber et al., 2010). The sensitivities of various groups of wildlife can be summarized as:

- Birds (more uniform than mammals) 100 Hz to 8–10 kHz; sensitivity at 0.10 dB
- Mammals <10 Hz to 150 kHz; sensitivity to −20 dB
- Reptiles (poorer than birds) 50 Hz to 2 kHz; sensitivity at 40–50 dB
- Amphibians 100 Hz to 2 kHz; sensitivity from 10 dB to 60 dB

As mentioned, turbine blades at normal operating speeds can generate significant levels of noise. How much noise? Based on a propagation model of an industrial-scale 1.5 MW wind turbine at 263 ft hub height, positioned approximately 1,000 ft apart from neighboring turbines, the flowing decibel levels were determined for peak sound production. At a distance of 300 ft from the blades, 45–50 dBA were detected; at 2,000 ft, 40 dBA; and at 1 mile, 30–35 dBA (Kaliski, 2009). Declines in densities of woodland and grassland bird species have been shown to occur at noise thresholds between 45 dB and 48 dB, respectively; while the most sensitive woodland and grassland species showed declines between 35 dB and 43 dB, respectively. Songbirds specifically appear to be sensitive to very low sound levels equivalent to those in a library reading room (~30 dBA) (Foreman and Alexander, 1998). Given this knowledge, it is possible that effects to sensitive species may be occurring at ≥1 mile from the center of a wind facility at periods of peak sound production.

As pointed out earlier, noise does not have to be loud to have negative effects. Very low-frequency sounds including infrasound (sound lower in frequency than 20 Hz) are also being investigated for their possible effects on both humans and wildlife. Wind turbine noise results in a high infrasound component (Salt and Hullar, 2010). Infrasound is inaudible to the human ear but this unheard sound can cause human annoyance, sensitivity, disturbance, and disorientation (Renewable Energy World, 2010). For birds, bats, and other wildlife, the effects may be more profound. Noise from traffic, wind, and operating turbine blades produce low-frequency sounds (<1–2 kHz; Dooling, 2002, Lohr et al., 2003). Bird vocalizations are generally within the 2–5 kHz frequency range (Dooling and Popper, 2007) and birds hear best between 1 kHz and 5 kHz (Dooling, 2002). Although traffic noise generally falls below the frequency of bird communication and hearing, several studies have documented that traffic noise can have significant negative impacts on bird behavior, communication, and ultimately on avian health and survival (e.g., Lohr et al., 2003, Lengagne, 2008, Barber et al., 2010). Whether these effects are attributable to infrasound effects or a combination of other noise factors is not yet fully understood. The fact is little is known about the combination effect of traffic noise and wind turbine noise. However, given that wind-generated noise including blade turbine noise produces a fairly persistent, low-frequency sound similar to that generated by traffic noise (Lohr et al., 2003; Dooling, 2002), it is plausible that wildlife effects from these two sounds could be similar. It is also plausible that wildlife effects from these two sounds combined could be detrimental to the wildlife of all kinds. Based on experience, this book supports this view.

Some may feel that the combination of road noise and wind turbine noise-causing wildlife effects is plainly a stretch. Although the author has studied, observed, measured, and monitored this phenomenon, the truth is little is known about the effects of noise related to road noise and wind turbines combined; moreover, at present, little to nothing on the subject is reported in the peer-reviewed literature.

Let's get back to why some people feel the statement that a combination of road noise and wind turbine noise has a wildlife effect is plainly a stretch. This point of view seems to be prevalent for those who do not travel the east–west US Interstates from East Coast areas to the Southwestern, Pacific Northwest or and the north–south route Indiana I-65, and others. On many roads (Interstates), it is common to drive in any direction and to view hundreds of tall wind turbines off in the far distance ahead and to the right and left of the highway. On the other hand, some of these wind turbines are very close to the shoulder of the roads. However, again, many wind turbine farms are viewed off in the vast, far distance as one travels the highways. It is this perception of distance on some outlying, remote hillside or ridge in some unoccupied wildness or high plains area that gives the unknowing viewer this wrong view of combined noise from road traffic and wind turbines. They ask if the wind turbines are off in some remote corner of nowhere, how can road noise contribute to and be added to normal wind turbine noise? The reality is very easy; so is the answer. Remember, service roads built to perform maintenance and preventive maintenance (i.e., inspection of components, servicing items that require some type of activity on a regular basis—retorquing bolts—and replacing consumable items at or before a specified age (i.e., replacing filters or changing the oil in the gearbox)) are accessed by light and heavy trucks and other vehicles on a routine basis—no matter the location. The larger the wind farm, the more access—the more traffic; the more noise. All access vehicles, including helicopters used to transport parts and personnel, produce noise; in some cases, a lot of noise.

The counterargument, of course, is that wind turbines run on their own and require little to no operation and maintenance (O&M). If this is the case (and it is not), then there is very little traffic on their associated service roads and, thus, little noise added to wind turbine noise. In the first place, wind turbines do not operate by themselves; they are normally operated remotely from a plant operations room by a human operator but they can also be operated within the turbine nacelle. Their operation is also monitored by Supervisory Control and Data Acquisition (SCADA) communication system designed to alert the appropriate service personnel through computer warning and automated telephone calls.

Based on experience, the major causes of downtime were O&M works and faults. The occurrence of O&M and faults was variable. Some sites have a downtime of no more than 43 h per month and others as much as 127 h (NREL, 2000). O&M downtime includes all troubleshooting, inspections, adjustments, retrofits, and repairs performed on the turbines. Faults generally require no more than a reset and most can be performed remotely. Increased downtime for O&M and fault reasons means more traffic to and from the turbine farm.

Fast forwarding to wildlife effects resulting from wind turbine noise alone, it is important to point out that a bird's inability to detect turbine noise at close range may also be problematic. Note that the threshold for hearing in birds is higher than that for humans at all frequencies and the overlap in the discernible frequencies between species indicates that birds do not filter out other species by simply being unable to detect them (i.e., birds can hear songs of other species). In their environment, birds must be able to discriminate their own vocalizations and those of other species apart from any background noise (Dooling, 1982). Calls are important in the isolation of species, pair bond formation, precopulatory display, territorial defense, danger, advertisement of food sources, and flock cohesion (Knight, 1974).

For the average bird in a signal frequency of 1–4 kHz, noise must be 24–30 dB above the ambient noise level in order for a bird to detect it. As noted above, turbine blade and wind noise frequencies generally fall below the optimal hearing frequency of birds. Additionally, by the inverse square law, the sound pressure level decrease by 6 dB with every doubling of distance. Therefore, although the sound level of the blade may be significantly above the ambient wind noise level and detectable by birds at the source, as the distance from the source increases and the blade noise level decreases toward the ambient wind noise level, a bird may lose its ability to detect the blade and risk colliding with the moving blade.

Some researchers have attempted to explain avian collisions with turbine blades on birds' inability to divide their attention between surveying the ground for prey and monitoring the horizon and above for obstacles; i.e., they are so busy searching the ground that they do not notice the turbines. This hypothesis derives from substituting our knowledge of human vision for that of avian vision. Humans are foveate animals; we search the visual world with a small area of the retina known as the fovea, which is our area of sharpest vision, like someone searching a dark room with a narrow-beam searchlight. This results from our very low ratio (approximately 1:1) of photoreceptors to ganglion cells in the macular region of the retina. Outside the macular region, the ratio of receptors to ganglion cells increases progressively to 50:1–100:1, and our visual acuity drops sharply. Birds and many other animals, on the other hand, have universal macularity, which means that they have a low ratio of receptors to ganglion cells (4:1–8:1) out to the periphery of the retina. They maintain good acuity even in peripheral vision (Hodos et al., 1997). In addition, raptors possess the specialization of two foveal regions: one for frontal vision and one for looking at the ground. Moreover, birds have various optical methods for keeping objects at different distances simultaneously in focus on the retina (Hodos et al., 1997). Because of these considerations, failure to divide attention seems like an unlikely hypothesis.

MOTION SMEAR

A bird approaching a moving blade under high-wind conditions may be unable to see the blade due to motion smear—reduced visibility of the blades, especially at the tips—and may not hear the blade until it is very close—if it is able at all to hear it at all (Dooling, 2002). As an object moves across the retina with increasing speed, it becomes progressively blurred; this phenomenon is known as "motion smear," "motion blur," or "motion transparency" and is well known in human psychophysical research. It results because the human visual system is sluggish in its response to temporal stimulation; i.e., the visual system (in humans) summates signals over periods of about 120 ms in daylight (Burr, 1980; Bex et al., 1995).

The phenomenon of motion smear is apparent at the tips of wind turbine rotor blades as the observer (bird, human, or camera) approaches the turbine. Motion smear is not apparent in the central regions of the rotors. Even though the central regions and the tips are rotating at the same number of revolutions per (RPM), the absolute velocity of the blades is much higher in the peripheral regions. The higher velocity of the blade tip has placed it in the temporal-summation zone, in which the retina is sluggish in its ability to resolve temporarily separated stimuli, whereas the lower velocities of the more central portions are below the transition point between blur and non-blur; the individual blades can be seen more or less clearly. Moreover, the absolute velocity of the blade in the visual world is not critical; rather, it is the absolute velocity of the image of the blade that sweeps across the retina that is the critical variable. For reasons that will be explained later, as the observer approaches the turbine, the retinal image of the blades increases in velocity until the retina can no longer process the information. This results in motion smear or motion transparency—the blade becomes transparent to the view. A solution to avian collisions with wind turbines must consider the causes of motion smear and consider whether blade patterns could minimize this effect.

The Theory of Motion Smear

One of the characteristics of motion smear is that it eliminates the high spatial frequencies from visual patterns, which is why they appear to go out of focus and become virtually transparent (Steinman and Levinson, 1990). High spatial frequencies are those Fourier components of a visual object that are found at edges and corners and in fine detail. The print on this page, for example, is made up mainly of high spatial frequencies. If they are removed by optical blur or refractive error, the text becomes transparent and, in the worst case, virtually disappears.

Motion smear causing bird collisions with seemingly slow-moving turbines seems paradoxical given that acute vision that most birds, especially raptors, possess. However, analysis indicates that as the eye approaches the rotating blades, the retinal image of the blade increases in velocity until it is moving so fast that the retina can't keep up with it. At this point, the retinal image becomes a transparent blur that the bird probably interprets as a safe area to fly through, with disastrous consequences (NREL, 2000).

The Law of the Visual Angle

Figure 9.4 shows how objects of different sizes and different distances can form the same size image within the eye. The angle (A') inside the eye is the same as the angle (A) from the eye to each of the objects. These angles, called "visual angles," are the conventional units to describe object size because they are directly related to retinal-image size, which is the only relevant variable for these purposes. Thus, a small object close to the eye can cast the same size retinal image as a large object seen from a much farther distance.

Another concern involves the effect of ambient noise on communication distance and an animal's ability to detect calls. For effects to birds, this can mean (1) behavioral and/or physiological effects, (2) damage to hearing from acoustic overexposure,

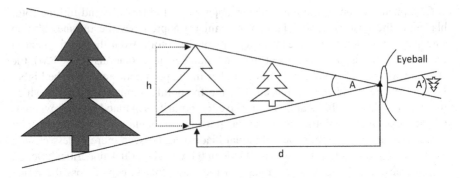

FIGURE 9.4 The law of the visual angle. Objects of different sizes and distances that subtend the same angle will cast the same size image on the retina. Angles A and A′ are the same.

and (3) masking of communication signals and other biologically relevant sounds (Dooling and Popper, 2007). Of the 49 bird species whose behavioral audibility curves and/or physiological recordings have been determined, Dooling and Popper (2007) developed a conceptual model for estimating the masking effects of noise on birds. Based on the distance between birds and the spectrum level, bird communication was predicted to be "at-risk" (e.g., at ~755 ft distance, where the noise was 20 dB), "difficult" (i.e., at ~755 ft, where the noise was 25 dB), and "impossible" (e.g., at ~755 ft, where the noise was 30 dB). While clearly there is variation between species and there is no single noise level where one-size-fits-all, this masking effect of turbine blades is of concern and should be considered as part of the cumulative impacts analysis of a wind facility on wildlife. It must be recognized the noise in the frequency region of avian vocalizations will be most effective in masking these vocalizations (Dooling, 2002).

Barber et al. (2010) assess the threats of chronic noise exposure, focusing on grouse communication calls, urban bird class, and other songbird communications. They determined that while some birds were able to shift their vocalizations to reduce the masking effects of noise, when shifts did not occur or were insignificant, masking could prove detrimental to the health and survival of wildlife (Barber et al., 2010). Although much is still unknown in the real world about the masking effects of noise on wildlife, the results of a physical model analyzing the impacts of transportation noise on the listening area (i.e., the active space of vocalization in which animals search for sounds) of animals resulted in some significant findings. With a noise increase of just 3 dB—a noise level identified as "just perceptible to humans"—this increase corresponded to a 50% loss of listening area for wildlife (Barber et al., 2010). Other data suggest noise increases of 3 dB to 10 dB correspond to 30% to 90% reductions in alerting distances (i.e., the maximum distance at which a signal can be heard by an animal and is particularly important for detecting threats) for wildlife, respectively (Barber et al., 2010). Impacts of noise could thus be putting species at risk by impairing signaling and listening capabilities necessary for successful communication and survival.

Swaddle and Page (2007) tested the effects of environmental noise on pair pref-erence selection of Zebra Finches. They noted a significant decrease in females' preference for their pair-bonded males under high environmental noise conditions. Bayne et al. (2008) found that areas near noiseless energy facilities had total passer-ine (i.e., birds of the order Passeriformes which includes more than half of all bird species) density 1.5 times greater than areas near noise-producing energy facilities. Specifically, white-throated sparrows, yellow-rumped warblers, and red-eyed vireos were less dense in noisy areas. Habib et al. (2007) found a significant reduction in ovenbird pairing success at compressor sites (averaging 77% success) compared to noiseless well pads (92%). Quinn et al. (2006) found that noise increases perceived predation risk in chaffinches, leading to increased vigilance and reduced food intake rates, a behavior which could, over time, result in reduced fitness. Francis et al. (2009) showed that noise alone reduced nesting species richness and led to a dif-ferent composition of avian communities. While they found that noise disturbance ranged from positive to negative, responses were predominately negative.

Schaub et al. (2008) investigate the influence of background noise on the forag-ing efficiency and foraging success of the greater mouse-eared bat, a model selected because it represents an especially vulnerable group of gleaning bats (predators of herbivorous insects) that rely on their capacity to listen for prey rustling sounds to locate food. Their study clearly found that traffic noise, and other sources of intense, broadband noise deterred bats from foraging in areas where these noises were pres-ent presumably because these sounds mashed relevant sounds or echoes the bats use to locate food.

Although there are few studies specifically focused on the noise effects of wind energy facilities on birds, bats, and other wildlife, scientific evidence regarding the effects of other noise sources (as mentioned, transportation) is widely documented. The results show, as documented in various examples above, that varying sources and levels of noise can affect both the sending and receiving of important acoustic signaling and sounds. This also can cause behavioral modifications in certain spe-cies of birds and bats such as decreased foraging and mating success and overall avoidance of noisy areas. The inaudible frequencies of sound may also have negative impacts on wildlife.

To this point, we have focused on the wildlife effects of wind turbines and wind turbine farms on birds. Little is known about the effects of noise related to wind tur-bines on invertebrates, fish, reptiles, amphibians, and various mammals. As stated earlier, some correlation and association have been made between road noise and wind turbine noise in producing wildlife effects. Additional extrapolation in making the connection between road-generated and wind turbine-generated noise is avoided here because not only is the jury still out; the fact is the jury has yet to be formed.

However, given the mounting evidence regarding the negative impacts of noise—specifically low-frequency levels of noise such as those created by wind turbines on birds, bats, and other wildlife, it is important to take precautionary measures to ensure that noise impacts at wind facilities are thoroughly investigated prior to development. Noise impacts to wildlife must be considered during the landscape site evaluation and constructions processes. In an attempt to meet this need, the US Fish and Wildlife

Service (USFWS, 2013) has developed the Eagle Conservation Plan Guidance (ECPG). Few would argue against the point that of all of America's wildlife, eagles hold perhaps the most revered place in our national history and culture. The United States has long imposed special protections for its bald and golden eagle populations. Now, as the nation seeks to increase its production of domestic energy, wind energy developers and wildlife agencies have recognized a need for specific guidance to help make wind energy facilities compatible with eagle conservation and the laws and regulations that protect eagles. The ECPG provides specific in-depth guidance for conserving bald and golden eagles in the course of siting, construction, and operating wind energy facilities. As research specific to noise effects from wind turbines further evolves, these findings should be added to ECPG guidelines and utilized to develop technologies and measures to further minimize noise impacts to wildlife (USFWS, 2013).

IMPACT ON HUMAN HEALTH

The average person would probably look upon an operational wind turbine or a massive wind farm and have various thoughts about what he or she perceives. For any number of reasons, maybe the observer's view would be positive or maybe it could be negative—it's all in the eyes of the beholder (literally, the perception is subjective). But the eyes are simply the windows to received neuron-transmitted thoughts to our brain cells and thus to our thought process. One thought the observer probably would not have, however, is that a wind turbine could not affect his or her health; the observer would be wrong, of course.

The logical question is: What is health? For our purposes in this text, we use the World Health Organization's (WHO) definition of health.

> Health is a state of complete physical, mental, and social well-being and not merely the absence of disease or infirmity.
>
> **—WHO (2011)**

While it is true that from a practical and well-documented viewpoint, the operation of a wind turbine or wind turbine farm doesn't directly impact human health, it is also true that factors such as stress and loss of sleep contribute to health problems for some residents living close to these installations. Stress can be generated by frustrated residents having to put up with noise pollution, visual impact, and loss of land value; at the same time, loss of sleep can be experienced by people living close to wind turbines as a result of noise pollution.

Observers have complained about turbine noise annoyance and unpleasant sounds that include rhythmic modulation of low-frequency noise (which may be more annoying than steady noise) and increasing sound pressure levels, resulting in increased levels of annoyance. Interestingly, the annoyance was reported more frequently when turbines were visible and also when the observer reported a negative impact on the surrounding landscape.

It has been widely reported that wind turbines are creating sounds and vibrations that can be sensed by people up to ten miles away. Magee (2014) points out that

Nina Pierpont MD PhD is reporting that people who live within two kilometers of wind turbines are reporting sickness that can be traced to the presence of these. Low-frequency noise and infrasound (sound that is less than 20 Hz) appear to be the problem.

(p. 227)

The problem that Pierpont and others report is commonly called wind turbine syndrome, which is the disruption or abnormal stimulation of the inner ear's vestibular system caused by turbine infrasound and low-frequency noise. Symptoms of wind turbine syndrome include:

- Sleep problems
- Headaches
- Dizziness
- Exhaustion, anxiety, anger, irritability, and depression
- Problems with concentration and learning
- Tinnitus (ringing in the ears)

Along with the turbine noise annoyance generated by mechanical and aerodynamic factors—i.e., the feeling of resentment displeasure, discomfort, dissatisfaction, or offense which occurs when noise interferes with someone's thoughts, feeling, or daily activities (WHO, 2011)—there have been complaints about rhythmic light flicker causing intermittent shadows known as shadow flicker or flickering shadows. Schworm and Filipov (2013) reported that a woman from Kingston, Massachusetts, stated that the problem with shadow flicker begins in the late afternoon. Stripes of shadow whip across her living room, kitchen, and bedroom, a pulse of flashing light and dark that can continue more than an and makes Reilly want to lose her mind. You can't stay in your room. "You get a headache," the woman said. "You can't live your life."

Another increasing complaint being heard concerning wind turbine noise generation is related to high levels of low-frequency noise over years of exposure. This problem is called vibroacoustic disease (VAD). The clinical progression is insidious, and lesions are found in many systems throughout the body. This disease or syndrome is commonly classified as mild, moderate, or severe as described below.

Vibroacoustic disease syndrome:

- *Mild* (—one to four years)—Slight mood swings, indigestion, heartburn, mouth/throat infections, bronchitis.
- *Moderate* (—four to ten years)—Chest pain, definite mood swings, back pain, fatigue, skin infections (fungal, viral, and parasitic, inflammation of the stomach lining, pain and blood in the urine, conjunctivitis, allergies).
- *Severe* (>ten years)—Psychiatric disturbances, hemorrhages (nasal, digestive, conjunctiva mucosa), varicose veins, hemorrhoids duodenal ulcers, spastic colitis, decrease in visual activity, headaches, several joint pain, intense muscular pain, neurological disturbances.

WIND TURBINE NOISE IMPACT BASED ON OBSERVATION
AND RECENT SCIENTIFIC RESEARCH

Between the 2010–2019 timeframe, while traveling from the US East Coast and visiting large wind farms in Indiana, Iowa, Oregon, Washington, California, and Oklahoma, I conducted a personal research project to determine for myself if the large wind turbines were and are responsible for bird kills. While it is true that I found remains of dead birds here and there, they were in lower numbers than what I had expected to find—numbering less than a dozen in total. Not only was I surprised by the scant number of dead birds that I found but it was not that long after visiting the first wind farm in Indiana and walking the site for three days still looking for bird remains but actually I found myself almost hypnotized by the rotating, whirling turbines themselves. I found that those rotating, whirling turbine blades are as hypnotic as waterfall and fire gazing. By the end of the second day after walking the Indiana wind farm site, my perception of the hypnotic effect ebbed somewhat when I realized that maybe it was not the rotating, whirling turbine blades that hypnotized, entranced, literally mesmerized me.

No. It was something else. It was something that was quite apparent but at the same time subtle besides the sighting of those rotating blades during the day and watching their lighted blade tips swoop along making a lighted circle in the darkened sky. What was it? Well, it was the noise—that swishing and whooshing sound, at least at my first detection, recognition, and perception. I thought it strange that it took me a couple of days to recognize that which was so apparent, so evident, so obvious, and so perceptible.

At each of the several sites I visited and studied, I basically expected to hear and as time went by to pay attention to—to take notice. I found via an experiment that the sounds from the wind turbines were different depending upon where I stood. Close up to one of the towers, the swishing and whooshing sound was obvious but as I moved off into the distance from the rotating wind turbine blades, the sound morphed into a hush, a hushing sound—not totally quiet but instead a muted shush.

So, in the transition from dead birds to haunting sounds, I came to wonder how the local residents (in places where there were residents near the wind farms) felt about the wind turbines and if they had any negative issues with them. What I found at every site where people were close by was basically the same complaint: "I don't like the sound … so irritating, infuriating, grating, and/or absolutely annoying."

Is it the and/or absolutely annoying? I wanted to answer this question myself, one way or another, so I did in a non-scientific manner. For example, not everyone complained about the noise but for those who did using various descriptors, some of them not printable, the common response was that they are "annoying"—the sounds.

Anyway, I changed my focus from searching for dead birds to using my calibrated sound meter to measure the sound levels at each site I visited. Interestingly, I was not able to detect or measure any sound level greater than 40 decibels on the (A) (dB(A)) scale of the meter. This is a noise level that is apparent but is usually drowned out by background noise from vehicles, airplanes, and farm machinery—no big deal, or so I thought.

Well, what those local folks reported to me was that they did not like the sound levels from the wind farms because they are/were *annoying* vs. *a health issue* is a common complaint that fit right into the data and information that several researchers found during their investigations of environmental issues occurring during wind farm operations.

From the scientific researchers' investigations, I discovered that researchers concluded that wind farms are sited properly, wind turbines are not related to adverse health effects. Research has shown that this claim is supported and made by a number of government health and medical agencies and legal decisions (MassDEP and MDPH, 2012; Krough, 2011; Thorne, 2011; Merlin et al., 2014). As I reviewed the evidence, it showed that while noise from wind turbines is not loud enough to cause hearing impairment and is not related to adverse effects, wind noise can be a source of annoyance for some people and that annoyance may be associated with certain reported health effects (e.g., sleep disturbance), especially at sound pressure levels >40 dB(A). Note that in all of my sound meter measurements at all of the sites I visited, I only found two sites where the decibel levels exceeded 40 dB(A) and those sites were either located next to an airport or an Interstate Highway where traffic was constant and sometimes heavy and contributed or augmented ambient noise levels.

Noise-related annoyance has been described as a feeling of displeasure evoked by noise; thus, the reported correlation between wind turbine noise and annoyance was expected and has been routinely linked to a variety of sources such as air, rail, and road traffic (Berglund and Lindvall, 1995; Laszlo et al., 2012; WHO, 2011). Research data indicate that noise-related annoyance from these more common sources is prevalent in many locations.

It is generally accepted by many experts that annoyance is considered to be the least severe likely effect of community noise exposure (WHO, 2011; Babisch, 2002). Moreover, it has been theorized that sufficiently high levels of annoyance could lead to negative demonstrative responses such as anger, disappointment, anxiety, and depression and psychosocial symptoms such as tiredness, stomach discomfort, and stress (WHO, 2011; Fields et al., 1998, 2001; Ohrstrom, 2004; Ohrstrom et al., 2006).

Keep in mind, however, that noise annoyance is known to be strongly affected by attitudinal factors such as fear of harm connected with the source and personal evaluation of the source as well as expectations or beliefs of residents (Fields, 1993; Guski, 1999; Miedema and Vos, 1999). For wind turbines, this has been revealed in studies that have shown that subjective variables like evaluations of visual impact, beautiful vs. ugly, attitude to wind turbines, benign vs. intruders, and personality traits are skewed and more clearly related to annoyance and health effects than the noise itself (Pedersen and Persson, 2004, 2007; Pederson and Larsman, 2008; Pederson et al., 2009; Taylor et al., 2012). Based on these research findings and my own interviews with residents living near wind turbine sites and my personal observations of attitudinal characteristics of those associated with the wind turbine industry, it appears that the adverse effects exhibited by some people who live near wind turbines are a response to stress and annoyance.

Again, based on personal observation of residents living near dozens of wind farms throughout the United States, the tension and displeasure appear to be driven by multiple environmental and personal factors and are not specifically caused by any unique characteristic of wind turbines. I also noticed that those people who benefit financially from wind turbine operations had significantly decreased levels of annoyance compared to individuals that received no economic benefit, despite exposure to similar, if not higher, sound levels. This is a classic example of two neighbors both of which live next to a factory whose smokestacks pollute the air in the surrounding area. One neighbor who does not work at the factory constantly complains about the foul air. However, the other neighbor who works at the factory smiles and states that the smoke and odor are signs of prosperity and the smell of money. With regard to those who work at wind farms and residents who do not, the research of Pederson et al. (2009) resulted in the same conclusion.

I was also able to substantiate and attest to the growing body of research that suggests that nocebo effects may play a role in a number of self-reported health impacts related to the presence of wind turbines. Note that nocebo effects are the opposite of placebo effects—placebo effects occur whenever a positive outcome is anticipated, and *nocebo* effects result due to negative expectations. Studies have indicated that negative attitudes and worries of individuals about perceived environmental risks have been shown to be associated with adverse health-related symptoms such as headache, nausea, dizziness, agitation, and depression, even in the absence of an identifiable cause (Boss, 1997; Henningsen and Priebe, 2003; Petrie et al., 2001). The circulation of negative information and priming of expectations, psychogenic factors, have been shown to impact self-assessments following exposure to wind turbine noise (Chapman et al., 2013; Crichton et al., 2014). Thus, it is important to consider the role of mass media in influencing public attitudes about wind turbines and how this may alter the response and perceived health impacts of wind turbines in the community. Deignan et al. (2013), for example, showed that newspaper coverage of the probable health effects of wind turbines in Ontario has tended to emphasize "fright factors" about wind turbines. Specifically, Deignan et al. stated that 94% of articles provided "negative, loaded or fear-evoking" descriptions of "health-related signs, symptoms or adverse effects of wind turbine exposure" and 58% of the articles suggested that the effects of wind turbines on human health were "poorly understood by science."

Based on evidence garnered through extensive research by specialists in human health related to wind turbine development and operation, Knopper et al. (2014) have suggested a series of best practices. However, note that the author of this work agrees with the researchers that subjective variables, attitudes, and expectations are strongly attendant to annoyance and have the potential to facilitate other health complaints via the nocebo effect. Thus, it is possible that a segment of the population may remain annoyed (no matter what; based on other health aspects) even when noise limits are enforced.

Having said all this, it is recommended that when building a wind farm and positioning each wind turbine that the selected positions of the wind turbines should not be based on setbacks alone but instead should also be sound-based. In other words, a sound survey should be conducted at the proposed site prior to construction so that the wind turbines, with regard to noise, will be less offensive while operating to the local population. The idea is to install the wind farm in such a manner that the turbine-generated noise level will not exceed ambient noise level in the area—the rule of thumb should be to make sure that the operating wind turbines noise is <40 dB(A) to keep the annoyance level non-determinable when mixed with the ambient noise level. After construction and installation, sound-level monitoring should regularly be conducted to ensure the maintenance of acceptable noise levels. If the noise levels are found to be greater than 40 dB(A), the neighbors should be consulted and informed to ensure that community support is maintained. The goal is to keep wind turbine noise and local ambient noise below 55 dB(A) (Knopper et al., 2014).

DID YOU KNOW?

The wind turbines which capture wind energy are as high as 20 story buildings and have 3 blades which are 60-m long.

ENVIRONMENTAL IMPACTS OF POWER TRANSMISSION LINES*

In the preceding paragraphs, we have set the stage for a more detailed discussion that follows, describing the environmental impact of power transmission lines. Basically, energy transmission involves three stages of implementation: site evaluation, project construction, and transmission operation.

ENERGY TRANSMISSION SITE EVALUATION IMPACTS

Energy site evaluation phase activities are generally temporary and conducted at a smaller magnitude than those at the construction and operation phases. Potential impacts from these activities are presented below, by the type of affected resource. The impacts described are for typical site evaluation activities, such as limited ground clearing, vehicular and pedestrian traffic, borings for geotechnical surveys, and positioning of equipment. If excavation or access road construction is necessary at this stage, impacts to resources would be similar in character, but lesser in magnitude, to those for the construction phase.

Route and access road selection that avoids major environmental impacts is ideal. Therefore, additional activities that could occur during this phase are field surveys for recording significant resources present in the potential project area (e.g., threatened and endangered species, wetlands, archaeological sites). These surveys are typical of the short and limited disturbance that occur during the site evaluation phase. The following impacts may result from the site evaluation activities.

---

* Material adapted from *Tribal Energy and Environmental Information Clearing House* accessed 03/16/14 @ http://teeic.anl.gov/er/trnasmission /imp-act/siteeval/index.cfm.

Air Quality

Impacts on air quality during surveying and testing activities would be limited to temporary and local generation of vehicle and boring equipment emissions and fugitive dust and would neither likely cause an exceedance of air quality standards nor have any impact on climate change.

Cultural Resources

The amount of surface and subsurface disturbance is minimal during the site evaluation phase. Cultural resources buried below the surface are unlikely to be affected; while material present on the surface could be disturbed by vehicular traffic, ground cleaning, and pedestrian activity (including the collection of artifacts). Surveying and testing activities could affect areas of interest to Native Americans depending on the placement of equipment and/or level of visual intrusion. Surveys conducted during this phase to evaluate the presence and/or significance of cultural resources in the area would assist developers in routing and designing the product to avoid or minimize impacts to these resources.

Ecological Resources

Impacts on ecological resources (vegetation, wildlife, aquatic, biota, special status species, and their habitats) would be minimal and localized during surveying and testing because of the limited nature of the activities. The introduction or spread of some nonnative invasive vegetation could occur as a result of vehicular traffic, but this would be relatively limited in extent. Surveys conducted during this phase to evaluate the presence and/or significance of ecological resources in the area would assist developers in routing and designing the project to avoid or minimize impacts to these resources (e.g., wetlands, migratory birds, and threatened and endangered species).

Water Resources

Minimal impact to water resources, local water quality, water flows, and surface water/groundwater interaction is anticipated. Very little water would likely be used or generated during the site evaluation phase. Any water needed could be trucked in from off-site.

Land Use

Site evaluation activities would likely result in temporary and localized impacts to land use. These activities could create a temporary disturbance in the immediate vicinity of a surveying or monitoring site (e.g., disturbing recreational activities or livestock grazing). Site evaluation activities are unlikely to affect mining activities, military operations, or aviation.

Soils and Geologic Resources

The amount of surface disturbance and use of geologic materials are minimal during the site evaluation phase, and soils and geologic recourse are unlikely to be affected.

Surveying and testing activities would be unlikely to activate geological hazards or increase soil erosion. Borings for soil testing and geotechnical surveys provide useful site-specific data on these resources. Surface effects from pedestrian and vehicular traffic could occur in areas that contain special (e.g., cryptobiotic) soils.

Paleontological Resources

The amount of subsurface disturbance is minimal during the site evaluation phase and paleontological resources buried below the surface are unlikely to be affected. Fossil material present on the surface could be disturbed by vehicular traffic, ground clearing, and pedestrian activity (including the collection of fossils). Surveys conducted during this phase to evaluate the presence and/or significance of paleontological resources in the area would assist develops in routing and designing the product to avoid or minimize impacts to these resources.

Transportation

No impacts on transportation are anticipated during the site evaluation phase. Transportation activities would be temporary and intermittent and limited to low volumes of heavy- and medium-duty construction vehicles (e.g., pickup trucks) and personal vehicles.

Visual Resources

Surveying and testing activities would have temporary and minor visual effects caused by the presence of workers, vehicles, and equipment.

Socioeconomics

The activities during the site evaluation phase are temporary and limited and would not result in socioeconomic impacts on employment, local services, or property values.

Environmental Justice

Site evaluation phase activities are limited and would not result in significant high and adverse impacts in any resource area; therefore, environmental justice is not expected to be an issue during this phase.

ENERGY TRANSMISSION CONSTRUCTION IMPACTS

Typical activities during the construction phase of an energy transmission project include ground clearing and removal of vegetative cover, grading, excavation, blasting, trenching, drilling, vehicular and pedestrian traffic, and project component construction and installation. Activities conducted in locations other than within the project right of way (ROW) include excavation/blasting for construction materials (such as sands and gravels), access road and staging area construction, and construction of other ancillary facilities such as compressor stations or pump stations.

Air Quality

Emissions generated during the construction phase include vehicle emissions; diesel emissions from large construction equipment and generators; VOC emissions from storage and transfer of fuels for construction equipment; small amounts of carbon monoxide, nitrogen oxides, and particulates from blasting activities; and fugitive dust from many sources such as disturbing and moving soils (clearing, grading, excavating, trenching, backfilling, dumping, and truck and equipment traffic), mixing concrete, storage of unvegetated soil piles, and drilling and pile driving. Air quality impacts could also occur if cleared vegetation is burned.

Cultural Resources

Potential impacts to cultural resources include:

1. Complete destruction of the resource if present in areas undergoing surface disturbance or excavation.
2. Degradation or destruction of near-surface cultural resources on- and off-site resulting from changing the topography, changing the hydrological patterns, and soil movement (removal, erosion, sedimentation).
3. Unauthorized removal of artifacts or vandalism as a result of human access to previously inaccessible areas.
4. Visual impacts resulting from large areas of the exposed surface, increases in dust, the presence of large-scale equipment, machinery, and vehicles for cultural resources that have an associated landscape component that contributes to their significance, such as a scared landscape or historic trail. Note that the accumulation of sediment mentioned above could serve to protect some buried resources by increasing the amount of protective cover.

Ecological Resources

Adverse impacts to ecological resources could occur during construction from:

1. Erosion and runoff
2. Fugitive dust
3. Noise
4. Introduction and spread of invasive nonnative vegetation
5. Modification, fragmentation, and reduction of habitat
6. Mortality of biota
7. Exposure to contaminants
8. Interference with behavioral activities

Site clearing and grading, coupled with the construction of access roads, towers, and support facilities, could reduce, fragment, or dramatically alter existing habitat in the disturbed portions of the project area. Wildlife in surrounding habitats might also be affected if construction activities (and associated noise) disturb normal behaviors, such as feeding and reproduction.

Water Resources

Water would be required for dust control, making concrete, and consumptive use by the construction crew. Depending on availability, it may be trucked in from off-site or obtained from local groundwater wells or nearby surface water bodies. Water quality can be affected by:

1. Activities that cause soil erosion
2. Weathering of newly exposed soils causing leaching and oxidation that can release chemicals into the water
3. Discharges of waste or sanitary water
4. Herbicide applications
5. Contaminant spills, especially oil

Applying sand and gravel for road construction, layout areas, foundations, etc., can alter the drainage near where the material is used. The size of the area affected can range from a few hundred square feet (for a support tower foundation) to a few hundred acres (for an access road). Surface and groundwater flow systems could be affected by withdrawals made for water use, wastewater and stormwater discharges, and the diversion of surface water flow for access road construction or stormwater control systems. Excavation activities and the extraction of geological materials may affect surface and groundwater flow. The interaction between surface water and groundwater may also be affected if the two are hydrologically connected, potentially resulting in unwanted dewatering or recharging.

Land Use

Impacts to land use could occur during construction if there were conflicts with existing land use plans and community goals; conflicts with existing recreation, education, religious, scientific, or other use areas; or conversion or cessation of the existing commercial land use of the area (e.g., mineral extraction). During construction, most land use impacts would be temporary, such as removal of livestock from grazing areas during periods of blasting or heavy equipment operations; curtailing hunting near work crews; or temporary effects to the character of a recreation area because of construction noise, dust, and visual intrusions. Long-term land use impacts would occur if existing land uses were not compatible with an energy transmission project, such as remote recreational experiences. Within forested areas, ROW clearing could result in the long-term loss of timber production.

Impacts to aviation are possible if the project is located within 20,000 ft (6,100 m) or less of an existing public or military airport or if proposed construction involves subjects greater than 200 ft (61 m) in height. The Federal Aviation Administration (FAA) must be notified if either of these two conditions occurs and would be responsible for determining if the project would adversely affect commercial, military, or personal air navigation safety. Similarly, impacts on military operations may occur if the project is located near a military facility and that facility conducts military testing and training activities that occur at low altitudes.

Soils and Geologic Resources

Surface disturbance, heavy equipment traffic, and changes to surface runoff patterns can cause soil erosion. Impacts of soil erosion include soil nutrient loss and reduced water quality in nearby surface water bodies. Impacts on special soils (e.g., cryptobiotic soils) could also occur.

Sands, gravels, and quarry stone would be excavated for use in the construction of access roads; concrete for foundations and ancillary structures; for improving ground surface for lay-down areas and crane-staging areas; and, as necessary, for backfill in pipeline trenches. Mining operations would disturb the ground surface, and runoff would erode fine-grained soils, increasing the sediment load farther down in steams and/or rivers. Mining on steep slopes and/or on unstable terrain without appropriate engineering measures increases the landslide potential in the mining areas. Possible geological hazards (earthquakes, landslides, avalanches, forest fires, geomagnetic storms, ice jams, mudflows, rock falls, flash floods, volcanic eruptions, geyser deposits, ground settlement, sand dune migration, thermals springs, etc.) can be activated by excavation and blasting for raw material, increasing slopes during site grading and construction of access roads, altering natural drainage patterns, and toe-cutting bases of slopes. Altering drainage patterns accelerates erosion and creates slope instability.

Paleontological Resources

Potential impacts to paleontological resources during construction include (1) complete destruction of the resource if present in areas undergoing surface disturbance or excavation; (2) degradation or destruction of near-surface fossil resources on- and off-site resulting from changing the topography, changing the hydrological patterns, and soil movement (removal, erosion, sedimentation); and (3) unauthorized removal of fossil resources or vandalism to the locality as a result of human access to previously inaccessible areas. The accumulation of sediment mentioned above could serve to protect some localities by increasing the amount of protective cover.

Transportation

Short-term increases in the use of local roadways would occur during the construction period. Heavy equipment would need to be continuously moved as construction professes along the linear project. Shipments of materials are not expected to significantly affect primary or secondary road networks but would depend on the ever-changing location of the construction site area relative to the material source. Overweight and oversized loads could cause temporary disruptions and could require some modification to roads or bridges (such as fortifying bridges to accommodate the size or weight). The weight of shipments is also a parameter in the design of access roads for grade determinants and turning clearance requirements.

Visual Resources

Potential sources of visual impacts during construction include visual constraints in the landscape from access roads and staging areas and conspicuous and frequent small-vehicle traffic for worker access and frequent large-equipment (trucks, graders, excavators, cranes, and possibly helicopters) traffic for project and access road construction. Project component installation would produce visible activity and dust in dry soils. Project construction may be progressive, persisting over a significant period of time. Ground disturbance (e.g., trenching and grading) would result in visual impacts that produce contrasts of color, form, texture, and line. Soil scares and exposed slop faces could result from excavation, leveling, and equipment movement.

Socioeconomics

Direct impacts would include the creation of new jobs for construction workers and the associated income and taxes generated by the project. As an example, the number of construction workers required for a 150-mile (241 km) length of the pipeline is only about 230 annual direct workers (fewer than 175 for an equivalent length of a transmission line or a petroleum pipeline). Indirect impacts are those impacts that would occur as a result of the new economic development and would include things such as new jobs at business that support the expanded workforce or that provide project materials, and associated income and taxes. Construction of an energy transmission project may affect the value of residential properties located adjacent to the ROW (there are conflicting reports on whether that would be adverse, beneficial, or neutral).

Environmental Justice

If significant impacts were to occur in any of the resource areas and these were to disproportionately affect minority or low-income populations, these could be an environmental justice impact. Issues that could be of concern are noise, dust, visual impacts, and habitat destruction from construction activities and possible impacts associated with new access roads.

ENERGY TRANSMISSION OPERATIONS IMPACTS

Typical activities during the operation and maintenance phase include operation of compressor stations or pump stations, ROW inspections, ROW vegetation clearing, and maintenance and replacement of facility components. Environmental impacts that could occur during the operation and maintenance phase would mostly occur from long-term habitat change with the ROW, maintenance activities (e.g., ROW vegetation clearing and facility component maintenance or replacement), noise (e.g., compressor station, corona discharge), the presence of workers, and potential spills (e.g., oil spills).

Air Quality

Vehicular traffic and machinery would continue to produce small amounts of fugitive dust and exhaust emissions during the operation and maintenance phase. These emissions would not likely cause an exceedance of air quality standards nor have any impact on climate change. Trace amounts of ozone would be produced by corona effects from transmission lines (e.g., less than 1.0 part per billion which is considerably less than air quality standards). Routine venting of pipelines and breakout tanks (for liquid petroleum products and crude oil) would also cause localized air quality impacts.

Cultural Resources

Impacts during the operations and maintenance phase could include damage to cultural resources during vegetation management and other maintenance activities, unauthorized collection of artifacts, and visual impacts. This threat is present once the access roads are constructed and the ROW is established, making remote areas more accessible to the public. Visual impacts resulting from the presence of the aboveground portion of a pipeline, transmission lines, and associated facilities could impact cultural resources that have an associated landscape component that contributes to their significance, such as a sacred landscape or historic trails.

Ecological Resources

During operations and maintenance, adverse impacts to ecological resources could occur from:

1. Disturbance of wildlife from noise and human activity
2. ROW maintenance (e.g., vegetation removal)
3. Exposure or biota to contaminants
4. Mortality of biota from colliding with transmission lines or aboveground pipeline components

Ecological resources may continue to be affected by the reduction in habitat quality associated with habitat fragmentation due to the presence of the ROW, support facilities, and access roads. In addition, the presence of an energy transmission project and its associated access roads may increase human use of surrounding areas, which, in turn, could impact ecological resources in the surrounding areas through:

1. Introduction and spread of invasive nonnative vegetation
2. Fragmentation of habitat
3. Disturbance of biota
4. Collision and/or electrocution of birds
5. Increased potential for fire

Water Resources

Impacts on water resources during the operation and maintenance phase would be limited to possible minor degradation of water quality resulting from vehicular traffic and machinery operation during maintenance (e.g., erosion and sedimentation) or herbicide contamination during vegetation management (e.g., from accentual sills). However, a large oil pipeline spill could potentially cause extensive degradation of surface waters or shallow groundwater.

Land Use

Land use impacts would be minimal, as many activities could continue with the ROW (e.g., agriculture and grazing). Other industrial and energy projects would likely be excluded within the ROW. In addition, construction of facilities (e.g., houses and other structures) would be precluded within the ROW and roads would only be allowed to cross ROWs, not run along their length). Recreation activities (e.g., off-highway vehicle (OHV) use and hunting) are also possible, although restrictions may exist for the use of guns, especially for aboveground pipelines or transmission lines. The ROW and access roads may make some areas more accessible for recreation activities. Activities centered on solitude and scenic beauty would potentially be affected. Military operations and aviation could be affected by the presence of transmission lines. For example, transmission lines could affect military training and testing operations that may occur at low altitudes (e.g., military training routes).

Soils and Geologic Resources

Following construction, disturbed potions of the site would be revegetated and the soil and geologic conditions would stabilize. Impacts during the operation phase would be limited largely to soil erosion impacts caused by vehicular traffic and machinery operation during maintenance activities. Any excavations required for pipeline maintenance would cause impacts similar to those from construction but to a lesser spatial and temporal extent. Herbicide would likely be used for ROW maintenance. The accidental spills of herbicides or pipeline products would likely cause soil contamination. Except in the case of a large oil spill, soil contamination would be localized and limited in extent and magnitude.

Paleontological Resources

Impacts during the operations phase would be limited to the unauthorized collection of fossils. This threat is present once the access roads are constructed and the ROW is established, making remote areas more accessible to the public.

Transportation

No noticeable impacts to transportation are likely during the operation and maintenance phase. Lower volumes of heavy- and medium-duty pickup trucks, personal vehicles, and other machinery are expected to be used during this phase. Infrequent, but routine, shipments of component replacements during maintenance procedures are likely over the period of operation.

Visual Resources

The above potions of energy transmission projects would be highly visible in rural or natural landscapes, many of which have few other comparable structures. The artificial appearance of a transmission line or pipeline may have visually incongruous "industrial" associations for some, particularly in a predominately natural landscape. Visual evidence of these projects cannot be completely avoided, reduced, or concealed. Additional visual impacts would occur during maintenance from vehicular traffic, aircraft, and workers. Maintenance, replacement, or upgrades of project components would repeat the initial visual impacts of the construction phase, although at a more localized scale.

Socioeconomics

Direct impacts would include the creation of new jobs for operation and maintenance workers and the associated income and taxes paid. Indirect impacts are those that would occur as a result of the new economic development and include things such as new jobs at businesses that support the expanded workforce or provide project materials, and associated income and taxes. The number of project personnel required during the operation and maintenance phase would be about an order of magnitude less than during construction. Therefore, socioeconomic impacts related directly to jobs would be minimal. Potential impacts on the value of residential properties located adjacent to an energy transmission project would continue during this phase.

Environmental Justice

Possible environmental justice impacts during operation include the alteration of scenic quality in areas of traditional or cultural significance to minority or low-income populations. Habitat modification, noise impacts, and health and safety impacts are also possible sources of environmental justice impacts.

REFERENCES

Archer, C. and Jacobson, M.Z. 2004. *Evaluation of Global Wind Power.* Stanford, CA: Department of Civil and Environmental Engineering, Stanford University.

Babisch, W. 2002. The noise/stress concept, risk assessment and research needs. *Noise Health* 4:1–11.

Barber, J.R., Cooks, K.R., and Fristrup, K. 2010. The costs of chronic noise exposure for terrestrial organisms. *Trends Ecology and Evolution* 25(3):180–189. Accessed @ http://www.sciencedirect.com/.

Bayne, E.M., Habib, L., and Boutin, S. 2008. Impacts of chronic anthropogenic noise from energy-sector activity on abundance of songbirds in the Boreal Forest. *Conservation Biology* 22(5):1186–1193.

Berglund, B. and Lindvall, T., Editors. 1995. *Community Noise.* Stockholm Center for Sensor Research, Stockholm University and Karolinska Institute, Stockholm, Sweden.

Bex, P.J., Edgar, G.E., and Smith, A.T. 1995. Sharpening of drifting, blurred images. *Vision Research* 35:2359–2546.

Boss, L.P. 1997. Epidemic hysteria: A review of the published literature. *Epidemiologic Reviews* 19:253–431.

Burr, D. 1980. Motion Smear. *Nature* 284(5752):164–165.

Chapman, S., St. George, A., Waller, K., and Cakic, V. 2013. The pattern of complaints about Australian wind farms does not match the establishment and distribution of turbines: Support for the psychogenic, 'communicated disease' syndrome. *PLoS One* 8:e76584.10.

Crichton, F., Dodd, G., Schmid, G., Gamble, G., Cunby, T., and Petri, K.J. 2014. Can expectations produce symptoms from infrasound associated with wind turbines? *Health Psychology* 33:360410.

Deignan, B., Harvey, E., and Hoffman-Goetz, L. 2013. Fright factors about wind turbine and health. *Health, Risk & Society* 15(3):234–250.

Dooling, R.J. 1982. Auditory perception in birds, in *Acoustic Communications in Birds*, Kroodsma, D.E. and Miller, E.H. (Eds.) (volume 1, pp. 95–129). New York: Academic Press.

Dolling, R.J. 2002. *Avian Hearing and the Avoidance of Wind Turbines*. Washington, DC: National Renewable Energy Laboratory, NREL/TP-500-30844, 83 p. Accesed @ http://www.nrel.gov/wind/pdfs/30844.pdf.

Dooling, R.J. and Popper, A.N. 2007. *The Effects of Highway Noise on Birds*. Report to the California Department of Transportation, contract 43A0139. Sacramento, California, USA: California Department of Transportation, Division of Environmental Analysis.

Fields, J.M. 1993. Effect of personal and situational variables on noise annoyance in residential areas. *The Journal of the Acoustical Society of America* 93:685–695.

Fields, J.M., de Jong, R., Brown, A.K.\L., Flindell, I.H., Gjestland, T., Job, R.F.S., et al., 1998. Guidelines for reporting core information from community noise reaction surveys. *Journal of Sound and Vibration* 206:685–951.

Fields, J.M., de Jong, R.G., Gjestland, T., Flindell, I.H., Job, R.F.S., Kerra, S., et al., 2001. Standardized general-purpose noise reaction questions for community noise surveys research and recommendation. *Journal of Sound and Vibration* 242:641–740.

Foreman, R.T.T. and Alexander, L.E. 1998. Roads and their major ecological effects. *Annual Review of Ecological Systems* 29:207–231.

Francis, C.D., Ortega, C.P., and Cruz, A. 2009. Noise pollution changes avian communities and species interactions. *Current Biology*, in press, doi: 10.1016/j.cub.2009.06.052.

Guski, R. 1999. Personal and social variables as co-determinants of noise annoyance. *Noise Health* 1:45–56.

Habib, L., Bayne, E.M., and Boutin, S. 2007. Chronic industrial noise affects pairing success and age structure of ovenbirds *Seiurus aurocapilla*. *Journal of Applied Ecology* 44:176–184.

Henningsen, P. and Priebe, S. 2003. New environmental illnesses: What are their characteristics? *Psychotherapy and Psychosomatics* 72:231–410.

Hodos, W., Ghim, M.M., Miller, R.F., Sterheim, C.E., and Currie, D.G. 1997. Comparative analysis of contrast sensitivity. *Invest Ophlhalmol Vis Sci* 38:5634.

Kaliski, K. 2009. The effects of noise on wildlife. Accessed 11/11/2021 @ https://www.rosemonties.us/files/references/045808,pdf.

Knight, T.A. 1974. A review of hearing and song in birds with comments on the significance of song in display. *Emu* 74:5–8.

Knopper, L.D., Olison, C.A., McCallum, L.C., Aslund, M.L.W., Berger, R.G., Souweine, K., and McDaniel, M. 2014. Wind turbines and human health. *Frontiers in Public Health* 2:63.

Komanoff, C. 2003. Even wind power can't be invisible. *The Providence Journal* xi:12–14.

Krough, C.M.E. 2011. Industrial wind turbine development and loss of social justice? *Bulletin of Science, Technology & Society* 31:321–331.

Laszlo, H.E., McRobie, E.S., Stansfeld, S.A., and Hansell, A.L. 2012. Annoyance and other reaction measures to changes in noise exposure—a review. *Science of the Total Environment* 435:551–621.

Lengagne, T. 2008. Traffic noise affects communication behavior in a breeding anuran, *Hyla arborea*. *Biological Conservation* 141: 2023–2031.

Lohr, B., Wright, T.F., and Dooling, R.J. 2003. Detection and discrimination of natural calls in masking noise by birds: Estimating the active space of a signal. *Animal Behavior* 65:763–777.

Magee, S. 2014. *Health Forensics*. Seattle, WA: (Amazon) CreateSpace Independent Publishing Platform.

MassDEP and MDPH. 2012. *Wind Turbine Health Impact Study: Report on Independent Expert Panel*. Boston, MA: Department of Environmental Protection and Department of Public Health. Accessed 09/09/2021 @ http://www.mass.gov/dep/energy/wind_trubine_impact_study_pdf.

Merlin, T., Newton, S., Ellery, B., Milverton, J., and Farah, C. 2014. *Systematic Review of the Human Health Effects of Wind Farms*. Canberra: ACTL Nation Health and Medical Research Council.

Miedema, H.M.E. and Vos, H. 1999. Demographic and attitudinal factors that modify annoyance from transportation noise. *The Journal of the Acoustical Society of America* 105:3336–4410.

Molburg, J.C., Kavicky, J.A., and Picel, K.C. 2007. *The Design, Construction and Operation of Long-Distance High-Voltage Electricity Transmission Technologies*. Argonne, IL: Argonne National Laboratory.

NREL. 2000. *Review of Operation and Maintenance Experience in the DOE-EPR Wind Turbine Verification Program*. Golden, Colorado. National Renewable Energy Laboratory.

Ohrstrom, E. 2004. Longitudinal surveys on effects of changes in road noise. *The Journal of the Acoustical Society of America* 115:719–729.

Ohrstrom, E., Skanberg, A., Svensson, H., and Gidlof-Gunnarsson, A. 2006. Effects of road traffic noise the benefit of access to quietness. *Journal of Sound and Vibration* 295:40059.

Pasqualetti, M.J. 2002. Living with wind power in a hostile landscape, in *Wind Power in View*, Gipe, P. and Righter, R. (Eds.). Atlanta, GA: Academic Press.

Pedersen, E. and Persson, W.K. 2004. Perception and annoyance due to wind turbine noise—a dose-response relationship. *The Journal of the Acoustical Society of America* 116:3460–7010.

Pedersen, E. and Persson, W.K. 2007. Wind turbine noise, annoyance and self-reported health and well-being in different living environments. *Occupational and Environmental Medicine* 64:480–610.

Pedersen, E. and Larsman, P. 2008. The impact of visual factors on noise annoyance among people living in the vicinity of wind turbines. *Journal of Environmental Psychology* 28:379–8910.

Pedersen, E., von deb Berg, R., Bakker, R., and Bouma, J. 2009. Response to noise from modern wind farms in The Netherlands. *The Journal of the Acoustical Society of America* 126:634–4310.

Petrie, K.J., Silversten, B., Hysing, M., Broadbent, E., Moss-Morris, R., Ericksen, H.R., et al., 2001. Thoroughly modern worries: The relationship of worries about modernity to reported symptoms, health and medical care utilization. *Journal of Psychosomatic Research* 51:395–411.

Quinn, J.L., Whittingham, M.J., Butler, S.J., and Cresswell, W. 2006. Noise, predation risk compensation and vigilance in the chaffinch Fringilla coelebs. *Journal of Avian Biology* 37:601–608.

Renewable Energy World, Editors. 2010. Measuring wind turbine noise. Are decibel levels the most important metric for determining impact? *Renewable Energy News*, November 22.

Righter, R.W. 2002. *Exoskeletal Outer-Space Creations*. Atlanta, GA: Academic Press.

Saito, Y. 2014. *Machines in the Ocean: The Aesthetics of Wind Farms*. Accessed 02/26/14 @ http://www.contempasethetics. org/newvolume/pages/article.php?articleID=247.

Salt, A.N. and Hullar, T.E. 2010. Responses of the ear to low frequency sounds, infrasound and wind turbines. *Hearing Research* 268: 12–21.

Schaub, A., Ostwald, J., and Siemers, B.M. 2008. Foraging bats avoid noise. *The Journal of Experimental Biology* 211:3174–3180.

Schworm, P. and Filipov, D. 2013. Flickering shadows from wind turbines draw complains. *The Boston Globe*, April 05.

Steinman, R.M. and Levinson, J.Z. 1990. The role of eye movement in detection and contrast and spatial detail. Accessed 10.31/2021 @ https://pubmed.ncbi.nim.nih.gov.

Swaddle, J.P. and Page, L.C. 2007. High levels of environmental noise erode pair preferences in zerbra finches: Implications for noise pollution. *Animal Behaviour* 74:363–368.

Taylor, J., Eastwick, C., Wilson, R., and Lawrence, C. 2012. The influence of negative oriented personality traits on the effects of wind turbine noise. *Personality and Individual Differences* 54:338–431.

Thorne, B. 2011. The problems with "noise numbers" for wind farm noise assessment. *Bulletin of Science, Technology & Society* 31:262–901.

United States Fish and Wildlife Service (USFWS). 2005. *A Fine line for Birds: A Guide to Bird Collisions at Power Lines.* Washington, DC. Accessed @ http://birds.fws.gov/imbd.

United States Fish and Wildlife Service (USFWS). 2013. *Eagle Conservation Plan Guidance: Module 1—Land-based Wind Energy.* Washington, DC: USFWS.

United States Fish and Wildlife Service (USFWS). 2014. *The Effects of Noise on Wildlife.* Accessed 02/18/14 @www.fws.gov/windenergy/docs/noise.pdf.

Van den Berg, G.P. 2004. Effects of the wind profile at night on wind turbine sound. *Journal of Sound and Vibration* 277:955–970.

WGA. 2009. *Western Renewable Energy ones Phase1 Report.* Denver, CO: Western Renewable Energy Zones Initiative.

WHO. 2011. *Burden of Disease from Environmental Noise: Quantification of Healthy Life Lost in Europe.* Copenhagen: WHO Regional Office for Europe.

10 Wind Turbine Safety

INTRODUCTION

On April 19, 2011, US Department of Labor's Occupational Safety and Health Administration (OSHA) cited a wind farm servicing company for six willful safety violations in an incident that occurred in Odell, Illinois. As a result of this incident, Outland Renewable Services was issued the six citations for safety violations after a wind farm technician suffered severe burns from an electrical arc flash on October 20, 2010. The US Department of Labor's Occupational Safety and Health Administration issued the citations following an investigation at the Iberdrola Streator Cayuga Ridge South Wind Farm near Odell. The company, a servicing and maintenance provider in the wind tower industry, faces proposed penalties of $378,000.

With regard to this incident, Secretary of Labor Hilda L. Solis said,

> Green jobs are an important part of our economy, and sectors such as wind energy are growing rapidly. Outland's management was aware of the potentially hazardous conditions to which its workers could have been exposed and showed intentional disregard for employee safety by ignoring OSHA's requirements for isolating energy sourced during servicing operations. Employers must not cut corners at the expense of their workers' safety.

Outland Renewable Services was issued the citations for exposing maintenance technicians to electrical hazards from the unexpected energization of transformers in three wind turbine towers. A willful violation is one committed with intentional knowing or voluntary disregard for the law's requirements, or with plain indifference to worker safety and health.

On the day of the tragic incident, Outland Renewable Services failed to ensure technicians working in wind farm towers affixed their own energy isolation devices—also known as personal lockout and tagout devices—on the tower turbine switch gear at ground level. Not following lockout–tagout procedures created the possibility for other workers to energize transformers in the turbine towers, upon which technicians were working at a distance of approximately 350 ft above ground. Anyway, as a result of non-compliance with OSHA's lockout–tagout standard, the injured worker suffered third-degree burns to his neck, chest, and arms, and second degree burns to the face as a result of an arc flash that occurred when a transformer was unexpectedly energized by another worker.

According to OSHA, the egregious violations that were committed in this case fall under the requirements of OSHA's Severe Violators Enforcement Program. Initiated in 2010, the program is intended to focus on employers that endanger workers by committing willful, repeat, or failure-to-abate violations in one of the following circumstances: a fatality or catastrophe; industry operations or processes that

DOI: 10.1201/9781003288947-12

expose workers to severe occupational hazards related to the potential release of highly hazardous chemicals; and all per-instance citation (egregious) enforcement actions (OSHA, 2011).

WIND ENERGY FATALITIES/INCIDENTS

Wind energy workers are exposed to hazards that can result in fatalities and serious injuries. Many incidents involving falls, severe burns from electrical shocks and arc flashes/fires, and crushing injuries have been reported to OSHA. Some examples are given below:

- On August 29, 2009, at 08:30 hours a 33-year male lineman was shocked as he grasped a trailer ramp attached to a low boy trailer containing an excavator. The excavator was being operated in anticipation of being off-loaded from the trailer. The trailer was parked on a rural aggregate road adjacent to an access road for a wind turbine generator. The excavator operator rotated the upper works of the machine prior to moving the machine from the trailer. During the rotation, the boom contacted a 7,200-volt primary rural power line. The power line was approximately 12 ft from the road with the trailer parked approximately 2 ft from the road edge. The injured worker had entry wounds in the hands and exit wounds in the feet. He was transported by EMS, treated and admitted for observation at a local hospital. He was discharged approximately 24 hours later and returned to work the following day.
- On May 10, 2009, the victim was working in the bottom power cabinet of a wind turbine. He was checking the electrical connections and came into contact with a bus bar and arc flash erupted, causing injury to the victim. Afterward the victim was taken to a hospital by their technician and was met by the ambulance on the way. After arriving at the hospital, he was later transferred by medi-vac to another hospital in Oklahoma City and was treated for injuries. On June 12, 2009, the company was notified by a representative of the hospital that the victim was deceased.
- On November 11, 2005, worker #1 and two coworkers were removing and replacing a broken bolt in the nacelle assembly of a wind turbine tower that was approximately 200 ft above the ground. They were heating the bolt with an oxygen–acetylene torch when a fire started. Worker #1 retreated to the rear of the nacelle, away for the ladder access area. While the two coworkers were able to descend the tower, worker #1 fell approximately 200 ft to the ground, struck an electrical transformer box, and died.
- At approximately 11:40 a.m. on June 17, 1992, a worker attempted to descend an 80 ft. ladder that accessed a wind turbine generator. The worker slipped and fell from the ladder and died. The victim was wearing his company-furnished safety belt, but the safety lanyards were not attached. Both lanyards were later discovered attached to their tie-off connection at the top of the turbine generator.

- A site supervisor was replacing a 480-V circuit breaker serving a wind turbine. He turned a rotary switch to what he thought was the open position in order to isolate the circuit breaker. However, the worker did not test the circuit to ensure that it was de-energized. The worker had placed the rotary switch in a closed position, and the circuit breaker remained energized by a back feed from a transformer. Using two plastic-handled screwdrivers, he shorted two contacts on the breaker to discharge static voltage buildup. This caused a fault and the resultant electric arc caused deep flash burns to the worker's face and arms and ignited his shirt. The worker was hospitalized in a burn unit for four days.

CASE STUDY 10.1: WIND TURBINE FATALITY*

The Oregon Department of Consumer and Business Services, Occupational Safety and Health Division (Oregon OSHA) fined Siemens Power Generation Inc. a total of $10,500 for safety violations related to an August 25, 2007, wind turbine collapse that killed one worker and injured another.

"The investigation found no structural problems with the tower," said Michael Wood, Oregon OSHA administrator.

> This tragedy was the result of a system that allowed the operator to restart the turbine after service while the blades were locked in a hazardous position. Siemens has made changes to the tower's engineering controls to ensure it does not happen again.

The event took place at the Klondike III Wind Farm near Wasco, where three wind technicians were performing maintenance on a wind turbine tower. After applying a service brake to stop the blades from moving, one of the workers entered the hub of the turbine. He then positioned all three blades to the maximum wind resistance posting and closed all three energy isolation devices on the blades. The devices are designed to control the mechanism that directs the blade pitch so that workers don't get injured while they are working in the hub. Before leaving the confined space, the worker did not return the energy isolation devices to the operational position.

As a result, when he released the service brake, wind energy on the out-of-position blades caused an "overspeed" condition, causing one of the blades to strike the tower and the tower to collapse, the Oregon OSHA investigation found.

Chadd Mitchell, who was working at the top of the tower, died in the collapse. William Trossen, who was on his way down a ladder in that tower when it collapsed, was injured. The third worker was outside the tower and unharmed.

---

* Department of Consumer Business & Services, 2008. Oregon OSHA releases findings in wind turbine collapse. Salem, Oregon: Oregon OSHA.

During the investigation, Oregon OSHA found several violations of safety rules:

- Workers were not properly instructed and supervised in the safe operation of machinery, tool, equipment, process, or practice they were authorized to use or apply. The technicians working on the turbine each had less than two months' experience, and there was no supervisor on site. The workers were unaware of the potential for catastrophic failure of the turbine that could occur as a result of not restoring energy isolation devices to the operational position.
- The company's procedures for controlling potentially hazardous energy during service or maintenance activities did not fully comply with Oregon OSHA regulations. Oregon OSHA requirements include developing, documenting, and using detailed procedures and applying lockout or tagout devices to secure hazardous energy in a "safe" or "off" position during service or maintenance. Several energy isolation devices in the towers, such as vales and lock pins, were not designed to hold a lockout device, and energy control procedures in place at the time of the accident did not include the application and removal of tagout devices.
- Employees who were required to enter the hub [a permit-required confined space (PRCS)] or act as attendants to employees entering the hub had not been trained in emergency rescue procedures from the hub.

As shown in Case Study 10.1, wind energy employers need to protect their workers from workplace hazards. Workers should be engaged in workplace safety and health and need to understand how to protect themselves from these hazards. Even though the wind energy industry is a growing industry, the hazards are not unique and OSHA has many standards that cover various worker on-the-job activities and exposures. The hazards (along with controls) that workers in wind energy may face are provided below.

WIND TURBINE HAZARDS AND APPLICABLE OSHA STANDARDS AND CONTROLS

As mentioned, working around, with, or on wind turbines and associated equipment presents many hazards to the installers, operators, and maintenance personnel. The primary hazards with wind turbine operations/maintenance and with the wind-power industry in general along with applicable OSHA standards and controls include:

- Falls from the tower or nacelle
- Confined space entry
- Fire

- Lockout/tagout (LOTO)
- Medical and first aid
- Crane, derrick, and hoist safety
- Electrical safety
- Machine guarding
- Respiratory protection

Each of these hazards and controls is briefly discussed below. (Note: many of the hazards discussed below apply, in one way or the other, to other renewable energy sources.)

FALLS*

Workers who erect and maintain wind turbines can be exposed to fall hazards. Wind turbines vary in height but can be over 100 ft tall (some are 400+ ft tall). Exposure to high winds may make work at high elevations even more hazardous. OSHA has different fall protection requirements for construction (installation of towers) and general industry (maintenance).

During installation, workers may need to access individual turbine sections to weld/fit individual sections together, run electrical or other lines, and install/test equipment—often at heights greater than 100 ft. Construction workers on wind farms when exposed to fall distances of 6 ft or more must be protected from falls by guardrail systems, safety net systems, or a personal fall arrest system.

CONFINED SPACE ENTRY†

Wind energy employers need to look at the spaces that workers enter to determine if they meet OSHA's definition of a confined space. By definition, a confined space:

- Is large enough for an employee to enter fully and perform assigned work
- It is not designed for continuous occupancy by the employee
- Has a limited or restricted means of entry or exit

Some confined spaces have recognized hazards, such as low oxygen environments, which can pose a risk for asphyxiation, or accumulation of hazardous gases. These confined spaces are called permit-required confined spaces and require additional safety precautions.

Wind energy employers also need to look at the hazards of the confined spaces to determine whether those spaces are "permit-required" confined spaces (PRCS). By definition, a PRCS has one or more of these characteristics:

---

* Based on information from OSHA's Green Job Hazards: Wind Energy. Accessed 2011 @ http://www. osha.gov/dep/greenjobs/windenergy_falls.html; F.R. Spellman 2013. *Safe Work Practices for Green Energy Jobs.* Lancaster, PA: DES*Tech* Publishers, Inc.

† Based on information from OSHA's *Green Job Hazards: Wind Energy—Confined Spaces.* Accessed 11/01/11 @ http://www.osha.gov/dep/greenjobs/windenergy_confined.html.

- It contains or has the potential to contain a hazardous atmosphere.
- It contains a material with the potential to engulf someone who enters the space.
- It has an internal configuration that might cause an entrant to be trapped or asphyxiated by inwardly converging walls or by a floor that slopes downward and tapers to a smaller cross section.
- It contains any other recognized serious safety or health hazards.

If workers are expected to enter permit-required confined spaces, the employer must develop a written permit space program and make it available to workers or their representatives. The permit space program must detail the steps to be taken to make the space safe for entry.

The configuration of all nacelles (i.e., the cover housing that houses all of the generating components in a wind turbine, including the generator, gearbox, drive train, and brake assembly) will classify them as confined spaces and during the maintenance activities inside the nacelles, workers may be exposed to hazards from electrical motors, gears, etc. Those hazards may classify a nacelle to be a PRCS. Technicians working in nacelles should make sure to perform air sampling (such as for low oxygen levels or other hazardous gases) prior to entering a nacelle. It is recommended that the technician should always carry a portable gas monitor in their toolkit and make sure that it is maintained properly.

Fires*

Wind turbines may have fire hazards because of the electrical pars and the combustible materials such as insulation or the material of construction used in the turbine housing (nacelle) or lubricants invoking in its operation.

Wind energy employees need to be trained about fire hazards at the worksite and about what to do in a fire emergency. This plan should outline the assignments of key personnel in the event of a fire and provide an evacuation plan for workers on the wind turbines. Where employers require workers to use portable fire extinguishers, the worker must be trained in the general principle of fire extinguisher use and the hazards involved with incipient stage firefighting.

Workers should be made aware that while fighting initial fires, toxic gases can be generated and oxygen can be depleted inside nacelles, and they can be exposed to such gases or can be asphyxiated from lack of oxygen.

If the employer chooses to use a fixed extinguishing system inside nacelles, then the freezing point of the extinguishing medium and the safety of workers (exposure to toxic gases and depletion of oxygen) including emergency escape method should be taken into consideration.

In addition to the fire extinguishing mechanism (whether the use of fire extinguishers or a fire extinguishing system or both), fire detection systems and emergency

---

* Based on information from OSHA's *Green Job Hazards: Wind Energy—Fires.* Accessed 11/01/11 @ http://www.osha.gov/dep/greenjobs/windenergy_fire.html.

alarm systems should be installed inside nacelles to give an early warning to workers to escape. If such systems are installed, they must be maintained in operable condition, see 29 CFR 1910.160(c) and 1910.165(d).

Workers should know exactly what to do and how to escape in a fire emergency. Wind turbines should be provided with quick escape descent devices for workers in escape in the event of a fire or other emergency.

Lockout/Tagout*

Lockout/tagout (LOTO) refers to specific practices and procedures to safeguard employees from the unexpected energization or startup of machinery and equipment, or the release of hazardous energy during service or maintenance activities.

Approximately 3 million workers work on equipment and face the greatest risk of injury if lockout/tagout is not properly implemented. Compliance with the lockout/tagout standard prevents an estimated 120 fatalities and 50,000 injures each year. Workers injured on the job from exposure to hazardous energy lose an average of 24 workdays for recuperation. In a study conducted by the United Auto Workers (UAW), 20% of the fatalities (83 of 414) that occurred among their members between 1973 and 1995 were attributed to inadequate hazardous energy control procedures, specifically lockout/tagout procedures. Wind turbines have lots of internal machinery and equipment, including blades that need to be maintained. Workers performing servicing or maintenance may be exposed to injuries from the unexpected energization, startup of the machinery or equipment, or relapse of stored energy in the equipment. Wind farm employers must implement lockout/tagout procedures outlined in OSHA standards (see 29 CFR 1910.269(d) and 29 CFR 1910.147).

The following are some of the significant requirements of a lockout/tagout procedure required under lockout/tagout program.

- Only authorized employees may lockout or tagout machines or equipment in order to perform servicing or maintenance.
- Lockout devices (locks) and tagout devices shall not be used for any other purposes and must be used only for controlling energy.
- Lockout and tagout devices (locks and tags) must identify the name of the worker applying the device.
- All energy sources to equipment must be identified and isolated.
- After the energy is isolated from the machine or equipment, the isolating device(s) must be locked out or tagged out in a safe or off position only by the authorized employees.
- Following the application of the lockout or tagout devices to the energy isolating devices, the stored or residual energy must be safely discharged or relieved.

---

* Based on information from OSHA's *Green Jobs Hazards: Wind Energy—Lockout/Tagout.* Accessed 11/02/11 @ http:www.osha.gov/dep/greenjobs/windenergy_loto.html.

- Prior to starting work on the equipment, the authorized employee shall verify that the equipment is isolated from the energy source, for example, by operating the on/off switch on the machine or equipment.
- Lock and tag must remain on the machine until the work is completed.
- Only the authorized employee who placed the lock and tag must remove his/her lock or tag, unless the employer has a specific procedure as outlined in OSHA's lockout/tagout standard.

MEDICAL AND FIRST AID*

Wind farms are normally located in remote locations, away from a hospital or other emergency treatment facilities. This is a major concern if a worker gets hurt—how will they be treated quickly? Wind energy employers should determine the estimates of emergency medical service response times for all their wind farm locations of all times of the day and night at which they have workers on duty, and they should use that information when planning their first aid program. The employers must ensure that medical personnel are available for advice and consultation and that someone who is trained is available to provide first aid.

Trained first aid providers must be available at all wind farms of any size, if there is no nearby clinic or a hospital. If a worker is expected to render first aid as part of his or her job duties, the work is covered by the requirements of the Occupational Exposure to Bloodborne Pathogens Standard (see 29 CFR 1910.1030). This standard includes specific training requirements.

DID YOU KNOW?

Blood-borne pathogens are infectious microorganisms in human blood that can cause disease in humans. These pathogens include, but are not limited to, hepatitis B (HBV), hepatitis C (HCV), and human immunodeficiency virus (HIV). Needle sticks and other sharp-related injuries may expose workers to blood-borne pathogens. Workers in many occupations including first aid team members, housekeeping personnel in some industries, nurses, and other healthcare personnel may be at risk of exposure to blood-borne pathogens.

OSHA's Electric Power Generation, Transmission, and Distribution standard requires that workers be trained in cardiopulmonary resuscitation (CPR), because a worker who may be exposed to an electric shock may experience a sudden cardiac arrest. In such adverse situations, automated external defibrillators (AEDs) can also assist in preventing a potential death. AEDs should be provided at wind farms and workers should be trained in how to use them. This training can be done when CPR training is provided to workers.

---

* Based on information from OSHA's *Green Job Hazards: Wind Energy—Medical and First Aid.* Accessed 11/02/11 @ http://www.osha.gov/dep/greenjobs/windenergy_medical.html.

CRANE, DERRICK, RIGGING, AND HOIST SAFETY*

Cranes, derricks, rigging equipment, and hoists are used to move the large, heavy loads during wind turbine installation and maintenance. Fatalities and serious injuries can occur if cranes are not inspected and used properly. Many fatalities can happen when the crane boom, load line, or load contacts power lines and shorts electricity to the ground. Other incidents happen when workers are struck by the load, are caught inside the swing radius, or fail to assemble/disassemble the crane properly. There are significant safety issues to be considered, both for the operators of the diverse "lifting" devices and for workers who work near them. See OSHA's General Industry Standards at 29 CFR 1910.179 and 29 CFR 1910.180, and Construction Standard at 19 CFR 1926.1417.

Because wind turbines are installed in windy areas, the effects of wind speeds need to be taken into consideration for lifting activities. Stability can be an issue when the boom is high and the wind coming from the rear, front, or side of the crane can cause the load to sway away from the crane, increasing the radius and thus possibly decreasing the crane capacity.

An employer needs to determine the wind speeds at which it is not safe to continue lifting operations. Load charts do not generally take wind speeds into consideration. If the load chart or the operating manual does not have information on wind speeds and derating (i.e., operating below design limits to prolong life and ensure safety) information, the crane manufacturer should be consulted. The procedures applicable to the operation of the equipment, including rated capacities (load charts), recommend operating speeds, special hazard warnings, instructions, and operator's manual, must be readily available in the cab at all times for use by the operator [see 29 CFR 1926.1417(c)]. The maximum allowable wind speed and derating information need to be posted conspicuously in the cab or on the load chart.

Extremely cold weather conditions can have an impact on crane and lifting operations. When temperatures drop below 10°F, appropriate consideration should be given to crane hydraulics and possible derating of the crane.

Bad weather, such as rain, snow, or fog, can also have an adverse impact on lifting. Equipment and/or operations must be adjusted to address the effect of wind, ice, and snow on equipment stability and rate capacity [see 29 CFR 1926.1417(n)]. During thunderstorms, a crane boom can become a lightning rod. If there is an indication of possible thunderstorms, lifting activities should be suspended and the boom should be lowered to a safe position, and workers should leave the area. If the crane is struck by lightning, it should be thoroughly inspected prior to putting it back into service.

Heavy rain along with high-speed winds also can affect crane operations. Water can get into components such as brakes or clutches and render them inoperable. When these conditions exist, operators should wait until the components are dried out.

---

* Based on information from OSHA's *Green Job Hazards: Wind Energy—Crane, Derrick and Hoist Safety*. Accessed 11/02/11 @ http://www.osha.gov/dep/greenjobs/windenegy_crane.html.

ELECTRICAL SAFETY*

Workers in wind farms are potentially exposed to a variety of serious hazards, such as arc flashes (which include arc flash burn and blast hazards), electrical shock, falls, and thermal burn hazards that can cause injury and death. Wind farm employers are covered by the Electric Power Generation, Transmission, and Distribution standards and, therefore, are required to implement the safe work practices and worker training requirements of OSHA's Electric Power Generation, Transmission, and Distribution standard, 29 CF 1910.269.

Workers need to pay attention to overhead power lines at wind farms. The hazard is from using tools and equipment that can contact power lines and workers must stay at least 10 ft away from them, because they carry extremely high voltage. Fatal electrocution is the main hazard but burns and falls from elevators can occur at wind farms.

MACHINE GUARDING†

The production of a wind turbine involves thousands of parts—gears, blades, and many other such parts. Manufacturing wind turbines, therefore, will involve machines of various configurations and may expose workers to the hazards of moving parts of the machines if they are not safeguarded properly.

Additionally, the moving parts associated with the turbine if not guarded properly may have the potential to cause severe workplace injuries, such as crushed fingers or hands, amputations, burns, or blindness. Employers must ensure that the workers are protected from the machine hazards, and workers should make sure that the rotating parts and points of operation machines are properly guarded prior to using them.

RESPIRATORY PROTECTION‡

Manufacturing turbine blades involve operations like buffing and resurfacing, which may expose workers to harmful gases, vapors, and dusts. Workers must be protected from inhalation hazards through the use of ventilation. If the ventilation alone is not adequate, then workers may also need to use appropriate respirators.

The use of respirators may give a false sense of security and workers should understand the limitations of the respirators. For example, during heavy exertion, the respirator seal is often compromised, which allows the chemical to enter the breathing zone (without being filtered) through the gaps between the respirator and the face. In such situations a worker who is not adequately trained may think that he or she is being protected. It is, therefore, essential that workers be provided training

---

* Based on information OSHA's *Green Job Hazards: Wind Energy—Electrical.* Accessed 11/03/11 @ http://www.osha.gov/dep/greenjobs/windenergy_electical.html.
† Based on information from OSHA's *Green Job Hazards: Wind Energy—Machine Guarding.* Accessed 11/03/11 @ http://www.osha.gov/dep/greenjobs/windenergy_machineguarding.html.
‡ Based on information from OSHA's *Green Job Hazards: Wind Energy—Respiratory Protection.* Accessed 11/02/11 @ http://www.osha.gov/dep/greenhobs/windenergy_respiratory.html.

in the proper use of respirators and their litigations. In addition, they must be trained on the proper storage and maintenance of respirators.

THE BOTTOM LINE ON WIND POWER

Wind energy jobs display the pros and cons associated with the double-edged sword syndrome: they provide the benefits related to moving us toward energy independence; they create new jobs; and they improve environmental conditions. However, the other edge of the sword is fraught with safety hazards to the workers. Workers may be exposed to the same conventional hazards found in most workplaces—such as slips, trips, and falls, confined spaces, electrical, fire, and other similar hazards. Additionally, workers may be exposed to new hazards which may not have been previously identified.

REFERENCE

OSHA. 2011. News Release:11–527-CHI. Washington, D.C.: U.S. Department of Labor.

Index

Printed in the United States
by Baker & Taylor Publisher Services